SEEPAGE, DRAINAGE, AND FLOW NETS

SEEPAGE, DRAINAGE, AND FLOW NETS

THIRD EDITION

HARRY R. CEDERGREN

A Wiley-Interscience Publication

JOHN WILEY & SONS, INC.

New York • Chichester • Weinheim • Toronto • Singapore • Brisbane

Library of Congress Cataloging-in-Publication Data

Cedergren, Harry R.
   Seepage, drainage, and flow nets / Harry R. Cedergren. – 3rd ed.
     p.   cm.
   "A Wiley-Interscience publication."
   Bibliography: p.
   Includes index.
   ISBN 0-471-61178-6 (cloth)
   ISBN 0-471-18053-X (paper)
   1. Seepage. 2. Drainage. 3. Soil percolation. I. Title.
TC176.C4   1988   89-18670
624.1'5136—dc19

To My Wife, EVELYN,
for her assistance in the development of this book

PREFACE

This book has been written primarily for students, practicing engineers, geologists, and others who are looking for practical methods for the solution of seepage and drainage problems. Although it has a fundamental approach, it is not a mathematical treatise on seepage theory and its use does not require proficiency in higher mathematics. Theory and mathematics are included only to the extent needed to promote an understanding of seepage and drainage phenomena. Emphasis is on practical applications rather than complex theories. Thus Darcy's law and the graphical flow net are given considerable emphasis, for these two fundamental tools offer solutions to countless everyday seepage and drainage problems.

When these techniques are used with experience and judgment they can help designers understand the *basic nature* of seepage and drainage problems and how to develop the best kinds of systems to prevent damage from harmful water actions.

In the past, when designers relied on "intuition" and "rules-of-thumb" for decisions about the design of overflow weirs, earth dams, pavements, and other facilities exposed to the detrimental actions of water, failures and inefficiencies were all too common. With the development of the "rational" and "experimental" methods by leaders such as K. Terzaghi and A. Casagrande, designers began to have a better understanding of the true nature of soil masses and earth foundations and how to design safe, economical engineering facilities. But, the flow of water in natural formations and in drainage materials intended to control it has somehow remained a mystery to many. This has often led to the selection of inadequate designs and completely "wrong" kinds of drainage materials. A large retaining wall in Portland, Oregon toppled after heavy rains, presumably because the "pervious" backfill behind the wall contained so many fines that it was in reality little more permeable than the soil it was supposed to drain. And, most of the world's pavements are breaking up prematurely because they are in fact poorly drained. In both of these examples, designers made the same mistake—not thinking of drains as *conveyors of water.* This book presents simple methods for analyzing the *discharge needs* of drains for all kinds of civil engineering works needing protection from water, and many illustrations of the practical application of these principles. After nearly five decades of working with earth projects I am fully convinced that seepage analysis and control are among the most important challenges facing civil engineers.

Although many projects are being made possible by drainage or other seepage control, the need still exists for better understanding and broader application of sound drainage principles. Helping to achieve these goals is a primary aim of this book.

Part I contains chapters on permeability, seepage fundamentals, and flow net construction. In presenting basic information about seepage and permeability, an effort has been made to relate fundamentals to commonly observed phenomena. Thus the chapter on permeability compares seepage through soils and porous aggregates with flow through pipes and open channels. Emphasis is also placed on the extremely wide variations in permeability that exist and on the even wider variations in velocities and quantities of seepage through soils and porous media. These wide variations tend to obscure the true nature of seepage in earth masses. Chapter 4 on flow nets provides instructions and step-by-step examples that should help beginners and advanced students alike to develop skill in flow-net construction.

Chapter 5 on filter and drain design describes basic design concepts and emphasizes the great advantages of *composite drains* for the control of seepage in many kinds of civil engineering works. Other chapters in Part II present typical seepage analyses and drainage designs for earth dams and levees (Chapter 6), foundation dewatering and stabilization by drainage (Chapter 7), slope improvement by drainage (Chapter 8), drainage of roads, airfields, and other surface facilities (Chapter 9), the improvement of miscellaneous structures by drainage (Chapter 10), and the protection of groundwaters from contamination (Chapter 11).

This edition contains numerous references to performance records of projects with water problems in many countries, and new ideas on seepage and drainage that have been developed by researchers and others throughout the world (see Chapter references). I have also drawn on my own experiences in the time that has elapsed since the writing of the earlier editions. Practical ways to obtain realistic large-scale evaluations of permeabilities from spreading or receding groundwater mounds are presented in Chapter 2 (Sec. 2.8); a practical method for correcting for semiturbulent-to-turbulent flow in coarse gravels and open-graded drainage layers is given in Chapter 3 (Sec. 3.7).

Chapter 6 (Sec. 6.6) describes new methods for installing slurry trench cut-off walls for rehabilitating water-endangered dams, or for other purposes, to previously unprecedented depths of 400 feet and greater. Chapter 7 has new information about the synthetic "wick" drains and "fin" drains that are being used in the place of "sand" drains and aggregate drains for numerous purposes. Documented information is given in Chapter 9 proving that the conventional un-drained pavements which fail prematurely from water damage are extremely uneconomical in the long-run when compared with well drained pavements using systems described in this book (see "Economics of Drained and Undrained Pavements," Sec. 9.4).

Because of the mounting contamination dangers to water supplies around the world, a new chapter has been added to this edition (Chapter 11), with substantial new information about this problem, and how to use basic seepage principles to better understand it, and how to try to keep it under control. Simple calculations with Darcy's law explain why serious groundwater contamination in so many areas went on so long without being noticed.

Up-to-date examples of a variety of seepage and drainage problems are given throughout the book, some replacing examples that appeared in the earlier editions.

I thank the many agencies and persons whose published papers, technical reports, and books have furnished material that has assisted in the development of this book. Whenever a source is known, credit has been given. If any omissions have occurred, they are unintentional, and I should appreciate being informed of them.

HARRY R. CEDERGREN

Sacramento, California

CONTENTS

SYMBOLS

The symbols used in this book closely follow those recommended in *Soil Mechanics Nomenclature,* Manual of Engineering Practice No. 22, published by the American Society of Civil Engineers in 1941. Descriptive subscripts are employed for clarity whenever possible. Care must be taken to use consistent units in all numerical computations.

A = area, dimension
a = area, constant, dimension
a_v = coefficient of compressibility
B = width
b = dimension, a constant
c = a constant, a distance, unit cohesion
c_h = horizontal coefficient of consolidation
c_v = coefficient of consolidation
C = a constant, total cohesion, centigrade, factor
D = depth, thickness, zone of influence of a drain
D_a = thickness of water-bearing strata at r_a
D_e = thickness of water-bearing strata at edge of hole
D_{10} = effective grain size
D_{15} = 15% size (15% is finer)
D_{50} = 50% size (50% is finer)
D_{60} = 60% size (60% is finer)
D_{85} = 85% size (85% is finer)
d = diameter, distance

d_e = diameter of zone of influence of a drain well, equivalent diameter of a drain
d_p = diameter of a pipe
d_w = diameter of a well
E = horizontal earthquake force
EOS = equivalent opening size
e = void ratio, volume of voids
e_a = volume of air-filled voids
e_o = initial void ratio
Δe = change in void ratio
F = resultant force, seepage force, a factor, Fahrenheit
F_v = vertical seepage force
G = specific gravity
G_a = air space ratio (drainage)
G_s = factor of safety
g = acceleration of gravity, longitudinal slope
H = thickness, distance, height, horizontal force
h = hydrostatic head
h_a = average hydrostatic head
h_b = head loss in blanket drain
h_c = height of capillary rise, constant head
h_d = head loss in a drain
h_m = head midway between two wells
h_t = critical uplift pressure
h_w = head losses at a well
Δh = increment of head, potential drop
i = hydraulic gradient
K = permeability (physicist's coefficient)
k = Darcy's coefficient of permeability (engineer's coefficient)
k' = semiturbulent-to-turbulent permeability coefficient
k_d = permeability of drainage layer
k_e = permeability of embankment
k_f = permeability of foundation
k_f = permeability of filter (Chapter 5 only)
k_g = permeability of grouted zone
k_h = horizontal permeability
k_s = permeability of soil
k_t = permeability at t degrees
k_v = vertical permeability
k_w = permeability of the well backfill
k_1, k_2 = permeability of zone 1, zone 2
k_{20} = permeability at 20 C
$\bar{k}$ = effective coefficient of permeability of stratified soil
L = length, distance

ΔL = increment of length, distance

l = distance, length

Δl = increment of distance

M = overturning moment

m = hydraulic mean radius

m_v = coefficient of volume change

N = normal component, an integer, number of wells discharging into a drain

n = porosity, an integer, well diameter ratio

n_d = number of equipotential drops in a flow net

n_e = effective porosity

n_f = number of flow channels in a flow net

n_{f-1} = number of flow channels in zone 1 in a flow net

n_{f-2} = number of flow channels in zone 2 in a flow net

P_{85} = 85% pore size of filter cloth or fabric (85% of openings are smaller)

P_{95} = 95% pore size of filler cloth or fabric (95% of openings are smaller)

P = resultant pressure

ΔP = differential pressure

p = pressure per unit of area

p_a = atmospheric pressure

Δp = increment of pressure

Q = seepage quantity

ΔQ = increment of seepage quantity

q = seepage quantity

q_d = discharge per drain

Δq = increment of seepage quantity

R = radius, hydraulic radius, resultant

Re = Reynold's number

r = radius

r_a = average radius

r_e = effective radius

r_o = average radius of a hole

r_s = remolded zone around a drain well

r_w = radius of a well

S = slope

$S\%$ = percentage of saturation, degree of saturation

s = a ratio, shearing resistance, drawdown, slope

T = time, time factor, basic time lag, tangential component, transmissibility factor

T_r = time factor for radial drainage to wells

T_v = time factor for vertical drainage

t = time, thickness (groundwater usage)

t_{50} = time for 50% drainage

Δt = increment of time

$-U$ = negative pore pressure

U = degree of consolidation (consolidation ratio), excess pore pressure, uplift force

$\overline{U}$ = average degree of consolidation

$\overline{U_r}$ = degree of consolidation for radial drainage

u = excess hydrostatic pressure

$\bar{u}_r$ = average excess pore pressure (radial flow)

$\bar{u}_z$ = average excess pore pressure (vertical flow)

u, v, w = discharge components

V = volume, vertical force

v = velocity, discharge velocity

v_d = discharge velocity

v_{dd} = velocity of drawdown of a water surface

v_s = seepage velocity

v_{sl} = velocity of a moving saturation line

W = weight

W_0 = weight of water, submerged weight

W' = effective weight

X = distance

x, y, z = distances, cartesian coordinates

y = an exponent

z = position head (hydraulics, drainage)

α = angle, factor

β = angle

γ = unit weight

γ_0, γ_w = unit weight of water

γ' = unit submerged weight

μ = viscosity

ϕ = angle of internal friction, coefficient of friction

ω = angle

SEEPAGE, DRAINAGE, AND FLOW NETS

PART I

BASIC CONSIDERATIONS

CHAPTER ONE

INTRODUCTION

1.1 NEED FOR CONTROL OF SEEPAGE

The Power of Water

Water is essential for life on our planet; however, too much in the wrong places under the wrong conditions can have detrimental effects. Throughout recorded history there is evidence that mankind has feared and respected the destructive power of water. Out in the open, in the form of tides and floods, it is one of the most powerful forces of nature. Hidden in rock crevices and soil pores, under the downward pull of the force of gravity it exerts unbelievable forces that tear down mountainsides and destroy engineering works. Railroad engineers, highway people, dam designers and builders, and many others have long known of the great importance of controlling water in pores and cracks in earth and rock formations. When groundwater and seepage are uncontrolled, they can cause serious economic losses and take many human lives. Controlled, they need not be feared.

Economic and Social Importance of Drainage

Civilization as we know it would not be possible without drainage. Many ancient civilizations that depended on irrigation for crop production disappeared from the face of the earth, in large measure because of diminishing crop production caused by lack of drainage.

Leading designers of earth dams, gravity dams, drydocks, basements, and other below-water structures, know that good drainage often opens the way to major savings in the cost of projects, energy, and vital natural resources.

The development of effective drainage systems to prevent harmful accumulations of excess water is one of the most important activities of civil engineers. Efficient drainage systems can improve the safety of earth dams, levees, storage reservoirs, concrete dams, and countless other structures. Furthermore, a well-drained structure is almost always more economical than its undrained counterpart. Utilization of the seepage and drainage analysis methods in this book can increase safety, reduce maintenance needs of facilities, and lower the long-term costs of many engineering works. The worldwide significance of good drainage is illustrated by the summary in Table 1.1 (estimated losses caused by poor drainage are for 1988).

TABLE 1.1 Illustration of World-Wide Importance of Good Drainage

Area of activity	Benefits of good drainage	Losses caused by poor drainage
Agriculture	Increased crop production, healthier crops; longer growing season; fewer problems from working on wet ground	Sour, waterlogged, unproductive land, accumulations of harmful salts, reduction in food production, destruction of many irrigated lands; demise of many ancient civilizations; (annual loss is probably more than $200 billion, worldwide)
Building foundations	Stable, dry structures below the water table; thinner basement slabs that result in savings in dollars and natural resources	Wet basements; uplifted, cracked floor slabs; heavy maintenance and repair bills
Mining	Dry, stable waste-disposal piles (dams)	Unstable mine tailing "dams" with some failures that cause loss of life (Buffalo Creek disaster); inefficient operation
	Stable, open-pit mine slopes	Slope failures, losses in time and production, damaged equipment, increased hazards to workmen, repair costs to correct slipouts
Water impoundment structures	Safe, stable dams, reservoirs, etc.; more efficient use of natural resources in their construction; fewer repairs	Unsafe, unstable structures whose failures often cause heavy losses in life and property (about 90% of all dam failures have been blamed on uncontrolled saturation and seepage)

TABLE 1.1 *(Continued)*

Area of activity	Benefits of good drainage	Losses caused by poor drainage
Pavements	Long-lasting, smooth-riding, "zero maintenance" pavements for highways, airfields, parking lots, etc., that require the least use of energy, fuel, and vital construction materials	Premature deterioration and destruction of important pavements nearly everywhere in the civilized world; costly repairs and replacements with great inconvenience to users and large waste of energy and natural resources; annual cost is probably at least $40 billion, worldwide.
Slopes of highways, railways, etc.	Stable, safe slopes with minimum failures and disrupted service	Slides, slipouts, and total collapse often putting important transportation and other facilities out of service, with loss of life, property damage, and large repair costs.
Playfields, athletic fields, stadiums, etc.	Long-lasting, trouble-free performance	Premature failure from deteriorating environmental actions, such as frost heave; and traffic actions.

Methods for Controlling Water in Structures

Powerful though the forces of water may be, civil engineers have long known that engineering works involving water can usually be made safe by (1) keeping the water out of places where it can cause harm or (2) controlling by drainage methods that which does enter. In most cases combinations of these two fundamental methods are used.

In civil engineering works such as reservoirs, basements, and underground structures damage from structural flooding by groundwater and seepage is often prevented by the use of watertight membranes. Cutoff trenches, impervious membranes or blankets, and grout curtains are used to reduce seepage in foundations of dams and levees. The membrane principle is also employed in the hulls of ships and submarines to prevent structural flooding. A steel hull is about as impervious a membrane as can be found, but even the boat designer provides pumps to remove extraneous water or leakage. The submarine "Thresher" is an example of an underwater ship that failed because of structural flooding.

Civil engineers, in common with boat designers, have learned that under most conditions where water is encountered in their projects they cannot depend wholly on keeping it out. Basements below the water table that depend only on watertight membranes for dryness can become wet when the membranes develop lapses. Grout curtains of even the best quality are only partially effective in controlling seepage under dams (Sec. 6.2). In nearly every kind of civil engineering construction involving seepage and gravitational water *drainage methods* are depended on, at least in part, to control water. When the amount of water is small, the demands on drainage systems are not great, and successful projects are the rule rather than the exception; but when large amounts of water must be removed, poorly designed systems have often become overloaded, and considerable deterioration, loss, and even failure has occurred. When sound drainage principles are followed, these problems can be almost entirely eliminated. The need to control seepage is emphasized in the next paragraphs.

The Consequences of Uncontrolled Seepage

Most children have read of the little Dutch boy who held his finger in a leak in a dike until help arrived. He knew that a small trickle could develop into a torrent that would flood his homeland. Failures by piping or erosion are the most easily understood. They represent one of the two basic kinds of failure caused by uncontrolled seepage. The second kind, which may be less obvious to the layman, is caused by seepage forces and pressures in soil and rock. In brief, most failures caused by groundwater and seepage can be classified in one of two categories.

1. Those that take place when soil particles migrate to an escape exit and cause piping or erosional failures.
2. Those that are caused by uncontrolled seepage patterns that lead to saturation, internal flooding, excessive uplift, or excessive seepage forces. The mere presence of free water in pavements and other surface facilities can lead to severe deteriorating actions plus serious damage under traffic impacts.

Some common types of failure due to lack of control of groundwater and seepage are listed in Table 1.2 under category 1 or 2 as just defined. Category 1 includes failures of dams, levees, reservoirs, and the like, caused by the migration of soil particles induced by a variety of defects.

Category 2 in Table 1.2 includes failures of dams, drydocks, and retaining walls caused by excessive saturation, seepage forces, and uplift pressures. In this category are listed the deterioration and failure of pavements from internal flooding, the uplifting of canal linings after drawdown, failures of fills and foundations caused by seepage forces, and uplift pressures in trapped water, landslides, and similar cases. Water that soaked into the base of Mt. Toc in

TABLE 1.2 Examples of the Consequences of Uncontrolled Seepage.

Category 1 Failures caused by migration of particles to free exits or into coarse openings	Category 2 Failures caused by uncontrolled saturation and seepage forces

1. Piping failures of dams, levees, and reservoirs, caused by:
  a. Lack of filter protection
  b. Poor compaction along conduits, in foundation trenches, etc.
  c. Gopher holes, rotted roots, rotted wood, etc.
  d. Filters or drains with pores so large soil can wash through
  e. Open seams or joints in rocks in dam foundations or abutments
  f. Open-work gravel and other coarse strata in foundations or abutments
  g. Cracks in rigid drains, reservoir linings, dam cores, etc. caused by earth movements or other causes
  h. Miscellaneous man-made or natural imperfections
2. Clogging of coarse drains, including French drains
	1. Most landslides, including those in highway or other cut slopes, reservoir slopes, etc., caused by saturation
	2. Deterioration and failure of roadbeds caused by insufficient structural drainage
	3. Highway and other fill foundation failures caused by trapped groundwater
	4. Earth embankment and foundation failures caused by excess pore pressures
	5. Retaining wall failures caused by unrelieved hydrostatic pressures
	6. Canal linings, basement and spillway slabs uplifted by unrelieved pressures
	7. Drydock failures caused by unrelieved uplift pressures
	8. Dam and slope failures caused by excessive seepage forces or uplift pressures
	9. Most liquefaction failures of dams and slopes caused by earthquake shocks

Italy led to the catastrophic landslide into Vaiont Reservoir late in 1963 (Fig. 8.6). The failure of Malpaset Dam in France in 1961 was caused by excessive water pressures in weak rock seams.

Altogether, the losses from uncontrolled saturation and seepage amount to many billions of dollars a year. Although the catastrophic types of failure— earth slopes, mountainsides, dams, and reservoirs and the like—cannot all be prevented, good drainage is one of the best ways, if not the best, to prevent many of the failures that take heavy toll of life and cause great property damage.

Recognition of the conditions that can lead to deterioration or failure of engineering works subjected to groundwater and seepage is an important step in their prevention. Some typical cases are described.

1.2 EXAMPLES OF DRAINAGE FAILURES CAUSED BY PIPING

Whenever soil particles can move freely to an escape exit there is the possibility that the condition may become progressively worse, with complete failure of an engineering structure ultimately occurring. Dangerous passageways for piping include large pores in natural gravel, conglomerate, open joints in bedrock, cracks caused by earthquakes or crustal movements, open joints in pipes, open voids in coarse boulder drains including French drains, gopher holes, cavities formed in levee foundations by rotting roots or buried wood, and the like. A notable disaster caused by piping was the failure of the St. Francis Dam in California, which failed quite suddenly in 1928 when a highly erodible sedimentary rock formation washed out, undermining portions of the concrete dam (Fig. 1.1).

Another example of a piping failure is a 100-ft-high earth dam on a loess (windblown silt) foundation, designed with a deep internal drain and cutoff trench for the control of underseepage. A 10 ft wide drainage blanket at the downstream side of the foundation trench was constructed with 6-in. minus stream gravel. The blanket, placed with center dump trucks, became badly segregated, thus producing a sloping zone of 6-in. boulders in direct contact

FIG. 1.1 The 205-foot-high St. Francis Dam after its failure in 1928. (From Report of the Commission appointed by Governor C. C. Young to investigate the causes leading to the failure of the St. Francis Dam near Saugus, California, California State Printing Office, Sacramento, 1928).

with compacted loess fill. During the first filling of the reservoir, several hundred cubic yards of loess washed out through the foundation drain. The dam was saved from destruction by quickly emptying the reservoir and making extensive repairs.

Sometimes failures are cause by conditions not anticipated by designers. On December 14, 1963, the Baldwin Hills reservoir in Los Angeles failed by piping and released about a thousand acre-feet of stored water that took five lives and caused $15,000,000 in property damage, showing that uncontrolled seepage from even a small-capacity reservoir can be extremely destructive. The State Engineering Board of Inquiry, appointed to investigate the failure, reported that, "Foundation displacement resulted in rupture of the reservoir lining and consequent entry of water under pressure into a pervious and erodible fault. Erosion in the fault and adjacent foundation proceeded rapidly, causing uncontrolled leakage through the east abutment of the main dam." Continuing, the report states that, "Evidence indicates that the earth movement which triggered the reservoir failure was caused primarily by land subsidence which had been experienced in the vicinity for many years."

1.3 EXAMPLES OF FAILURES CAUSED BY UNCONTROLLED SATURATION AND SEEPAGE FORCES

When water accumulates in earth and rock foundations or in constructed embankments or roadbeds faster than it is drained away, deterioration and failure may occur. Drainage systems, both natural and man-made, on which the maintenance of a proper balance between inflow and outflow depends, must have sufficient capacity to keep up with demands. Some examples of failures caused by inadequate drainage follow.

Most of the major slope failures occur in saturated earth or rock, many taking place during or shortly after an earthquake (Sec. 8.1) or other sudden change in conditions affecting stability. When water soaks into soil and rock formations, it frequently lowers their strength by reducing *cohesion*. The presence of free water in earth slopes also increases their vulnerability to *liquefaction* under shocks (Sec. 8.2).

Some important slides occur in masses that are too large to be stabilized effectively by man, but stabilization measures can greatly reduce the dangers of serious slides in many slopes. Lowering of saturation levels in slopes by drainage is often one of the best means of preventing landslides during earthquakes.

Uncontrolled saturation is also causing pavement systems throughout the developed parts of the world to break up in a third or less of the time they could perform satisfactorily if well drained. Even though the detrimental effects of water trapped in slow-draining pavements and bases have been known for centuries and the need for good internal drainage has also been recognized, hardly any modern pavement designers provide good internal drainage in the pavements they design. After the modern experimental and analytical methods for designing pavements became established (Cedergren, 1982), a high level of confidence in the "newfound" methods led most designers to believe there is no need for good internal drainage. But, during the times free water in pavements is pounded on by heavy vehicles and is acted on by harmful environmental conditions, rates of deterioration can be even hundreds to thousands of times greater than when little or no free water is present (Highway Research Board, 1952, 1955, 1962). As a consequence, thousands of miles of "rationally designed" pavements are in need of major repairs or restorations after as little as 5 to 8 years (see Fig. 1.2).

Good internal drainage systems, designed as outlined in Chapter 9 and described in greater detail in a recent book (Cedergren, 1987), can ensure pavements that will last literally forever with only nominal maintenance and an occasional thin overlay. A key element of these good drainage systems is a layer of highly permeable drainage material (ranging in size from about 1/4 in. up to a maximum of about 3/4 to 1 1/2 in.) protected with filters so they cannot become clogged by intrusion of adjacent fine materials (See Fig. 9.7).

Drainage systems with high water-removing capabilities are needed for essentially all kinds of Civil Engineering works needing protection from the harmful effects of water. When such facilities are *not* provided with good drainage, severe damage and even failure can occur, as illustrated next.

In 1956 an earth dam was constructed with a horizontal drainage blanket

FIG. 1.2   Eight-year-old state highway on a 10-ft-high fill in an eastern state; the asphalt concrete is badly damaged by traffic and surface water infiltration aggravated by bad drainage (Sept. 1970).

and an inclined chimney drain for the control of seepage through the dam and its moderately pervious foundation and abutments. Several months after the reservoir was filled, heavy seepage emerged on the downstream slope, saturating it to one-third of the height of the dam. Reconstruction measures saved this dam from possible failure. Tests on the drainage aggregates revealed that the blanket drain and chimney drain had been constructed with graded filter aggregates containing up to 8% of minus 200 material. The discharge capacity of the drains was too low to remove the seepage without excessive buildup of head. Although this dam appeared to be modern and sound in design, failure to specify drainage aggregate with the required degree of permeability nullified the good intentions of the designers.

Even smaller amounts of fines in a well-graded mixture of sand and gravel can produce permeabilities of low levels. Butterfield (1964) tells about the material the U.S. Corps of Engineers used in the least permeable section of the Howard A. Hanson Dam on the Green River in the State of Washington. The sand and gravel mixture used in the *impervious core* of this dam had a fines content (passing a No. 200 sieve) of about 3%, which ensured a coefficient of permeability of not more that 1×10^{-3} ft/min or 1.4 ft/day. For this major dam a class of material with even fewer fines than were allowed in the drain of the dam described in the previous paragraph has successfully restricted seepage. How can we expect a material that serves as an effective barrier to seepage also to serve as a good conveyor?

Here is another example of an engineered structure suffering from uncontrolled saturation. In 1959, a concrete retaining wall 60 ft high toppled over after a period of heavy rainfall. An investigation of the failure disclosed that a "permeable" silty gravel backfill layer, which was supposed to prevent large hydrostatic pressure from building up behind the wall, actually contained as much as 10% of minus 200 mesh silt and clay. Although the drainage material contained gravel up to 3 in. in diameter, and to the workmen must have appeared to be excellent drainage material, it was in fact only slightly more permeable than the adjacent soil. Supposedly, the wall had been "designed" to withstand severe seepage conditions, yet lack of sufficient permeability in the dirty gravel drainage material contributed to its failure.

1.4 HOW CAN SEEPAGE FAILURES BE PREVENTED?

Historical Notes

In the foregoing paragraphs examples have been given of the consequences of uncontrolled seepage. Groundwater and seepage may be encountered almost anywhere in the world and unless their influences are taken into account in designing and building engineering works untimely deterioration or failure will take place. Man has tried to cope with these problems from the early days of history but has often been baffled by the mysterious ways in which water gets into places where it can cause damage. Much of the responsibility for these problems rests with the physical property *permeability* which governs the rate of flow of water through the pores and interstices of soil and rock (Chapter 2). The ability of various kinds of earth materials to conduct water varies over a range of more than 10 billion times. Furthermore, natural earth formations almost always contain hidden inconsistencies that can exert a major influence on saturation and seepage.

Before the twentieth century, designing and building in soil and rock was generally considered to be more of an art than a science. It was widely believed that theory, science, and mathematics had no place in the design of engineering works constructed of earth materials. The proportions of overflow weirs, the shapes of earth dams, the selection of "drain rock" for roads, dams, and other purposes; and similar matters pertaining to earthworks and drainage were developed by empirical "rules of thumb," by the "intuition" or "judgment" of designers, or simply by copying something that had succeeded in the past. Unfortunately these procedures largely disregarded *reasons* for the success or failure of a project and often led to uneconomical designs which offered no assurance of even marginal factors of safety against failure.

Darcy's fundamental experiments with seepage phenomena (Sec. 2.3) furnished a simple explanation for the mechanics of seepage through homogeneous soil and porous aggregate drainage systems. The historic work of Terzaghi (1925) opened the way to the scientific and experimental approach to earthwork design that is being applied with increasing vigor by engineers throughout the world. The lucid explanations of practical seepage theory by Dr. Arthur Casagrande (1937, 1961) represent major steps forward in the design of earth dams and other water-impounding works. Many others in many lands have made important contributions, which collectively remove much of the guesswork from the design and construction of earth and rock structures capable of resisting the damaging forces of nature, including those of seepage and groundwater.

The great benefits of the rational approach to the design of earthworks and drainage systems must be evident from the comparatively few failures that take place when these methods are properly employed.

When rational and experimental methods are used properly in the design of projects, sound designs can be developed for extremely difficult site conditions.

Fallacy of Using "Standard" Designs

The fallacy of resorting to "standard" types of treatment, without full regard for individual soil and groundwater conditions, may be illustrated by reference to popular trends in the selection of drainage aggregates for engineering works.

Early in the history of road building, after about 1750, road builders employed coarse rock in "French drains" and macadam bases, following leaders of the times like Pierre Marie Tresaguet and John McAdam who instinctively must have realized that coarse stone and gravel are relatively good conductors of seepage. Those who understood soil behavior placed a few inches of dry stone screenings on subgrades to prevent soil from working up into base courses (Oglesby, 1975). Unfortunately a tendency developed to place coarse materials directly on wet sand, silt, or clay subgrades without an intervening protective layer. As a consequence of this disregard for physical conditions coarse stone and gravel bases and drains often became clogged with fine soil and the overlying roadways soon deteriorated from lack of drainage.

After about 1930 the pendulum swung far to the opposite extreme with the widespread use of *well-graded* (having a wide range of sizes) filter materials that provide a high degree of protection against soil migration but are often so impermeable that they are virtually useless as conveyers of water. In keeping with this trend there was a tendency for some to consider concrete sands and comparable filter aggregates as "universal" drainage materials, satisfactory for nearly every purpose. Because of this unfounded belief, miles of highways and roads and other works are deteriorating prematurely from lack of drainage.

In contrast with the optimistic view of some engineers of the suitability of concrete sands, agricultural experts look on beach sands, which usually are considerably more permeable than concrete sands, as being too impermeable for some agricultural uses. This view is expressed in the *Sunset Garden Book,* second edition, 1964, p. 97: ". . . beach sand is unsatisfactory (as a rooting medium) because it contains salt and *is too fine to permit free drainage* and thorough aeration."

The analysis methods presented in this book point up the fact that although well-graded aggregates can provide excellent filter protection they have limited value as conveyors of seepage. An economical answer to many drainage situations is a *graded filter,* composed of a fine filter layer in contact with the soil and a layer of clean-washed pea gravel or other uniform-sized gravel or stone of high permeability protected when necessary by another fine filter layer. An alternate, made possible by recent improvements in synthetic filter fabrics (geotextiles), is the use of open-graded aggregate with high water-removing capability protected with suitable fabric filters (see Sec. 5.7).

Degree of Accuracy To Be Expected

Most experienced seepage and drainage engineers regard seepage theory as a means of predicting the general *order of magnitude* of problems and for developing appropriate *types of solutions.* When the physical factors that influence a seepage or drainage situation are known only approximately (which is often the case), fairly simple and approximate seepage calculations can point the way to the right kind of solution and help the designer avoid gross mistakes (Sec. 3.1). Although it must be recognized that refined theoretical seepage analyses can be of considerable value, the marvels of electronic computers and other sophisticated methods may tend to lull a person into a false sense of overall accuracy that does not exist.

Although a seepage analysis method may have a high degree of theoretical accuracy and repeatability, a solution to a problem is no more accurate than the physical data used in the problem; for example, if the permeabilities of soil formations are known with a probable accuracy of plus or minus 50%, no theoretical solution to a seepage or groundwater problem involving these materials can be more accurate than plus or minus 50%, *even though the analysis is exactly repeatable.*

A sound approach to every seepage and drainage situation requires the greatest possible care in assigning permeabilities and other physical data as realistically as possible and the use of reasonable care in carrying out an appropriate theoretical or mathematical solution. Water will find its way into every significant pipe, seam, openwork layer, and interconnected conducting feature in a structure or foundation, *whether the designer knows about it or not.* Water will be blocked or partly blocked by impermeable barriers. If the designer's knowledge of the section being analyzed is based on limited numbers of test holes and small individual permeability tests, he will have a smaller chance of predicting actual seepage conditions for a completed work than if he had a broader, more complete picture of the water-transmitting properties of the formations. Experience and common sense must play an important part in the selection of appropriate methods of analysis and in the assessment of the probable overall degree of accuracy of the final solution.

Need for Observation of Completed Works

Because of the wide variation in seepage conditions that can develop in earthworks and their foundations and the difficulties of making accurate predictions, observation of the *actual* behavior of completed projects should always be required (Sec. 6.1). This need for observation of completed works is implied by the following statement of Dr. Terzaghi in an address to the First International Conference on Soil Mechanics and Foundation Engineering in 1936: "In soil mechanics the accuracy of computed results never exceeds that of a crude estimate, and the principle function of theory consists of teaching us *what and how to observe in the field.*" This admonition by the recognized originator of modern soil mechanics does not weaken the argument in favor of rational methods for the analysis of seepage and seepage control; instead, it strengthens it. Widespread experience is proving that when these methods are used with reservation and judgment and followed up by the observation of completed works they point the way to economical, safe designs.

Among the more important observations to be made of structures and slopes influenced by seepage and groundwater are general seepage patterns, hydrostatic pressures, and seepage quantities. Any unexpected or unexplained change in conditions is a cause for concern. Seepage quantities can frequently be measured by installing small weirs at suitable locations; saturation levels and hydrostatic pressures may be measured with small-diameter observation wells or *piezometers*. The U.S. Bureau of Reclamation (1974), Sherard et al. (1963), and others describe a variety of wells and piezometers that are used for this purpose. A number of applications of piezometers and other devices for monitoring seepage behavior in dams, earth slopes, and a variety of civil engineering facilities are discussed in Chapters 6, 7, 8, 10, and 11.

The need for careful observation and instrumentation of earth dams and other water-impounding works is becoming widely recognized and put into practice. Nearly every important dam in the United States and in many other countries is being equipped with extensive systems of piezometers and other instruments for observing behavior. Wadhwa and Srinivasai (1960) reported, for example, that after obtaining valuable information from piezometer systems in Hirikud and Gangapur Dams in India the Central Water and Power Commission in New Delhi planned to have piezometers installed in all earth and masonry dams more than 100 ft high. Observation wells and piezometers are also widely used for measuring seepage conditions in earth slopes, cofferdams, drydocks, retaining walls, and many other structures.

1.5   DEGREE OF CONSERVATISM NEEDED

It is obvious that the consequences of inadequate control over groundwater and seepage can vary from minor failures that involve only small economic losses to major failures that will take a heavy toll of life and property. A logical question of basic philosophy is "How conservative should designers of structures involving water try to be?" Some designers believe that if they do not build a failure now and then it will be proof that they are too conservative; for example, it is customary to design highway cut slopes with the expectation that some slides will occur. Highway slides do not usually endanger lives; hence the degree of conservatism is primarily a matter of economics and convenience to the public. But, if dams fail from inadequate drainage (or from any other cause) enormous amounts of property can be destroyed and large numbers of people killed.

In arriving at the degree of conservatism required in the design of engineering works, two dominating factors stand out.

1. How serious are the consequences of a failure?
2. What is the level of uncertainty involved?

Each case should be considered individually, but certain guidelines are in order. If major loss of life and property are likely to result, engineering works must be made unquestionably safe for all conditions that can be anticipated during the life of the works. If loss of life is not a problem and only minor economic losses are to be expected, there is less need for conservatism. The less sure a designer is that he has detected every important detail that can have harmful effects on his project, the more conservative he must try to be.

Sometimes a high level of conservatism can be obtained for little or no added cost. When this is possible, it seems foolish to accept less; for example, the capacity of drainage layers to remove seepage and groundwater can be increased hundreds or thousands of times (often even at a reduction in cost) by using drains constructed with highly permeable, open-graded aggregates protected with suitable filters (see Chap. 5). When a designer has the choice between a drainage system having little or no known margin of safety and one with hundreds or thousands of times greater discharge capacity, it seems unwise to accept the poorer of the choices, particularly when one considers that the real needs of drainage systems can never be known precisely before they are required to operate.

A second example of getting full value in drainage is the design of embankment dams capable of withstanding severe earthquakes as well as the normal static forces. Earth dams with large well-drained downstream sections are inherently more stable than homogeneous dams (Sec. 6.4). It therefore seems unwise ever to accept *undrained* dams in any situation in which a failure could have serious consequences.

1.6   SCOPE OF THIS WORK

The primary objectives of this book are twofold: (1) to show the *benefits* of the analytical and experimental approach to the design of earthworks involving seepage and seepage control measures and (2) to present and illustrate practical methods of analysis and design. Emphasis is placed on designing structures for individual conditions rather than relying on "standard" types of design.

Although the basic fundamentals of seepage are contained in the pages of several good books on soil mechanics, they are sometimes buried under complex mathematics. Two basic seepage principles, *Darcy's law* and the *flow net,* have considerable practical value in the solution of everyday seepage and drainage problems. These methods are given special emphasis in this book and are illustrated by many examples of their use in the design of earth dams, drainage systems, and other practical applications.

Chapters are presented on permeability, seepage principles, and filter design. A chapter on flow net construction contains step-by-step instructions that describe proved methods for the construction of flow nets, from the simplest to the more complex. The balance of the book is devoted to the analysis of seepage and the design of practical drainage systems for the protection of earth dams and levees, foundations, slopes, roads, airfields, and other surface facilities, and a variety of structures subjected to groundwater and seepage.

Numerous charts and flow nets are to be found throughout. Although these solutions have a certain value in solving specific seepage and drainage situations, their primary purpose is to illustrate *methods* of analysis and design. It is hoped the reader will be encouraged to use the fairly simple methods as guides to the development of his own solutions to individual types of problems.

1.7 CONCLUSIONS

Drainage and other forms of seepage and groundwater control have become increasingly important in the performance of many kinds of engineering projects. Although the principles of seepage and drainage are comparatively straightforward, appropriate solutions to individual situations are not always self-evident. The rational and experimental methods of designing earthworks subjected to groundwater and seepage offer a sound basis for adapting individual projects to specific site conditions. Methods developed in this book for the analysis of seepage problems, when used with experience and judgment, can help to remove much of the guesswork from drainage design. These methods can aid in obtaining maximum life and safety for a great many kinds of engineering works at the least overall cost.

REFERENCES

Butterfield, Glen R. (1964), "Sand and Gravel Core at Howard A. Hanson Dam," U.S. Committee on Large Dams (USCOLD), *Newsletter,* May 1964.

Casagrande, A. (1937), "Seepage Through Dams," Harvard University Publication 209, reprinted from *Journal of the New England Water Works Association,* June 1937.

Casagrande, A. (1961), "Control of Seepage Through Foundations and Abutments of Dams," *Geotechnique,* Vol. 11, September 1961, pp. 161–181.

Cedergren, H. R. (1982), "Overcoming Psychological Hangups is Greatest Drainage Challenge," Proceedings, 2nd Int. Conf. on Geotextiles, Las Vegas, Nevada, Oct. 1-6, 1982, Vol. I, pp. 1-6.

Cedergren, Harry R. (1987), "Drainage of Highway and Airfield Pavements," [Original Pub. by John Wiley & Sons, Inc., N. Y. (1974)]; Updated printing by Robert E. Krieger Pub. Co., Inc., Melbourne, Florida, 289 pages.

Highway Research Board, HRB (1952), "Special Report 4, Final Report on Road Test One-MD," Washington, D.C., 1952.

Highway Research Board, HRB (1955), "Special Report 22," 1955.

Highway Research Board, HRB (1962), "Summary Report (Report 7) the AASHO Road Test," National Advisory Committee, HRB, Special Report 616, Publication Number 1061, 1962.

Oglesby, Clarkson H. (1975), *Highway Engineering,* 3rd ed., Wiley, New York, pp. 561–562.

Sherard, James L., R. J. Woodward, S. F. Gizienski, and W. A. Clevenger (1963), *Earth and Earth-Rock Dams,* Wiley, New York, pp. 384–399.

Terzaghi, K. (1925). *Erdbaumechanik auf bodenphysikalischer Grundlage,* Vienna, Deuticke, 399 pp.

U.S. Bureau of Reclamation (1974), *Earth Manual,* Denver, 2nd ed., pp. 650–699.

Wadhwa, H. L., and B. Srinivasai (1960), "Field Behavior of Some Earth Dams in India," *Proceedings,* First Asian Regional Conference, International Society of Soil Mechanics and Foundation Engineering, New Delhi, February 1960, Paper I(b) (iii).

CHAPTER TWO

PERMEABILITY

2.1 GENERAL CONSIDERATIONS

Permeability, or *coefficient of permeability, k* as used in this book, is one of the most intriguing properties of materials that engineers, hydrogeologists, and groundwater professionals must deal with. Because it can vary over a range of many billions of times, its real significance can be hard to grasp, and many misconceptions exist regarding what is a "permeable" material, what is a good drainage material, what is an "impermeable" material, and so on. A prime objective of this chapter is to enable readers to have a greater practical understanding of this very important property of earth and rock masses, and materials used for engineering drainage.

Laboratory and field tests should always be made to establish important properties of all materials used in drainage systems. However, a simple examination of a proposed mineral aggregate with the unaided eye can give an indication of general levels of permeability. For example, if a k of 5,000 ft/day is needed, one should be able to see openings between particles at least 1/20 inch in diameter; if 20,000 ft/day is required, visible openings should be at least 1/10 inch in diameter; and for 100,000 ft/day, at least 1/4 inch. The making of even such a crude "eye test" can help designers avoid serious mistakes in selecting types of aggregates to be used in drainage systems.

A *permeable* material is one that is capable of being penetrated or permeated by another substance, usually a gas or liquid. Thus dry cement is permeable to air, and an air permeability test is a useful means of obtaining an indirect measure of its fineness of grind, since the speed of flow of air through it can be related to the size of the pore spaces between the particles. Likewise, soils and aggregates and jointed, cracked, or vesicular rocks are often permeable to air and water. Many materials allow the movement of fluids by a diffusion process, but that is not within the meaning of *permeability* as used in soil mechanics. In the study of soil mechanics a material is considered permeable if it contains interconnected pores, cracks, or other passageways through which water or gas can flow. A rock may be virtually impervious, yet contain cracks or joints that make a formation highly permeable to the flow of water. The permeability of most rock abutments and dam foundations is determined almost entirely by the joint and crack patterns, and many clays are extremely resistant to the flow of water, yet shrinkage cracks or interbeds of silt or sand may increase their permeabilities thousands of times.

As explained in Section 2.2, the permeability coefficient used in this book is the engineers' coefficient, or Darcy's coefficient, which is defined as a discharge velocity.

The availability of vast underground supplies of groundwater and petroleum depends on, among other factors, the *permeabilities* of soil and rock formations. Economic considerations, in addition to engineering, have been responsible for a great deal of study and research in the flow of fluids in porous media. Extensive bibliographies, such as V. C. Fishel's (listed by Wenzel, 1942) and those prepared by Johnson (1963), Bittinger and Maasland (1963), and Richter (1963), reflect the scope and magnitude of the work that has been done and are excellent references for those interested in thoroughly pursuing the study of seepage and permeability. Todd (1959) provides extensive lists of references on groundwater movement, permeability, flow through porous media, and related subjects. In this chapter permeability is discussed primarily with engineering applications in mind. The treatment necessarily is rather brief.

Many of the design and construction problems associated with hydraulic structures and other engineering works involving drainage are caused by imbalances of permeabilities of earth and rock masses. Frequently water enters spaces behind walls and under pavements, canal linings, and the like, more readily than it can escape, thus creating conditions detrimental to safety and performance. Water that becomes trapped in earth and rock masses contributes to landslides and is a serious threat to stability during earthquakes. Undetected joints or strata of high permeability in the foundations or abutments of dams create serious leakage and uplift problems (i.e., see Fig. 6.1).

No other engineering property of construction materials is so variable as permeability. Coarse gravels, open joints, and seams allow water to flow rapidly and in large quantity; but, on the other extreme, fat clays are so impervious that the rate of flow through them and the drainage of water out of them are infinitesimal. The coefficient of permeability of natural earth deposits and man-made embankments can vary from a high of more that 100,000 ft/day (35 cm/sec) for clean gravels to a low of around 0.000,003 ft/day (1×10^{-9} cm/sec) for fine clays, a range of several billion times. In contrast, variations in other properties of common engineering materials are comparatively modest (Fig. 2.1).

Permeability can vary so widely that its physical significance is often difficult to comprehend, and the *velocities* of seeping water are subject to even greater variations than permeability. The general range of possible seepage velocities in a wide variety of soils and aggregates is indicated in Fig 2.2, which also gives some calculated velocities in pipes and open channels. All are for a hydraulic gradient of 0.01 and are related in terms of the "effective diameter" of flow channels. Flow in large open channels often attains velocities of 40 ft/sec and in small pipes, several feet per second. In contrast, flow through a clean aggregate with an "effective" pore diameter of 0.2 in. (one-fifth the D_{10} size of 1.0 in.) can have a velocity of about 0.05 ft/sec and in fine clay soils, under 1×10^{-8} ft/sec, which is only a fraction of an inch per year.

Further practical comparisons between flow in open channels and pipes and seepage through soils and aggregates are made in Table 2.1. Recognition of this vast range of possible behavior helps to develop an understanding of the physical meaning of permeability and an appreciation of possible rates of seepage through soils, rocks, and porous media, compared with flow in pipes and channels.

Figure 2.3 illustrates vividly the dramatic differences in flow that can be caused by variations in permeability. Both photos show field permeability tests being made on base courses for highway pavements. Figure 2.3*a* is a view of

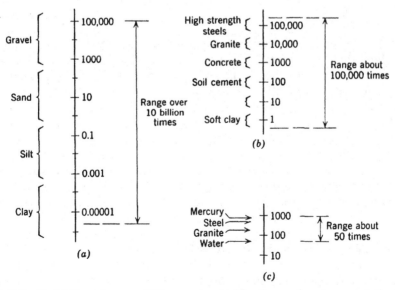

FIG. 2.1 Variability of permeability compared with other engineering properties (log scale). (*a*) Permeability, ft/day. (*b*) Strength, lb/in.² (*c*) Unit weight, lb/ft.³

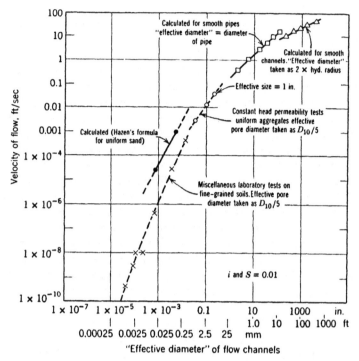

FIG. 2.2   Comparison of flow in porous media with flow in pipes and smooth channels (log-log plot).

a test of one of the conventional, well-graded aggregate bases used widely throughout the world. After 100 minutes the water level in this hole has fallen about ⅛ in. In sharp contrast is the test underway in Fig. 2.3b. Here a special subbase of open-graded AC (protected with a filter layer on the subgrade) has a coefficient of permeability of nearly 30,000 ft/day (10 cm/sec). The difference between the two materials is a permeability range of about 1,000,000, which is a narrow sector of the overall range that is possible. Yet even this difference (which makes a comparison between the "state-of-the-art" and the "state-of-the-knowledge" of pavement design) represents a conspicuous difference in drainage capability. The test in Fig. 2.3a could have been kept supplied with an eyedropper; the one in Fig. 2.3b needed a 2-in. diameter hose!

2.2   COEFFICIENT OF PERMEABILITY

The coefficient of permeability as used in this book is the *engineers' coefficient,* or *Darcy's coefficient,* and is defined as the discharge velocity through

TABLE 2.1 Comparison Between Flow in Open Channels and Pipes and Seepage Through Soils and Aggregates

Flow medium	Typical coefficients of permeability, ft/day	Effective channel diameter	Hyd. gradient	Discharge velocity, ft/sec	Q, cfs	Q, gpm	Square feet needed for discharge of 2 in. pipe[a]
Smooth channel		80 ft = 2R	0.01	35	450,000		
Smooth pipe		8 ft = d	0.01	15	750		
		1 ft = d	0.01	4.5	3.5		
		2 in. = d	0.01	0.5	0.018	8	2 in. pipe
$1\frac{1}{2}$–1 in. gravel	140,000	0.2 in.	0.01	0.015	0.015[b]	6.8[b]	1.2
1–$\frac{1}{2}$ in. gravel	50,000	0.1 in.	0.01	0.005	0.005[b]	2.3[b]	3.5
$\frac{3}{8}$"–#4 gravel	8,000	0.03 in.	0.01	0.0008	0.0008[b]	0.36[b]	22
Coarse sand	800	0.01 in.	0.01	0.0001	0.0001[b]	0.045[b]	180
Fine sand, or graded filter agg.	1.0	0.002 in.	0.01	1×10^{-6}	1×10^{-6b}	4.5×10^{-4b}	18,000
Silt	0.001	0.00024 in.	0.01	1×10^{-9}	1×10^{-9b}	4.5×10^{-7b}	18,000,000
Fat clay	0.00001	0.00004 in.	0.01	1×10^{-11}	1×10^{-11b}	4.5×10^{-9b}	2,000,000,000

[a] $S = 0.01$.
[b] Q per square foot area.

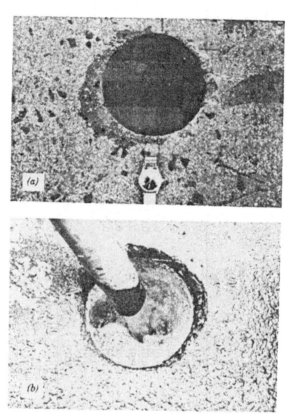

FIG. 2.3 Field percolation tests in pavements illustrate wide variations in permeability. (*a*) Gravel base under a deteriorating PCC pavement in New England, water level has gone down 1/8 in. in 100 min. (*b*) Open-graded AC drainage layer under County road in California is taking 75 gpm flow. (From "Drainage of Highway and Airfield Pavements," Wiley, New York, 1974, p. 83; also updated printing by Robert E. Krieger Pub. Co., Inc., Malabar, Florida, 1987, p. 83.)

unit area under unit hydraulic gradient. It is a term in Darcy's law for laminar flow in porous media (Sec 3.2).

$$Q = kiAt \tag{2.1}$$

In Eq. 2.1 Q is the quantity of seepage in a cross section with an area A normal to the direction of flow, under a hydraulic gradient i, during a length of time t. The coefficient of permeability k is equal to the discharge velocity under a hydraulic gradient of 100%. By rearranging the terms, Eq. 2.1 provides the basis for many experimental determinations of permeability that measure seepage quantity:

$$k = \frac{Q}{iAt} \qquad (2.2)$$

Darcy's *discharge velocity* multiplied by the entire cross-sectional area, including voids e and solids 1, gives the seepage quantity Q under a given hydraulic gradient $i = \Delta h / \Delta l$ or h/L. It is an imaginary velocity that does not exist anywhere. The average *seepage velocity* v_s of a mass of water progressing through the pore spaces of a soil is equal to the discharge velocity $(v_d = ki)$ multiplied by $(1 + e)/e$ or the discharge velocity divided by the effective porosity n_e; hence permeability is related to seepage velocity by the expression

$$k = \frac{v_s n_e}{i} \qquad (2.3)$$

For any seepage condition in the laboratory or in the field in which the *seepage quantity,* the area perpendicular to the direction of flow, and the hydraulic gradient are known the coefficient of permeability can be calculated. Likewise, for any situation where the *seepage velocity* is known at a point at which the hydraulic gradient and soil porosity also are known, permeability can be calculated.

Experimentally determined coefficients of permeability can be combined with prescribed hydraulic gradients and discharge areas in solving practical problems involving seepage quantities and velocities. When a coefficient of permeability has been properly determined, it furnishes a very important factor in the analysis of seepage and in the design of drainage features for engineering works.

The coefficient of permeability as used in this book and in soil mechanics in general should be distinguished from the physicists' coefficient of permeability K, which is a more general term than the engineers' coefficient and has units of centimeters squared rather than a velocity; it varies with the porosity of the soil but is independent of the viscosity and density of the fluid. The transmissibility factor T represents the capability of an aquifer to discharge water and is the product of permeability k and aquifer thickness t.

The engineers' coefficient, which is used in practical problems of seepage through masses of earth and other porous media, applies only to the flow of water and is a simplification introduced purely from the standpoint of convenience. It has units of a velocity and is expressed in centimeters per second, feet per minute, feet per day, or feet per year, depending on the habits and personal preferences of individuals using the coefficient. In standard soil mechanics terminology k is expressed in centimeters per second.

Although coefficient of permeability is often considered to be a constant for a given soil or rock, it can vary widely for a given material, depending on a number of factors. Its absolute values depend, first of all, on the properties of water, of which viscosity is the most important. For individual materials and formations its value depends primarily on the dimensions of the finest pore spaces through which water must travel and on the size and continuity of cracks and joints in rocks and fissured clays. In short, the ease with which water can travel through soils and rocks depends largely on the following:

1. The viscosity of the flowing fluid (water).
2. The size and continuity of the pore spaces or joints through which the fluid flows, which depends in soils on
   (*a*) the size and shape of the soil particles,
   (*b*) the density,
   (*c*) the detailed arrangement of the individual soil grains, called the *structure.*
3. The presence of discontinuities.

These factors have an important influence on the coefficient of permeability of soil and rock formations and are discussed in the following section.

2.3   FACTORS INFLUENCING PERMEABILITY

Basic Considerations

A review of the physical factors that influence the flow of water through small pore spaces aids in the study of permeability and the factors influencing its magnitude.

When water flows through a pipe or open channel, the velocities near the edges are considerably smaller than those in the center of the flowing stream, but when water flows through homogeneous soils or other porous media under uniform gradients the average velocities are no greater *at the center of a formation* than at its edges. Variations in velocity do exist, but within the individual pores or cracks through which the water is flowing. This is a fundamental difference between ordinary flow in pipes and conduits and flow through porous media. Another important difference is that flow in pipes and conduits is almost always *turbulent,* whereas in soils and aggregates it is almost always *nonturbulent* or *laminar.*

Whenever a fluid is in motion, layers of the fluid slip and move in relation to other layers. The ease with which they slip depends on the *viscosity* of the fluid, which is the resistance or "drag" offered to motion. For water it has a small value, but machine oils are about 200 times more viscous than water at room temperature and glycerine and castor oil, about 1000 times. Although the viscosity of water, like that of most fluids, is reduced at high temperatures, the range is much narrower than for other fluids. Over the widest range in temperatures ordinarily encountered in seepage, viscosity varies about 100% (Fig. 2.4). Although this variation is not highly important, it causes variations

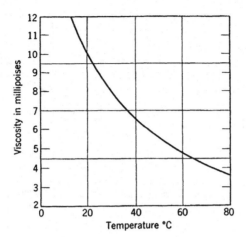

FIG. 2.4 Viscosity of water.

in permeability of a like amount; hence it is customary to standardize permeability values at 20°C or 70°F and make a correction if field temperatures are substantially different.

By application of laws of physics it can be shown that the resistance to *viscous* or *streamline* flow increases in direct proportion to viscosity, and the velocities attained by seeping fluids vary inversely with viscosity. When flow is viscous or laminar, layers of fluid are slipping past other layers and shearing resistance, s is related to viscosity according to *Newton's law of friction:*

$$s = \mu \frac{dv}{dr} \tag{2.4}$$

In this expression μ is the absolute viscosity in poises and dv is the change in velocity in a distance dr normal to the direction of flow.

Starting with Newton's law of friction, it can be shown that permeability should vary approximately with the squares of the sizes of pore spaces and particle sizes. The general validity of this conclusion has been verified by the testing of actual soils. It suggests that large variations in permeability should indeed be expected. It has been the basis for a number of indirect methods for estimating the permeability of clean filter sands (Eq. 2.9). The plot in Fig. 2.5 shows that the actual permeabilities of these soils increase at a rate somewhat more than the square of pore diameters (assuming pore diameter = $D_{10}/5$).

To obtain an interesting comparison between laminar and turbulent flow synthesized "permeabilities" of pipes and smooth channels have been added to the plot in Fig. 2.5. The discharge capacities of several pipes and channels were first calculated by using conventional hydraulic formulas. Artificial coefficients of "permeability" were then calculated from "k" = Q/SA, assuming

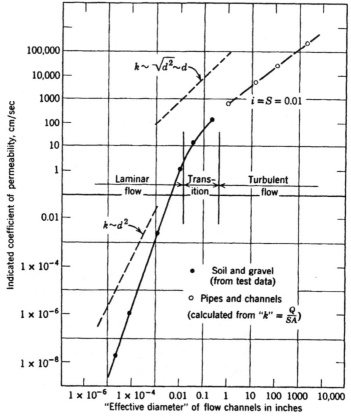

FIG. 2.5 Variations in "permeability" with size of flow channels.

momentarily that flow in pipes and channels follows the general Darcy relationship for laminar flow. It is seen that the synthetic "permeabilities" for turbulent flow conditions vary nearly with the first power of the sizes of flow channels, not with the square, as for laminar flow. This verifies the already known fact that turbulent flow consumes large amounts of energy in relation to laminar flow.

The actual pore channels through which water finds its way through soils are very tortuous and often semidiscontinuous and the hydraulics of flow through such channels is extremely involved. By making simplifying assumptions efforts have been made to calculate the permeabilities of simulated soils purely from theoretical considerations. Such efforts have on the whole been fruitless because of the many simplifying assumptions that have to be made and the great complexity of the differential equations that represent the highly oversimplified flow systems. The main value of such efforts has been in disclosing fundamental relationships that govern flow through minute pore spaces.

Darcy (1856) investigated this problem experimentally by using a simple apparatus to force water through small specimens of sand (Fig 2.6). Darcy's experiments demonstrated that the rate of flow q through the sand varies in direct proportion to the cross-sectional area A of the specimens and to the difference between the hydrostatic head at the two ends of specimens and is inversely proportional to the length L of the column of sand tested.

These relationships can be expressed as

$$q \sim \frac{A \, \Delta h}{L}$$

or

$$q = (\text{a constant}) \frac{A \, \Delta h}{L}$$

Darcy's experiments produced this simple relationship which has since become known as *Darcy's law*. Darcy's direct approach to the determination of permeability eliminated all of the difficulties of solving complex differential equations involving more or less uncertain assumptions. It provides a practical and powerful tool for the analysis of flow through porous media and for determining permeability. One of the common forms of Darcy's law (Sec. 3.2) is

$$Q = kiAt \tag{2.1}$$

The coefficient of permeability defined by Darcy is a statistical average factor that represents a definite cross section of soil. When determined by the careful experimental testing of a given mass of soil, it is representative of the

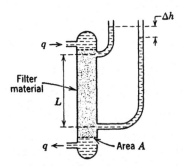

FIG. 2.6 Darcy's experiments. (See *Foundation Engineering,* edited by G. A. Leonards, McGraw Hill Book Co., New York, 1962, Fig. 2–17, p. 109.)

volume of soil tested. To be of use in practical problems it is usually necessary to estimate the permeability of extensive soil deposits from tests on limited volumes of soil. As a rule, the larger the mass represented by a test, the more reliable the results; however, tests on comparatively minute specimens can be extremely valuable when applied carefully and with the proper regard for limitations inherent to all such determinations.

According to Darcy's law, the velocity of flow and quantity of discharge through porous media are directly proportional to the hydraulic gradient. For this condition to be true flow must be laminar or nonturbulent.

The transition between laminar and turbulent flow is rather well defined by *Reynolds number* Re expressed as

$$\text{Re} = \frac{\gamma v d}{\mu}$$

where γ = density of the fluid, d = diameter of circular pipe, v = mean velocity, μ = coefficient of viscosity. Turbulence in round pipes exists when Re exceeds about 2000.

Taylor (1948) presents information indicating that flow through most soils is generally laminar; however, when the effective diameter of particles exceeds about 0.5 mm, flow becomes semiturbulent. Terzaghi investigated the range of validity of Darcy's law and concluded that the law is valid for a wide range of soils and hydraulic gradients. Muskat (1937) concludes that "in the great majority of flow systems of physical interest the flow will be strictly governed by Darcy's law. . . . " Although it is generally agreed that Darcy's law must be valid over a wide range of conditions, it has been difficult to establish criteria to judge its validity in any individual case. It seems wise to allow liberal margins of safety in all situations in which turbulence would create adverse conditions. Section 3.7, *Semiturbulent and Turbulent Flow* discusses a method of compensating for turbulence in highly permeable open-graded drainage layers for roadbeds, airfields, earth dams, and other structures.

In the preceding paragraphs fundamentals of flow and permeability have been discussed briefly. The influence of important physical factors on permeability is now described.

Influence of Grain Sizes

From theoretical considerations it has been shown that permeability can be expected to vary with the squares of the diameters of pore spaces and the squares of the diameters of soil particles. Typical permeabilities of soils and aggregates of several sizes are given in Fig. 2.7; variations in permeability with flow channel diameter are given in Fig. 2.5. The permeability of soils varies significantly with grain size and is extremely sensitive to the quantity, character, and distribution of the finest fractions.

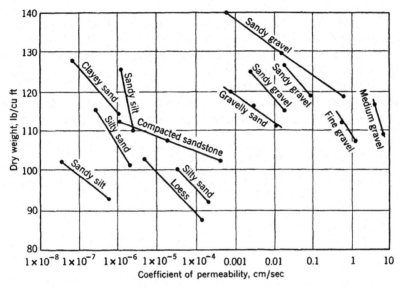

FIG. 2.7 Relation between coefficient of permeability and soil type and density (log scale).

The influence of fines on the permeability of manufactured washed filter aggregates grading from 1 in. to finer than the 100 mesh sieve is illustrated in Table 2.2. With no material finer than 100 mesh, the permeabilities ranged from 80 to 300 ft/day, but with 7% passing a No. 100 sieve, the permeability was 3 ft/day or less.

As a rule permeability should be determined by test, not from other properties such as grain size. The general soil type and grain size are useful, however, in indicating the approximate range of permeability to be expected. The gen-

TABLE 2.2 Influence of Percent of Minus 100 Fraction on Permeability of Washed Filter Aggregates (Typical)

Percentage Passing Number 100 Sieve	Permeability, ft/day
0	80 to 300
2	10 to 100
4	2 to 50
6	0.5 to 20
7	0.2 to 3

eral way in which the test permeability has been found to vary with the type of soil is illustrated in Fig. 2.7 which also shows that soil density influences permeability. This factor is discussed in more detail on the following pages.

Figure 2.8, a useful chart developed by A. Casagrande and R. E. Fadum, gives common ranges of permeabilities of soils varying from fine clays to gravels. It should not be considered a precise means of determining permeability—only a general guide for common soil types.

Influence of Particle Arrangement (Structure)

The arrangements of soil particles can influence permeability in two important ways:

1. By sorting or stratification.
2. By detailed orientation of particles and the balling up of fines or broad dispersion of the fines.

Natural soil deposits are always more or less stratified or non-uniform in structure. Water-deposited soils are usually constructed in a series of horizontal layers that vary in grain-size distribution and permeability. These deposits are generally more permeable in a horizontal than vertical direction (Sec. 3.3). Windblown sands and silts are often more permeable vertically than horizontally because of tubular voids believed to be left by rotted plant or grass roots. Many variations occur, depending on the way soil deposits are formed; consequently an understanding of the methods of formation of soils aids in evaluating their engineering properties. An outstanding example of stratification is openwork gravel contained in ordinary gravel or soil. The presence of openwork gravel can have tremendous influence on the watertightness of dam foundations and abutments.

Effect of Openwork Gravel

An indication of the presence of openwork gravel in a soil formation is given by the behavior of the water table. The extremely high permeabilities of these formations, which contain no fines, often permit rapid equalization of hydrostatic pressures, thus allowing the water table to rise and fall almost as quickly as an adjacent stream. When a water table hundreds of feet from a stream stands near the level of the stream and rises and falls with little lag time, there is a strong possibility that openwork gravel is present. Openwork seams cannot always be detected by foundation explorations, for they may be only an inch or two in thickness and not easily detected by normal sampling methods. Test pits or other open excavations may be needed to prove their presence. Carefully performed field pumping tests also may reveal their presence, although pervious strata can be clogged by drilling operations and remain undetected.

Coefficient of Permeability k in cm per sec (log scale)

	10^2	10^1	1.0	10^{-1}	10^{-2}	10^{-3}	10^{-4}	10^{-5}	10^{-6}	10^{-7}	10^{-8}	10^{-9}
Drainage	Good							Poor		Practically Impervious		
Soil types	Clean gravel	Clean sands, clean sand and gravel mixtures			Very fine sands, organic and inorganic silts, mixtures of sand silt and clay, glacial till, stratified clay deposits, etc.				"Impervious" soils, e.g., homogeneous clays below zone of weathering			
					"Impervious" soils modified by effects of vegetation and weathering							
Direct determination of k	Direct testing of soil in its original position—pumping tests. Reliable if properly conducted. Considerable experience required											
	Constant-head permeameter. Little experience required											
			Falling-head permeameter. Reliable. Little experience required				Falling-head permeameter. Unreliable. Much experience required		Falling-head permeameter. Fairly reliable. Considerable experience necessary			
Indirect determination of k	Computation from grain-size distribution. Applicable only to clean cohesionless sands and gravels								Computation based on results of consolidation tests. Reliable. Considerable experience required			

After A. Casagrande and R. E. Fadum

FIG. 2.8 Permeability and drainage characteristics of soils. (Table 6, p. 48, of *Soil Mechanics in Engineering Practice*, K. Terzaghi and R. B. Peck, Wiley, New York, 1948.)

The great influence of openwork gravel on leakage through natural soil formations is illustrated in Fig. 2.9. The sketch in Fig. 2.9*a* shows a soil cross section in a dam foundation as surmised from several drill holes. The mechanical analysis of samples secured at frequent intervals erroneously indicated that the deposit was composed of relatively uniform sandy gravels. Laboratory permeability tests on the disturbed samples produced coefficients of permeability of roughly 1×10^{-6} cm/sec (0.003 ft/day). By use of this coefficient the probable water loss beneath the proposed dam without any cutoff provisions was estimated to be 3 cu ft/day, which is an insignificant quantity. The design engineer had observed many openwork streaks along the banks of the river and had noted that the water table fluctuated rapidly with changes in the level of the river. He also noted that the water table was perfectly level for hundreds of feet away from the river. Field pumping tests were then made which indicated somewhat variable permeabilities but none approaching those of openwork gravels. On the grounds of the basic evidence, the dam was designed with a cutoff trench to bedrock. During its excavation streaks of openwork gravel were found throughout the foundation.

The idealized cross section as it really existed is given in Fig. 2.9*b*. A revised seepage computation based on a coefficient of permeability of 30 cm/sec (86,000 ft/day) for the openwork seams indicated that without the cutoff the theoretical leakage would have been in the order of 1,000,000 cu ft/day.

If openwork streaks or other important discontinuities in earth formations remain undetected, serious troubles from excessive seepage and hydrostatic pressures are likely to develop (see Fig. 6.1). This example illustrates the need for thorough, experienced examinations of the foundations of important hydraulic structures.

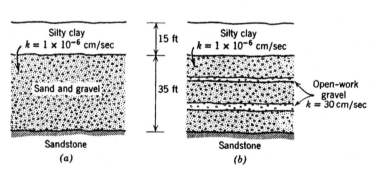

FIG. 2.9 Open-work gravel can be a dominant factor in underseepage. (*a*) Soil profile as surmised from drill holes. Estimated seepage under dam is 3 cu ft/day. (*b*) Soil profile revealed by cutoff trench. Estimated seepage under dam is 1,000,000 cu ft/day (without cutoff trench).

Influence of Dispersion of Fines

The detailed arrangement of soil particles can have a major influence on permeability and other soil properties; for example, if soils are compacted in a relatively dry state, a comparatively harsh permeable structure is usually formed. On the other hand, if liberal amounts of moisture are present, the particles tend to slide over one another into a relatively well-knit, smooth, impermeable type of structure. Engineers of bygone years undoubtedly were aware of the practical value of using liberal amounts of water in the placement of soils to secure good watertightness. The puddled and hydraulic cores of many old dams have a high degree of watertightness. Lambe (1960), Seed and Chan (1961), and others have done considerable research on the influence of the molding water content on the properties of compacted clays.

Cary et al. (1943) report extremely wide variations in the permeabilities of glacial till-gravel mixtures for the core of Mud Mountain Dam with small variation in placement water content. A mixture of 20% till and 80% gravel had permeabilities of around 5×10^{-4} cm/sec at a water content of 14% but only 5×10^{-7} cm/sec at 16%, a change of 1000 times with variation in water content of only 2%. Many other soils show similar but considerably smaller trends. The mixture tested by Cary contained about 1% montmorillonite, an extremely active clay, which if dispersed through soil-gravel mixtures can have an exceedingly important influence on the properties of the mixtures. Not only is the permeability drastically reduced, but the shearing strength can be greatly lowered.

Influence of Density

Density, also *void ratio* or *porosity,* of soil masses, though less important than grain size and soil structure, can have a substantial influence on permeability. The denser a soil and the smaller the pores, the lower its permeability. Often in the construction of reservoirs the soil at the bottom of the reservoir is compacted in place to improve watertightness. Highway engineers know that the construction of embankments over wet hillsides can entrap groundwater, partly through increased density and lower permeability of soil formations. In some cases serious slipouts have resulted (Fig. 8.5). In the construction of dams and levees thorough compaction of fill materials is required to obtain strong embankments and to ensure the best possible watertightness of impervious zones. Some typical plots illustrating the way density (or void ratio) is related to permeability are given in Fig. 2.7. From the loosest to the densest states represented on these plots permeabilities varied 2 to 1000 times; as a rule, however, the narrower the range of particle sizes, the less the permeability is influenced by density.

When blends of sand and gravel are thoroughly compacted with ample water, their permeabilities can be drastically reduced. Strohm et al. (1967) discovered that when well-graded mixtures of sand and gravel contained as little as 5% of fines (sizes smaller than a No. 200 sieve) high compactive efforts reduced the effective porosities nearly to zero and the permeabilities to less than 0.01% of those at moderate densities. These tests explain one of the reasons that blends of sand and gravel often used for drains are virtually useless as drainage aggregates if they contain more than insignificant amounts of fines.

In the preceding paragraphs variations in the permeability of remolded materials caused by variable compaction were discussed. Any factor that densifies soils reduces permeability. Studies of the rate of consolidation of clay and peat foundations are sometimes made by using initial coefficients of permeability of compressible formations. While the consolidation process is going on in foundations their permeabilities are becoming less. Generally, decreases in the permeabilities of clay foundations are rather moderate, but they can be large in highly compressible organic silts and clays and in peats. Modified calculation methods utilizing the changing permeability are needed in the analysis of highly compressible foundations. Some typical variations in permeability caused by consolidation are given in Fig. 2.10, a plot of consolidation pressure versus permeability.

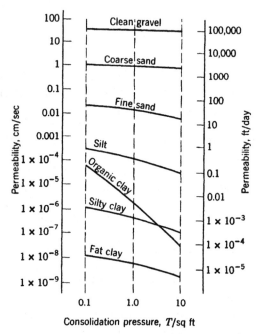

FIG. 2.10 Permeability versus consolidation pressure.

Influence of Discontinuities

Undetected joints, seams, or strata of openwork gravel can lead to serious trouble with hydraulic structures. The effect of openwork gravel on permeability has already been noted.

Compact clays often contain shrinkage or shear cracks that render such formations thousands of times more permeable than the clay between the cracks (Sec. 2.1). Likewise, jointed rocks often have mass permeabilities many times greater than the basic materials between joints. Major dam failures have been caused by some unknown seam or joint system that fed water under pressure into abutments or foundations or allowed piping to occur (E.N.-R., 1977). Careful and adequately deep explorations should always be made for important projects to locate any discontinuities or other features of importance.

The effective permeability of jointed rocks and soils can be determined effectively only by large-scale pumping tests or by careful examination of all strata and assessment of the probable influence of defects and joints. The permeability of many foundations is established almost entirely by the discontinuities, and in such formations tests on individual cores or chunks can be very misleading. Water-pressure tests in drill holes can be useful if the holes are thoroughly bailed before testing (Sec. 2.6) and other precautions are taken to minimize errors.

Londe (1972) points out that the permeability of jointed rocks depends on the widths of the fissures, and that slight changes in the widths of fissures greatly alter permeability. He says that "even a slight change in applied stresses might significantly alter the local permeability of the rock." His comprehensive study of the possible causes of failure of Malpasset Dam in France led to the conclusion that the gneiss in the left abutment of this thin arch dam became extremely impermeable in the zone compressed by the arch, which allowed full reservoir pressures to build up in the rock, causing it to blow out and allow the dam to fail. As may be seen in Table 2.3, the permeability of the Malpasset gneiss is extremely sensitive to pressure (as determined by laboratory tests), whereas other rocks tested were much less affected by changes in pressure. At a positive pressure of 10 tons/sq ft (10 kg/cm^2) the permeability is only 0.2% or 1/500th of that at a slight negative pressure.

Limestone, gypsum, or other water-soluble rock or mineral constituents can lead to the development of solution channels that get larger with time. This can raise the permeability substantially and lead to increased water losses through dam foundations and from reservoirs. When hydraulic structures are built on such formations, it is especially important that the seepage be collected and periodically measured throughout the life of the project. If seepage quantities increase with the passage of time at a given head, steps should be taken to reduce the permeability by grouting, constructing impervious cut-off walls, or other suitable methods (see also Sec. 6.6).

TABLE 2.3   Effect of Pressure on Fissures on the Permeability of Rocks

Kind of rock	Pressure (kg/cm² or tons/sq ft)	Permeability	
		cm/sec	ft/day
Oolithic	− 1.0	5×10^{-7}	1.4×10^{-3}
limestone	10.0	5×10^{-7}	1.4×10^{-3}
	50.0	5×10^{-7}	1.4×10^{-3}
Granite	− 1.0	2×10^{-9}	6×10^{-6}
(compact)	10.0	1×10^{-9}	3×10^{-6}
	50.0	0.8×10^{-9}	2.4×10^{-6}
Malpasset	− 1.0	1×10^{-6}	3×10^{-3}
gneiss	10.0	2×10^{-9}	6×10^{-6}
	50.0	0.4×10^{-9}	1×10^{-6}

(After Londe, 1972. Imperial College Rock Mechanics Research Report No. 17, "The Mechanics of Rock Slopes and Foundations," London, England, April, 1972.)

Influence of Size of Soil or Rock Mass

When the coefficients of permeability of earth masses are being determined in the development of projects in which seepage conditions will be changed by the project, it is important that the scope of the study be adapted to the size of the soil or rock masses that will influence seepage behavior. Therefore it is important that all informations influencing the seepage be investigated. If only small specimens are obtained from test holes (e.g., even if a detailed testing program is carried out), the answers will be representative of the overall mass only if the samples are representative of the mass. Neely (1974) has shown that the resulting values of permeability obtained from *in situ* tests were much larger than those obtained in laboratory tests on small samples because the laboratory specimens did not contain a representative proportion of joints and fissures in the rocky materials tested. Likewise, large-scale well-pumping tests are usually more dependable than large numbers of tests on undisturbed samples taken from drill holes and test pits.

If the groundwater or saturation level within formations at sites can be observed during seasonal or other changes in the inflow or outflow of the system, valuable clues to the mass permeability often can be obtained. Whenever possible, selected drill holes or test pits should be made into observation wells and water levels recorded for as long a time as practicable—hopefully during at least one annual wet and dry cycle. The rate of rise and fall of the water table with changes in natural or artificial inflows (rainfall, irrigation, and the like) and outflows (natural drainage seeps, well pumping rates, and the like) will point to the presence of openwork gravel or other highly permeable seams, strata, openings, and the like. Designers should always be on the alert for any large-scale groundwater changes that occur while a project's seepage needs are being investigated (including any that have been recorded in the past).

2.4 LABORATORY METHODS FOR DETERMINING PERMEABILITY

The coefficient of permeability of soil and rock masses can be determined by any controlled test in which the cross-sectional area, the hydraulic gradient, and the quantity of flow are known or can be approximated. The permeability can be computed from Darcy's law in the form

$$k = \frac{Q}{iAt} = \frac{q}{iA}$$

Laboratory permeability tests commonly used are the *constant head* and *falling head* types (Fig. 2.11).

The *constant head* permeability test is most applicable to permeable mate-

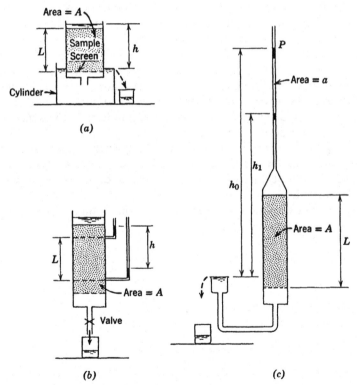

FIG. 2.11 Laboratory permeameters. (*a*) Constant head permeameter. (*b*) Constant heat permeameter. The arrangement in (b) eliminates errors due to filter skin at top or bottom of specimen. (*c*) Falling head permeameter. (Some of the above arrangements are from *Soil Mechanics in Engineering Practice*, K. Terzaghi and R. B. Peck, Wiley, New York, 1948, Fig. 14, p. 46.)

rials such as filter or drainage aggregates. Simple constant head permeameters are shown in Fig 2.11a and b. A specimen of the material is placed in a cylindrical mold and a continuous supply of water is fed through the sample. An overflow arrangement may be used to pass excess water into a sink or suitable drain. The water that passes through the sample in time t flows into a container, where it is collected and the rate q, calculated. The cross-sectional area is the area of the cylinder A, and the hydraulic gradient i is equal to the net head h divided by the length of the sample L; hence

$$k = Q \frac{L}{hAt} = q \frac{L}{hA} \tag{2.5}$$

The constant head permeability test is suitable only for relatively permeable soils and aggregates. If the permeability is low, the time becomes excessive and evaporation during the test introduces errors in the results. Low-permeability soils can be tested in the laboratory by the falling head test.

The *falling head* permeability test is run in apparatus with the general features shown in Fig. 2.11c. A specimen is placed in a tubular chamber of suitable diameter, usually a few inches, and connected with a suitable overflow arrangement and collection container. A small-diameter standpipe tube is connected to the top of the larger tube; its diameter is adjusted to the permeability of the material being tested. If the standpipe is too large, the rate of fall of the water level will be excessively slow; if the standpipe is too small, the rate will be too fast for accurate measurement.

In making a test with a falling head type of apparatus, the standpipe is filled to a level somewhat above point P in Fig 2.11c. When it is at P, a stopwatch is started and the time required for the water level to drop to one or more lower points is recorded.

To permit the calculation of permeability from a falling head test the relationships can be set up as follows:

$$dQ = -a \, dh = k \frac{h}{L} A \, dt$$

By transposing terms we have

$$k = \frac{aL}{A \, dt} \log_e \frac{h_0}{h_1} \tag{2.6}$$

This is the general equation for computation of permeability from a falling head test. To permit the use of common logarithms the equation is rewritten:

$$k = \frac{2.3aL}{A \, dt} \log_{10} \frac{h_0}{h_1} \tag{2.7}$$

Falling head tests can be made with flow downward or upward through the specimen. The water pressure of downward flow holds the soil in place; however, it may cause migration of fines and lead to clogging. Considerable care must be taken to prevent segregation during placement and to prevent the formation of a skin of fines at the top or bottom of the specimen.

Permeability coefficients can be calculated from falling head permeability tests by using Eq. 2.6 or 2.7 for any two positions of water in the standpipe at the beginning and end of a recorded time interval.

If readings are taken for relatively small differences in head in relation to the magnitude of the head, the permeability can be computed by using the average and constant head formulas. The volume of seepage Q for time t is the area of the standpipe a multiplied by the drop in head.

Test Precautions and Details. Care should be taken to avoid segregation during placement of soil in permeameters. If fines are permitted to segregate, thin impervious skins may form and distort the test results.

Frequently, as permeability tests are run, the measured permeability becomes progressively smaller. When this is the case, air from the test water is probably filling the voids in the soil, causing air locking. A considerable amount of air is usually present in ordinary tap water. The use of distilled water at higher than room temperature usually eliminates air locking.

Bouška (1973) tells about reducing the effects of air bubbles in laboratory permeability tests on soils and rocks by deaerating the water with a device that sprays the water in a partial vacuum and stores it in a partial vacuum. Many laboratories use deaerated water in permeability tests. The U.S. Department of the Army, Corps of Engineers (1986) says that when deaired water is used, there should be no significant decrease in permeability with time, but if it does occur, "a prefilter consisting of a layer of the same material as the test specimen, should be used between the deaired distilled water reservoir and the test specimen to remove the air remaining in solution."

When testing soils for permeability in the laboratory, it is necessary to hold the number of variables to a minimum. One minor variable, the viscosity of water, is standardized for performing tests at 20°C or by making a correction for tests performed at other temperatures. The correction is as follows.

$$
\begin{bmatrix} k \text{ at any} \\ \text{temperature } t, \\ \text{the test} \\ \text{temperature} \end{bmatrix} = \begin{bmatrix} k \text{ at another} \\ \text{temperature} \\ \text{(standard} \\ \text{temperature)} \end{bmatrix} \times \begin{bmatrix} \text{the ratio} \\ \text{between the} \\ \text{viscosities at} \\ \text{the two} \\ \text{temperatures} \end{bmatrix}
$$

or

$$
\frac{k_{20}}{k_t} = \frac{\mu_t}{\mu_{20}}
$$

and

$$k_{20} = k_t \frac{\mu_t}{\mu_{20}}$$

Over the maximum range of temperatures encountered in soils the maximum correction for viscosity is about 100% (Fig. 2.4).

The mineral content of the water used in permeability tests can influence the coefficients obtained. Some investigators have found that the kind of salts in the water had a marked influence on the permeabilities obtained on agricultural soils tested at relatively low densities. Permeabilities several hundred times greater have been obtained with tap water containing several hundred parts per million of mixed salts than with distilled water or waters containing a high percentage of sodium salts. Major changes in the rate of flow of water in a soil can take place because of changes in the soil caused by reactions between the salts in the soil and the salts in the percolating water.

If it is known that a nonstandard condition exists in a particular project, it may be desirable to perform permeability tests under conditions simulating the real conditions; however, as a rule permeability tests should be performed under standard conditions with deaerated pure water.

2.5   INDIRECT METHODS FOR DETERMINING PERMEABILITY

Many methods, both direct and indirect, have been devised for determining the permeabilities of soils and rocks. Two indirect methods are described here.

The permeabilities of clays and silts can be calculated from data recorded in laboratory consolidation tests by using the following relationship developed by Terzaghi (1943):

$$T_v = \frac{k}{\gamma_w m_v} \frac{t}{H^2} \tag{2.8}$$

in which T_v is the *time factor* for a given percent consolidation; k is the coefficient of permeability; γ_w, the density of water; m_v, the coefficient of volume change of the specimen; t, the time required to reach the given percentage of consolidation; and H, the longest drainage path. The coefficient m_v is equal to $a_v/(1 + e_0)$ in which a_v is the coefficient of consolidation de/dp, and e_0, the initial void ratio.

From dial readings the percentage of consolidation at any given time t can be calculated after the test is completed. The time factor T_v is determined from the *percentage of consolidation-time factor* curve for one-dimensional consolidation. The distance H is half the thickness of a consolidation specimen drained at the top and bottom.

Frequently the permeability of clays and silts is determined directly by using the consolidation test apparatus as a falling head permeameter. Values thus obtained are checked by indirect calculations, using Eq. 2.8 while being careful to use consistent dimensions. Some engineers believe that the calculated values are more dependable than the directly determined values, because calculated values are free of errors caused by piping along the sides of specimens.

The permeability of clean filter sand can be calculated from a number of formulas such as the following developed by Hazen (1911):

$$k(\text{cm/sec}) = C_1 D_{10}^2 \tag{2.9}$$

in which D_{10} is the *effective size* in centimeters and C_1 varies from about 90 to 120. A value of 100 is often used for C_1. Hazen's formula was developed by testing clean filter sands in a loose state. Even minute quantities of silt or clay can greatly diminish the permeability of sands.

Determination of permeability from Hazen's formula or one of the numerous others that have been developed should be looked on as approximations. As a rule the permeabilities of most soils should be determined by direct test methods.

2.6   FIELD METHODS THAT DEPEND ON PUMPING OR CHANGING THE HEAD IN HOLES

No matter how carefully laboratory permeability tests are made, they represent only minute volumes of soil at individual points in large masses. Their value in solving field seepage and drainage problems depends on how well they represent masses of materials that actually exist in the field. When used with careful consideration of field conditions, laboratory methods can be of considerable value. Nevertheless, in important projects it is often advisable to require field tests that measure the permeabilities in large masses of soil *in situ*.

Examination of the bibliographies listed in Sec. 2.1 will give an indication of the wide variety of methods that have been devised for determining the in-place permeabilities of water-bearing materials. In this chapter only a few of the methods that illustrate some of the fundamental approaches are described. Valuable references are to be found in publications of the U.S. Department of the Navy, Naval Facilities Engineering Command (1982), the U.S. Department of the Army, Corps of Engineers (1957), the U.S. Department of the Interior, Bureau of Reclamation (1960), (1966), (1973), (1974), and the U.S. Department of the Interior, Geological Survey (Wenzel, 1942).

Pumped Wells with Observation Holes—Steady State

A widely used field permeability test is the well-pumping test, in which water is pumped into or out of a well while water level readings are made in several nearby sounding wells. The test is continued until steady conditions are reached. In radial flow toward wells that penetrate the entire water-bearing formation Darcy's law and the Dupuit assumption provide the basis for deriving the *simple well formula.* Dupuit assumed that the hydraulic gradient at any point is a constant from the top to the bottom of the water-bearing layer and is equal to the slope of the water surface. This assumption introduces large errors near wells but is reasonably accurate at moderate distances.

The simple well formula may be derived as follows, with reference to Fig. 2.12. After development of a steady state the quantity of water flowing toward a well in unit time is

$$q = kiA$$

At a distance r from the well the area A through which the water is flowing is $2\pi rh$. Applying Dupuit's assumption that $i = dh/dr$

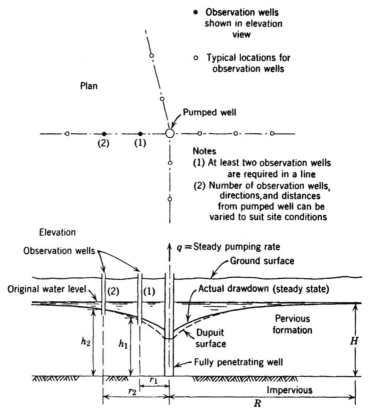

FIG. 2.12 Typical arrangements for determining soil permeability by well pumping test.

$$q = 2k\pi rh \frac{dh}{dr}$$

$$\frac{q\,dr}{r} = 2k\pi h\,dh$$

and

$$q\,\log_e r = \frac{2\pi kh^2}{2} + c = \pi kh^2 + c$$

At a radial distance R where drawdown is negligible $h = H$ and

$$q\,\log_e R = \pi kH^2 + c$$

Therefore

$$c = q\,\log_e R - \pi kH^2$$

Then

$$q\,\log_e r = \pi kh^2 + q\,\log_e R - \pi kH^2$$

and

$$q\,\log_e \frac{R}{r} = \pi k(H^2 - h^2)$$

hence

$$k = \frac{q\,\log_e (R/r)}{\pi(H^2 - h^2)}$$

In common logarithms

$$k = \frac{2.3q\,\log_{10}(R/r)}{\pi(H^2 - h^2)} = \frac{2.3q\,\log_{10}(r_2/r_1)}{\pi(h_2^2 - h_1^2)} \qquad (2.10)$$

Equation 2.10 is the simple well formula used in computing coefficients of permeability from well-pumping tests (a sample computation is given at the end of this section). By transposing terms it can be used in estimating the yield of wells, for

$$q = \frac{\pi k(H^2 - h^2)}{2.3\,\log_{10}(R/r)} = \frac{\pi k(h_2^2 - h_1^2)}{2.3\,\log_{10}(r_2/r_1)} \qquad (2.11)$$

The well formula is based on the following assumptions.

1. The pumping well penetrates the full thickness of the water-bearing formation.
2. A steady-state flow condition exists.
3. The water-bearing formation is homogeneous and isotropic and extends an infinite distance in all directions.
4. The Dupuit assumption is valid.

The reliability of well-pumping tests depends on how accurately the above assumptions are fulfilled. Pumping tests can be used for level or sloping water tables. If the water table is slightly sloping, the bottom boundary of the water-bearing strata may be drawn parallel to the original water table and the computation carried out the same as for a level water table.

The photo in Fig. 2.13 shows a well-pumping test discharge pipe with a flow of 150 gpm. Observation wells and the pumping well are not visible in this picture. Highly permeable gravels were known to be in this foundation for an earth dam east of Fairbanks, Alaska. With this rate of pumping, the water level in the test well was drawn down only 0.6 ft, confirming the presence of extremely permeable gravels. Later on, a moderate excavation was required very near this spot for a concrete structure to be founded on deep alluvial

FIG. 2.13 Discharge pipe for a well pumping test near Fairbanks, Alaska, with flow of 150 gpm.

deposits. To lower the water level approximately 10 ft to enable the work to be carried out in the "dry," a pumping rate of 40,000 gpm was estimated from the well pumping test results. The actual rate to allow the excavation and other work to be carried out was 38,000 gpm, showing that this pumping test had given dependable results.

A field-pumping test should have a pumping well of sufficient diameter to permit the insertion of a pump of suitable type and capacity to do the necessary work. Suction pumps can be used in tests in which the water table is not more than about 15 ft below the surface; however, setting the pump in a small pit can increase the depth to which it can lower the water table. For tests at locations in which the water table must be drawn down more than about 20 ft below the ground level the well must be capable of taking a deep-well pump. The pumping well should penetrate the full depth of the water-bearing strata being tested, although observation wells often need to penetrate only to the lowest level that the water table will reach in a test. Observation wells need a diameter only sufficient to permit the insertion of a suitable water-level measuring device. Holes 1 to 2 in. in diameter are usually adequate for sounding wells.

A simple but accurate way to measure the level in sounding wells is with an "M-scope," which is simply a pair of insulated wires connected in series with a battery and galvanometer. A double-wire cable is fed off a reel into an observation well. When the short uninsulated ends come in contact with the water surface, the galvanometer is activated. By carefully working the wires up and down after the water surface has been contacted, its position can be accurately established. The amount of drawdown is determined by measuring the depth to the water surface just before the start of pumping and periodically while the pumping is in progress. The depth to the water surface is measured from the top of the casing, although a tightly drawn wire or other stationary reference may be used. The "M-scope" is a satisfactory method of measuring drawdown in wells; however, many ingenious sounding devices have been developed for this purpose. The method chosen must give dependable, repeatable results.

The cost of well-pumping setup is relatively high, but normally observation wells are a comparatively small part of the total. At least four observation wells should be used; two is the minimum number in one radial line that permits a single computation of permeability (Fig. 2.12). In highly stratified formations multilevel piezometers are sometimes needed.

The customary procedure in making a field-pumping test is to make a number of permeability computations based on observation wells in various combinations and to use an average as representing a reasonable estimate of the permeability of the soil tested. To provide checks on the accuracy of the determinations the well is usually pumped to equilibrium under at least two rates of flow. The pump should have sufficient capacity to lower the water level in the nearest observation well as least 6 in. When observation wells extend outward in two or more directions from a pumped well, the drawdowns in the several directions may be averaged to reduce the number of permeability computations.

When the water table has an appreciable slope, observation wells more or less in the direction of natural flow are desirable. Uhliarik (1973) says that when use is made of drawdown measurements produced by well pumping tests greater accuracy is achieved by locating pairs of standpipes (wells) at equal distances from the well in the direction of axial cross (at right angles).

Example of Well-Pumping Test

Assumptions

1. Nonartesian conditions comparable to Fig. 2.12. Eq. 2.10 is applicable.
2. Steady pumping rate $= q = 5$ gpm.
3. $H = 60$ ft.
4. $r_1 = 20$ ft, $h_1 = 55$ ft.
5. $r_2 = 30$ ft, $h_2 = 58$ ft.

Computation. Because dimensions are in feet, q must be expressed in cubic feet; hence $q = $ (5 gpm) (1440/7.5) $= 960$ cu ft/day. By using Eq. 2.10

$$k = \frac{2.3q \, \log_{10} (R/r)}{\pi(H^2 - h^2)} = \frac{2.3q \, \log_{10} (r_2/r_1)}{\pi(h_2{}^2 - h_1{}^2)}$$

$$= \frac{2.3(960) \, \log_{10} (30/20)}{\pi(58^2 - 55^2)} = \frac{(2.3)(960)(0.176)}{(3.14)(339)}$$

$$= 0.37 \text{ ft/day}$$

Some formations contain thin layers of high permeability "sandwiched" between materials of much lower permeability. Fortunately an instrument is available to help detect and evaluate the influence of such layers in the foundations of dams and levees. This instrument is a special type of flowmeter developed by the U.S. Army Corps of Engineers (1954), Waterways Experiment Station. Figure 2.14 shows the meter ready for insertion in a pumping well near Fairbanks, Alaska. A calibration chart relates the speed of rotation of an impeller with the rate of flow of water in a well. By taking readings at small intervals throughout the depth of a well, the locations at which water is entering (and the quantities) and the *relative* permeabilities of the various strata penetrated by a well can be determined. This information can be valuable in designing seepage control measures and dewatering systems.

FIG. 2.14 Waterways Experiment Station flow meter ready for inserting into a pumping well near Fairbanks, Alaska. (Printed with permission of Alaska District, U.S. Corps of Engineers.)

Pumped Wells with Observation Holes—Nonsteady State

When field permeability tests are made with the method described for steady-state flow, pumping must be continued until the water levels in observation holes have approximately stabilized. Although true equilibrium may require extremely long periods of pumping, practical results usually can be obtained by pumping at a steady rate for periods that range from a few hours to a few days, depending largely on the permeability.

During the period in which the water table around a pumped well is lowering, water is draining out of the aquifer. Useful solutions to seepage conditions during the nonsteady period are furnished by basic differential equations (Glover, 1966).

When field permeability tests are made by measuring the drawdown in observation wells adjacent to a pumped well, the pumping is usually continued until the drawdowns are reasonably stabilized. If the pumping is continued a number of hours or days until the rate of change in drawdown is small, errors in permeability will be within moderate limits, although, from a rigorous viewpoint, drawdowns continue to increase as long as the pumping is continued.

Borehole Tests (USBR Method E-18)

Because complete well-pumping tests are costly, efforts are frequently made to estimate permeabilities of inplace soils and rocks by pumping into or out of drill holes without the use of observation wells. These procedures are used in exploration boreholes where they provide a physical index of the flow into or out of moderate volumes of inplace material at relatively little cost. They can furnish useful permeability information, but they must be applied with care because the results are not easily checked for accuracy and errors are possible. The most frequent causes of error are the following:

1. Leakage along casing and around packers.
2. Clogging due to sloughing of fines or sediment in the test water.
3. Air locking due to gas bubbles in soil or water.
4. Flow of water into cracks in soft rocks that are opened by excessive head in test holes.

When the stratum being tested is above the water table, it is necessary to perform pumping-in tests; when the layers are beneath the water table, tests can be either the pumping-in or the pumping-out type. The pumping-in tests are susceptible to plugging even when care is taken to avoid causes 2 and 3; hence the pumping-out type should be made whenever conditions permit. All kinds of field permeability tests should be conducted by experienced persons, because improperly performed tests can be highly misleading. The most serious errors are those that produce permeabilities that are too low. Because of the clogging problems, these tests may fail to detect pervious joints or seams in rock formations and highly pervious sand or gravel strata in alluvial formations. When they are used in the design of important projects, the results should be evaluated in relation to broad soils and geological features of the deposits being tested (Sec. 2.3).

The open-end and packer tests described in this section are the pumping-in type performed by the U.S. Bureau of Reclamation (1973) (1974), Designation E-18.

Open-End Tests.   Figure 2.15 shows the arrangements for testing through the open end of a casing which has been drilled to the desired depth and carefully cleaned out to the bottom of the casing. If the hole extends below the ground-

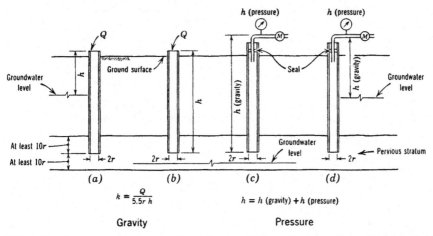

FIG. 2.15 An open-end pipe test for soil permeability which can be made in the field (U.S. Bureau of Reclamation.)

water level, as in Fig. 2.15a, it should be kept filled with water to minimize the squeezing of soil into the bottom of the casing. The test is made by maintaining a constant head by adding clear water through a measuring device. When tests are made above the water table *(b)* in Fig. 2.15, a smooth, consistent water level is seldom obtained, and surging of a few tenths of a foot at a steady rate of flow for about 5 min is considered a satisfactory test by the Bureau of Reclamation.

When desired, additional pressure head can be added to the gravity head, as shown in (c) and (d) of Fig. 2.15. The required data include the amount of head maintained during a constant rate of flow into the hole, the diameter of the casing, and elevations of the top and bottom of the casing. The permeability is calculated from the following relationship determined by electric analogy tests:

$$k = \frac{q}{5.5rh} \tag{2.12}$$

In Eq. 2.12 q is the constant rate of flow into the hole, r is the inside radius of the casing, and h is the differential head of water used in maintaining the steady rate. Any consistent units may be used in computing the permeability. The Bureau of Reclamation expresses k in feet per year and calculates its value from

$$k = C_1 \frac{q(\text{in gallons per minute})}{h(\text{in feet})} \tag{2.13}$$

TABLE 2.4   Variation in C_1 with Size of Casing

Size of casing	C_1
EX	204,000
AX	160,000
BX	129,000
NX	102,000

(From *Design of Small Dams*, U.S. Bureau of Reclamation, 2nd ed., 1973, p. 194.)

The factor C_1 varies with the size of the casing, as shown in Table 2.4.

For tests made below the water table the head h is the difference in feet between the groundwater level and the elevation of the sustained water level in the casing. In tests above the water table h is the depth of water in the hole. When pressure tests are made, the applied pressure head in feet of water (1 psi = 2.3 ft) is added to the gravity head to obtain the total head h [(c) and (d) in Fig. 2.15]. In all borehole tests care must be taken to avoid the use of amounts of head that will split formations and cause erroneously high rates of flow, thereby giving erroneously high permeabilites.

Packer Tests. Figure 2.16 shows the arrangements for making permeability tests in drill holes below the casing. If the formation is strong enough to remain open, tests can be made above or below the water table. These tests are used for testing bedrock with the number of packers necessary to isolate the section of hole being tested.

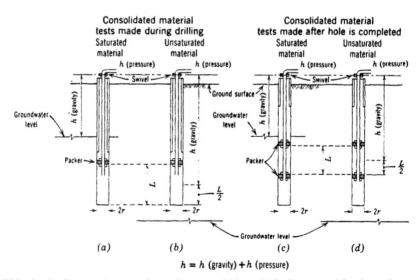

$$h = h \text{ (gravity)} + h \text{ (pressure)}$$

FIG. 2.16 The packer test for soil permeability. (U.S. Bureau of Reclamation.)

Permeabilities are calculated from the following formulas:

$$k = \frac{q}{2\pi Lh} \log_e \frac{L}{r}, L \geq 10r \qquad (2.14)$$

$$k = \frac{q}{2\pi Lh} \sinh^{-1} \frac{L}{2r}, \ 10r > L \geq r \qquad (2.15)$$

In Eqs. 2.14 and 2.15 q is the constant rate of flow into the hole under differential head h, L is the length of the section of hole being tested, r is the radius of the hole tested, $\log_e$ is natural logarithm, and $\sinh^{-1}$ is the arc hyperbolic sine. Throughout this book k is the coefficient of permeability.

The Bureau of Reclamation (1973) (1974) has found that Eqs. 2.14 and 2.15 are most valid when the thickness of the tested stratum is at least $5L$ and are more accurate in testing below the groundwater level than above. The Bureau calculates permeabilities in feet per year, with the formulas rewritten as

$$k = C_p \frac{q(\text{in gallons per minute})}{h(\text{in feet})} \qquad (2.16)$$

The term C_p varies with the size of the test hole and the length of test section as shown in Table 2.5.

In performing packer tests, it is customary to drill the hole, remove the core barrel or other tool, install the packer, perform a test, remove the packer, drill the hole deeper, reseat the packer, and test the newly drilled part of the hole.

TABLE 2.5 Variation in C_p with Diameter of Hole and Length of Test Section

Length of test section L (ft)	Diameter of test hole (Fig. 2.16)			
	EX	*AX*	*BX*	*NX*
1	31,000	28,500	25,800	23,300
2	19,400	18,100	16,800	15,500
3	14,400	13,600	12,700	11,800
4	11,600	11,000	10,300	9,700
5	9,800	9,300	8,800	8,200
6	8,500	8,100	7,600	7,200
7	7,500	7,200	6,800	6,400
8	6,800	6,500	6,100	5,800
9	6,200	5,900	5,600	5,300
10	5,700	5,400	5,200	4,900
15	4,100	3,900	3,700	3,600
20	3,200	3,100	3,000	2,800

(From *Design of Small Dams*, U.S. Bureau of Reclamation, 2nd ed., 1973, p. 196.)

TABLE 2.6 Shape Factors for Computation of Permeability from Variable Head Tests

Condition	Diagram	Shape factor, F	Permeability, k by variable head test (for observation well of constant cross section)	Applicability
(a) Uncased hole		$F = 16\pi DSR$	$k = \dfrac{R}{16DS} \times \dfrac{(h_2 - h_1)}{(t_2 - t_1)}$ for $\dfrac{D}{R} < 50$	Simplest method for permeability determination; not applicable in stratified soils; for values of S see Fig. 2.17
(b) Cased hole, soil flush with bottom		$F = \dfrac{11R}{2}$	$k = \dfrac{2\pi R}{11(t_2 - t_1)} \ln\left(\dfrac{h_1}{h_2}\right)$ for 6 in. $\leq D \leq$ 60 in.	Used for permeability determination at shallow depths below the water table; may yield unreliable results in falling head test with silting of bottom of hole
(c) Cased hole, uncased or perforated extension of length L		$F = \dfrac{2\pi L}{\ln (L/R)}$	$k = \dfrac{R^2}{2L(t_2 - t_1)} \ln\left(\dfrac{L}{R}\right) \ln\left(\dfrac{h_1}{h_2}\right)$ for $\dfrac{L}{R} > 8$	Used for permeability determinations at greater depths below water table
(d) Cased hole, column of soil inside casing to height L		$F = \dfrac{11\pi R^2}{2\pi R + 11L}$	$k = \dfrac{2\pi R + 11L}{11(t_2 - t_1)} \ln \dfrac{h_1}{h_2}$	Principal use is for permeability in vertical direction in anisotropic soils

Observation well or piezometer in saturated isotropic stratum of infinite depth

Observation well or piezometer in aquifer with impervious upper layer

Condition	Diagram	Shape factor, F	Permeability, k by variable head test	Applicability
(e) Cased hole, opening flush with upper boundary of aquifer of infinite depth		$F = 4R$	$k = \dfrac{\pi R}{4(t_2 - t_1)} \ln\left(\dfrac{h_1}{h_2}\right)$	Used for permeability determination when surface impervious layer is relatively thin; may yield unreliable results in falling head test with silting of bottom of hole
(f) Cased hole, uncased or perforated extension into aquifer of finite thickness: (1) $\dfrac{L_1}{T} \le 0.20$ (2) $0.2 < \dfrac{L_2}{T} < 0.85$ (3) $\dfrac{L_3}{T} = 1.00$ *Note.* R_0 equals effective radius to source at constant head		(1) $F = C_s R$ (2) $F = \dfrac{2\pi L_2}{\ln(L_2/R)}$ (3) $F = \dfrac{2\pi L_3}{\ln(R_0/R)}$	(1) $k = \dfrac{\pi R}{C_s(t_2 - t_1)} \ln\left(\dfrac{h_1}{h_2}\right)$ (2) $k = \dfrac{R^2 \ln(L_2/R)}{2L_2(t_2 - t_1)} \ln\left(\dfrac{h_1}{h_2}\right)$ for $\dfrac{L_2}{R} > 8$ (3) $k = \dfrac{R^2 \ln(R_0/R)}{2L_3(t_2 - t_1)} \ln\left(\dfrac{h_1}{h_2}\right)$	(1) Used for permeability determinations at depths greater than about 5 ft, for values of C_s see Fig. 2.17 (2) Used for permeability determinations at greater depths and for fine grained soils using porous intake point of piezometer (3) Assume value of $\dfrac{R_0}{R} = 200$ for estimates unless observation wells are made to determine actual value of R_0

The procedure is repeated as many times as desired. When the hole stands open without casing, the Bureau of Reclamation usually drills it to its final depth, fills the hole with water, surges it, and bails it out. Two packers on pipe or drill stems are set as shown in (c) and (d) of Fig. 2.16. When testing between two packers, it has been found best to start at the bottom and work upward.

Other Methods

In addition to pumped wells with observation holes and the borehole methods already described in this section, a variety of other steady and nonsteady seepage methods are used for determining *in situ* permeabilities. The Bureau of Reclamation (1974) also recommends a method (Designation E-19) by which the rate of flow of water out of uncased auger or bore holes is measured and permeabilities are calculated by using appropriate shape factors. The U.S. Department of the Navy, Naval Facilities Engineering Command (1982), has standard methods of performing variable head tests to estimate the permeability *in situ* by means of piezometer observation wells. Table 2.6 gives shape factors and formulas for the calculation of permeabilities from variable head tests in cased and uncased holes, and Fig. 2.17 summarizes methods of calculating permeabilities from these tests.

Hvorslev (1951) made a thorough study for the U.S. Corps of Engineers, Waterways Experiment Station, of methods of measuring hydrostatic pressures and of determining permeability in borings, observation wells, piezometers, and hydrostatic pressure cells. Whenever a piezometer or other device is installed, the initial hydrostatic pressure recorded by the device seldom equals the true surrounding porewater pressure. Water must flow to or from the hole or device until the pressure or head in the device equals that in the surrounding soil. A flow must also occur with a corresponding *time lag* when the surrounding pore pressure increases or decreases. The magnitude of the time lag is inversely proportional to the permeability of the soil and varies with the size and type of pressure measuring device. Table 2.7 is a tabulation prepared by Hvorslev which gives the approximate time lags for pore-pressure measuring devices of various kinds and dimensions. It is a useful guide to the selection of an appropriate type of installation for given conditions. Hvorslev (1951) recommends that actual time lags be determined by field experiments so that subsequent observations may be corrected for its influence. The time lag theory is also a practical device for determining the permeability of soils in the field.

If water is flowing into or out of a casing or other pore-pressure sensing device, the flow q may be expressed by the following simplified expression:

$$q = Fkh = Fk(z - y) \qquad (2.17)$$

In Eq. 2.17 F is a factor that depends on the shape and size of the intake or well point, h is the active head, z is the distance from a reference level to the outside piezometric level, and y is the distance from the reference level to the

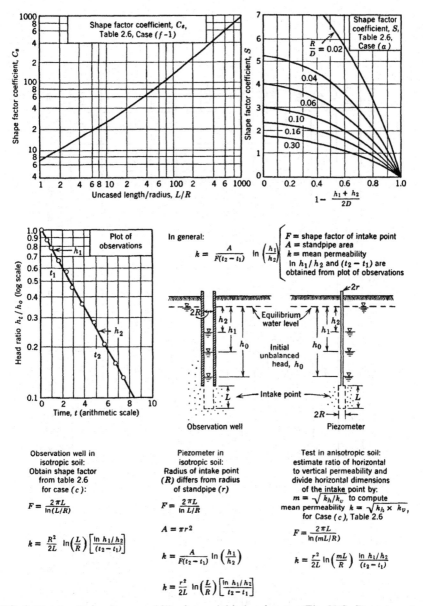

FIG. 2.17   Analysis of permeability by variable head tests. (The U.S. Department of the Navy, Naval Facilities Engineering Command, 1982.) (Shape factor S chart was developed from Fig. 10–10 *Soil Engineering*, M. G. Spangler, International Textbook Co., 2nd ed., 1963.)

TABLE 2.7 Approximate Hydrostatic Time Lags (From Hvorslev (1951), U.S. Corps of Engineers, W.E.S.)

Time lags	For 90% equalization = T_{90}										Basic time lag T
	Sand			Silt			Clay				
Approximate soil type	10^{-1}	10^{-2}	10^{-3}	10^{-4}	10^{-5}	10^{-6}	10^{-7}	10^{-8}	10^{-9}	10^{-10}	10^{-4}
Coefficient of permeability in cm/sec											
1 2-in. casing—soil in casing, $L = 3D = 6$ in.	6^m	1^h	10^h	4.2^d							193^d
2 2-in. casing-soil flush bottom casing	0.6^m	6^m	1^h	10^h							17^d
3 2-in. casing-hole extended, $L = 3D = 6$ in.		1.5^m	15^m	2.5^h	25^h	10^d					4.5^d
4 2-in. casing-hole extended, $L = 12D = 24$ in.			6^m	1^h	10^h	4.2^d	42^d				47^h
5 3/8-in. piezometer with well point diameter 1½ in., length 18 in.				3^m	30^m	5^h	50^h	21^d			130^m
6 3/8-in. piezometer with well point and sand filter, $D = 6$ in., $L = 36$ in.					12^m	2^h	20^h	8.3^d	83^d		51^m
7 1⅜-in. mercury manometer, single tube with porous cup point, $D = 1¼$ in., $L = 2½$ in.	*One-half of values for 1⅜-in. mercury U-tube manometer or 4½ in. bourdon gage*					2^m	20^m	3.3^h	33^h	14^d	52^s
8 1⅜-in. mercury manometer, single tube with well point, $D = 1½$ in., $L = 18$ in.							6^m	1^h	10^h	4.2^d	16^s
9 3-in. W. E. S. hydrostatic pressure cell in direct contact with soil								16^m	2.6^h	26^h	4^s
10 3-in. W. E. S. hydrostatic pressure cell in sand filter, $D = 6$ in., $L = 18$ in.									16^m	2.6^h	0.4^s

Symbols: s = seconds, m = minutes, h = hours, d = days. *Assumptions:* constant groundwater pressure and intake shape factor, isotropic soil, no gas, stress adjustment time lags negligible. The computed time lags have been rounded off to convenient values.

piezometer level in the transient state (Fig. 2.18). Equation 2.17 is also valid for anisotropic permeabilities if modified values of $\bar{F}$ and $\bar{k}$ are used. Considering the volume of flow during a time dt and assuming that friction losses in the pipe can be neglected, the following equation is obtained:

$$q\,dt = A\,dy \tag{2.18}$$

In Eq. 2.18 A is the cross-sectional area of the standpipe or an appropriate area representing the relationship between volume and pressure changes in a pressure cell or gage. By taking the expression for q from Eq. 2.17 a differential equation assumes the following form:

$$\frac{dy}{z-y} = \frac{Fk}{A}\,dt \tag{2.19}$$

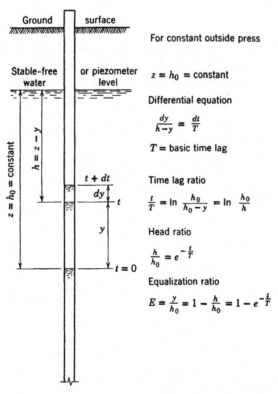

Ground surface

For constant outside press

Stable-free water | or piezometer level

$z = h_0 = $ constant

Differential equation

$$\frac{dy}{h-y} = \frac{dt}{T}$$

$T = $ basic time lag

Time lag ratio

$$\frac{t}{T} = \ln\frac{h_0}{h_0-y} = \ln\frac{h_0}{h}$$

Head ratio

$$\frac{h}{h_0} = e^{-\frac{t}{T}}$$

Equalization ratio

$$E = \frac{y}{h_0} = 1 - \frac{h}{h_0} = 1 - e^{-\frac{t}{T}}$$

$z = h_0 = $ constant

$h = z - y$

$t + dt$

dy

t

y

$t = 0$

FIG. 2.18 General conditions for time lag determination. (After Hvorslev, U.S. Corps of Engineers, W.E.S.)

The volume of flow required to equalize the pressure difference h is $V = Ah$. Hvorslev (1951) defines the *basic time lag* T as the time required for equalization of this pressure difference when the original rate of flow $q = Fkh$ is maintained. It then follows that

$$T = \frac{V}{q} = \frac{Ah}{Fkh} = \frac{A}{Fk} \qquad (2.20)$$

Equation 2.19 may be rewritten

$$\frac{dy}{z - y} = \frac{dt}{T} \qquad (2.21)$$

Equation 2.21 is the basic differential equation for the hydrostatic time lag.

In the field the basic time lag is determined by raising or depressing the head in a standpipe or other pressure measuring device and recording the head at a number of time intervals. A plot is then made (Fig. 2.19) of time on an arithmetic scale and the head ratio (h/h_0) on a log scale. The basic time lag is the time at which the head ratio equals 0.37. The *equalization ratio* is defined as $(1 - h/h_0)$; thus when the head ratio is 0.37 the equalization ratio is 0.63. An equalization ratio of 0.90, which corresponds to a time lag 2.3 times the basic time lag, is considered by Hvorslev to be adequate for many practical purposes. The basic time lag T corresponds to $h = 0.37h_0$; that is,

$$\log_e \left(\frac{h_0}{h} \right) = \log_e \left(\frac{h_0}{0.37h_0} \right) = \log_e 2.7 = 1.0$$

When the shape factor F of a pore-pressure measuring device is known, it is possible, theoretically, to calculate coefficients of permeability of soil in place by field measurements. Thus, if a constant head h_0 and a steady rate of flow q are maintained, Eq. 2.17 becomes

$$k = \frac{q}{Fh_c} \qquad (2.22)$$

and, under variable head conditions, the following equation can be derived:

$$k = \frac{A}{F(t_2 - t_1)} \log_e \left(\frac{h_1}{h_2} \right) \qquad (2.23)$$

When the basic time lag T is known, the coefficient of permeability can be obtained by rearranging Eq. 2.20 in the form

$$k = \frac{A}{FT} \qquad (2.24)$$

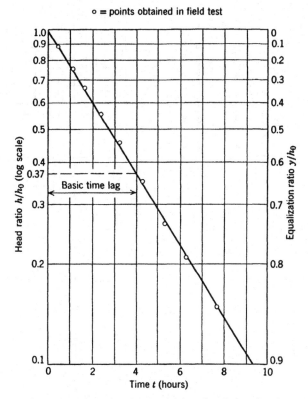

FIG. 2.19 Plot used for basic time lag determination.

Figure 2.20 is a summary of formulas compiled by Hvorslev (1951) for the determination of coefficients of permeability by constant head, variable head, and basic time lag tests. Diagrams in the lower part of the figure identify the various installations for which solutions are given.

2.7 FIELD METHODS THAT DEPEND ON SEEPAGE VELOCITIES

The field permeability testing methods described in Sec. 2.6 require the pumping of measured quantities of water into or out of test holes or measurements of changes in water levels or pressures. Methods described in this section make use of the principle that permeability is related to the *seepage velocity* (Sec. 2.2):

$$k = \frac{v_s n_e}{i} \tag{2.3}$$

Case	Constant head	Variable head	Basic time lag	Notation
				D = diam. intake, sample, cm d = diameter, standpipe, cm L = length, intake, sample, cm h_c = constant piez. head, cm h_1 = piez. head for $t = t_1$, cm h_2 = piez. head for $t = t_2$, cm q = flow of water, cm³/sec t = time, sec T = basic time lag, sec k_v = vert. perm. ground, cm/sec k_v' = vert. perm. casing, cm/sec k_h = horz. perm. ground, cm/sec k_m = mean coeff. perm. cm/sec m = transformation ratio $k_m = \sqrt{k_h \cdot k_v}$, $m = \sqrt{k_h/k_v}$ $\ln = \log_e = 2.3 \log_{10}$
a	$k_v = \dfrac{4 \cdot q \cdot L}{\pi \cdot D^2 \cdot h_c}$	$k_v = \dfrac{d^2 \cdot L}{D^2 \cdot (t_2 - t_1)} \ln \dfrac{h_1}{h_2}$ $k_v = \dfrac{L}{t_2 - t_1} \ln \dfrac{h_1}{h_2}$ for $d = D$	$k_v = \dfrac{d^2 \cdot L}{D^2 \cdot T}$ $k_v = \dfrac{L}{T}$ for $d = D$	
b	$k_m = \dfrac{q}{2 \cdot D \cdot h_c}$	$k_m = \dfrac{\pi \cdot d^2}{8 \cdot D \cdot (t_2 - t_1)} \ln \dfrac{h_1}{h_2}$ for $d = D$	$k_m = \dfrac{\pi \cdot d^2}{8 \cdot D \cdot T}$ for $d = D$	
c	$k_m = \dfrac{q}{2.75 \cdot D \cdot h_c}$	$k_m = \dfrac{\pi \cdot d^2}{11 \cdot D \cdot (t_2 - t_1)} \ln \dfrac{h_1}{h_2}$ for $d = D$	$k_m = \dfrac{\pi \cdot d^2}{11 \cdot D \cdot T}$ for $d = D$	
d	$k_h = \dfrac{q \cdot \ln\left[\dfrac{2mL}{D} + \sqrt{1 + \left(\dfrac{2mL}{D}\right)^2}\right]}{2 \cdot \pi \cdot L \cdot h_c}$	$k_h = \dfrac{d^2 \cdot \ln\left[\dfrac{2mL}{D} + \sqrt{1 + \left(\dfrac{2mL}{D}\right)^2}\right]}{8 \cdot L \cdot (t_2 - t_1)} \ln \dfrac{h_1}{h_2}$ $k_h = \dfrac{d^2 \cdot \ln\left(\dfrac{4mL}{D}\right)}{8 \cdot L \cdot (t_2 - t_1)} \ln \dfrac{h_1}{h_2}$ for $\dfrac{2mL}{D} > 4$	$k_h = \dfrac{d^2 \cdot \ln\left[\dfrac{2mL}{D} + \sqrt{1 + \left(\dfrac{2mL}{D}\right)^2}\right]}{8 \cdot L \cdot T}$ $k_h = \dfrac{d^2 \cdot \ln\left(\dfrac{4mL}{D}\right)}{8 \cdot L \cdot T}$ for $\dfrac{2mL}{D} > 4$	
e	$k_h = \dfrac{q \cdot \ln\left[\dfrac{mL}{D} + \sqrt{1 + \left(\dfrac{mL}{D}\right)^2}\right]}{2 \cdot \pi \cdot L \cdot h_c}$	$k_h = \dfrac{d^2 \cdot \ln\left[\dfrac{mL}{D} + \sqrt{1 + \left(\dfrac{mL}{D}\right)^2}\right]}{8 \cdot L \cdot (t_2 - t_1)} \ln \dfrac{h_1}{h_2}$ $k_h = \dfrac{d^2 \cdot \ln\left(\dfrac{2mL}{D}\right)}{8 \cdot L \cdot (t_2 - t_1)} \ln \dfrac{h_1}{h_2}$ for $\dfrac{mL}{D} > 4$	$k_h = \dfrac{d^2 \cdot \ln\left[\dfrac{mL}{D} + \sqrt{1 + \left(\dfrac{mL}{D}\right)^2}\right]}{8 \cdot L \cdot T}$ $k_h = \dfrac{d^2 \cdot \ln\left(\dfrac{2mL}{D}\right)}{8 \cdot L \cdot T}$ for $\dfrac{mL}{D} > 4$	

Piezometric head h (Log scale) — Time t (linear scale)

h_0 0.37 h_0 0.10 h_0 $t = T$ $t = 2.3 \cdot T$

Determination basic time lag T

Assumptions: Soil at intake, infinite depth and directional isotropy (k_v and k_h constant); no disturbance, segregation, swelling or consolidation of soil; no sedimentation or leakage; no air or gas in soil, well point, or pipe; hydraulic losses in pipes, well point or filter negligible. (After Hvorslev, U.S. Corps of Engineers, W.E.S.)

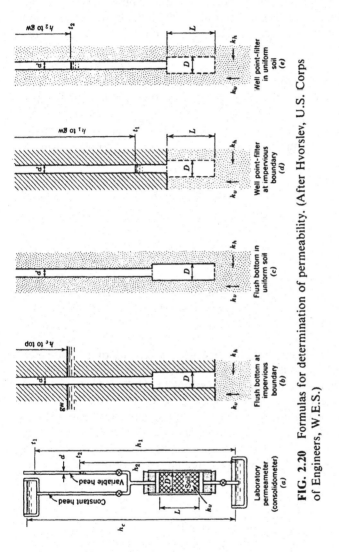

FIG. 2.20 Formulas for determination of permeability. (After Hvorslev, U.S. Corps of Engineers, W.E.S.)

Equation 2.3 permits the determination of the permeability of soil formations by methods that do not require pumping equipment. When, at a common point, the velocity of the flowing water and the hydraulic gradient are known, the permeability can be estimated. Frequently the hydraulic gradient of an existing water table can be estimated from wells in the area. If not, observation wells must be installed. The velocity of flow can be determined by a number of practical methods:

1. Insertion of an electrolyte into a test hole and the use of galvanometers to detect the time required for the electrolyte to pass a known distance through the soil.
2. Use of radioactivated charges and geiger counters or other instruments to detect the time required for the charge to travel a known distance through the soil.
3. Insertion of a dye such as fluorescein sodium into a test hole and observation of the time it takes to emerge in a nearby test pit or on a bank from which seepage is emerging.

The procedure used in these tests is illustrated in Fig. 2.21. An electrolyte or radioactive charge is inserted into the sloping water table in hole A, the time for the charge to reach hole B is measured with suitable instruments, and the seepage velocity is determined by dividing distance L by time t. The effective porosity n_e is determined from test data for the in-place soil; if no tests are available, it is estimated. The permeability is calculated from Eq. 2.3, $k = v_s n_e / i$.

When dyes or other tracers are used to measure the velocity of groundwater for permeability determinations or other purposes, the relatively sluggish nature of most groundwater should not be overlooked. Unless the formations contain extremely permeable strata, the time required for tracers to move even short distances can be very long. Also, the drill holes or test pits used for the tests must be placed along *streamlines* or *flow lines* (see Chapter 3) or negative results will be obtained. In 1939 I attempted to make tests in the Willamette Valley in Oregon by inserting dye in wells located only 8 or 10 ft from pumped

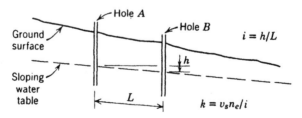

FIG. 2.21 Typical arrangement for determining soil permeability by measurement of seepage velocity.

TABLE 2.8   Calculation of Coefficient of Permeability of Cobbles from Dye Tests

Travel distance L (ft)	Head, H (ft)	Average i	Estimated n	Travel time	Calculated k (ft/day)
5200	17	0.0033	0.3	2.7 days	180,000
120	11	0.09	0.3	43 min	13,000
150	1.5	0.01	0.3	28 min	200,000
1000	4.0	0.004	0.3	0.8 days	95,000
920	18.4	0.02	0.3	95 min	200,000
700	17.5	0.025	0.3	100 min	120,000

test pits. In several tests no dye was detected in the pits and the tests were discontinued after 6 to 7 days. In 1971 I helped the Sacramento District of the U.S. Corps of Engineers to make fluorescein dye tests of complex cobble deposits in a gold-dredging field proposed for a cutoff trench for an earth dam. Positive and quick results were obtained with travel distances ranging from 120 ft to nearly a mile. Some of the permeabilities calculated from these tests (at six different locations) are listed in Table 2.8.

Table 2.9 gives the relationship between permeability, hydraulic gradient, and the rate of movement of seepage through porous media whose effective

TABLE 2.9   Relationship Between Permeability and Rate of Movement of Groundwater ($n_e = 0.25$)

Permeability		i	v_s (ft/day)	Time to move 1 ft	
ft/day	cm/sec			Days	Hours
0.01	3.5×10^{-6}	0.01	4×10^{-4}	2500	
		0.02	8×10^{-4}	1250	
		0.05	20×10^{-4}	500	
		0.10	40×10^{-4}	250	
1.0	3.5×10^{-4}	0.01	0.04	25	600
		0.02	0.08	12.5	300
		0.05	0.20	5	120
		0.10	0.40	2.5	60
100	0.035	0.01	4.0	0.25	6
		0.02	8.0	0.13	3
		0.05	20	0.05	1.2
		0.10	40	0.025	0.6
10,000	3.5	0.01	400	0.0025	0.06
		0.02	800	0.0013	0.03
		0.05	2000	0.0005	0.012
		0.10	4000	0.0003	0.006

porosity is 25%; for example, if water is flowing under a hydraulic gradient of 0.02 in a bed of washed concrete sand with a permeability of 1.0 ft/day, the possible seepage velocity is 0.08 ft/day, and the travel time for dye or other tracers is 300 hr/ft. Under the same hydraulic gradient seepage can travel through coarse pea gravel with a permeability of 10,000 ft/day at a rate of 800 ft in 24 hr.

2.8 FIELD METHODS THAT DEPEND ON OBSERVATION OF SPREADING OR RECEDING GROUNDWATER MOUNDS

Groundwater hydrologists often make use of mathematical models to study the behavior of groundwater systems under changing conditions and to predict probable behavior under future changes. Weber and Hassan (1972) outline the progress of model development in recent years and describe the procedures that have been used on major groundwater basins. Analog simulators, analog computers, and digital computers are among the available methods. I use simplified approximate computations with flow nets and Darcy's law for estimating soil permeabilities and transmissibilities from changing water tables. Computations are based on (1) measurements of the *rate of spread* of saturation or (2) *volumes of water* flowing into or out of soil systems, estimated from volumes of soil that become saturated or unsaturated during a known period of time.

Estimating *k* from Rate of Spread of Water—*Unconfined Flow*

When water is spreading from a sudden rise of rivers in flood stage, the mass permeabilities of soil formations can often be estimated from the *rate of spread* and rise of saturation measured in observation wells. If the spreading takes place in relatively uniform soil with unconfined flow, an estimate of permeabilities can be made with the use of *transient flow nets*. In Sec. 3.6 transient flow nets are used for estimating the rate of movement of saturation lines when the permeability is known (see Fig. 3.25 and Eq. 3.31). If the time of spread is known, permeability can be estimated:

From Darcy's law

$$v_{sl} = \frac{ki}{n_e} \quad \text{and} \quad k = \frac{v_{sl}n_e}{i}$$

(v_{sl} is the seepage velocity, k is the coefficient of permeability, i is the average hydraulic gradient, and n_e is the effective porosity).

Because *distance = velocity × time,* it follows that *time = distance/velocity* and that time Δt for an increment of distance Δl is $\Delta l/v_{sl}$ and the total time T for water to move a given distance L is equal to $\Sigma \Delta t$. Therefore

$$T = \sum \frac{\Delta l}{v_{sl}} = \sum \frac{\Delta l n_e}{ki} = \frac{n_e}{k} \sum \frac{\Delta l}{i}$$

and, if the travel time is known, k can be estimated from

$$k = \frac{n_e}{T} \sum \frac{\Delta l}{i} \tag{2.25}$$

To make a calculation the soil zone into which water has spread a known distance in a known time is divided into several increments and transient flow nets are drawn as shown in Fig. 3.25. According to the methods described in Sec. 3.6, the ratio of $\Delta l/i$ is determined for each of the increments and the total value of $\Sigma(\Delta l/i)$ is calculated. Then, by using Eq. 2.25, the average coefficient of permeability is calculated.

Estimating *k* from *Quantity of Water* Flowing Into or Out of Soil

When water is spreading from a rapidly rising river through *highly permeable strata* and rising into *moderately permeable upper strata,* as described in the example in Fig. 2.22, the permeability of the underlying strata often can be estimated from the rise of water in observation wells. The actual flow can be very complex and extremely difficult to analyze rigorously; however, practical estimates of permeability can often be obtained by the following procedure:

1. Plot a cross section, as in Fig. 2.22, to show the initial position of the water table (level 1) and its position after the river has been at level 2 for T days (rise in river is assumed to be instantaneous).
2. Estimate the *mean* travel distance L for the volume of water that has filled volume V in T days (L is the distance to the center of gravity of the cross-hatched [saturated] area in Fig. 2.22).

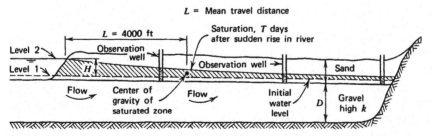

FIG. 2.22 Estimating k from quantity of water flowing into a soil system (moderately permeable sand above extremely permeable gravel). An approximate method.

Then calculate k of the underlying gravel from Darcy's law:

$$k = \frac{q}{iA} \qquad (2.26)$$

In this example

the total quantity of water in T days $= Q = Vn_e$,

the quantity per day $= q = Q/T$,

the cross-sectional area normal to flow $= A = D$ (times 1 ft).

The average hydraulic gradient i is assumed to be equal to H/L and

$$\begin{aligned}
H &= 5 \text{ ft}, \\
L &= 4000 \text{ ft}, \\
V &= 20{,}000 \text{ cu ft}, \\
n_e &= 0.35, \\
T &= 58 \text{ days}.
\end{aligned}$$

Then

$$Q = 20{,}000 \ (0.35) = 7000 \text{ cu ft},$$

$$q = \frac{Q}{T} = \frac{7000}{58} = 120.7 \text{ cu ft/day},$$

$$i = 5 \text{ ft}/4000 \text{ ft} = 0.00125.$$

Also assume that $A = 50$ sq ft/lin ft. Now

$$k = \frac{q}{iA} = \frac{120.7}{(0.00125)(50)} = 1930 \text{ ft/day}$$

This procedure was used for estimating the permeabilities of large volumes of soil in perhaps the biggest field permeability test ever performed. In a study of methods for designing levees to protect 90,000 acres of irrigated land in the Gila River valley in Arizona I learned that some 20 billion gal of water had soaked into the foundations for the proposed levees several years before, when a flood volume was released from an upstream flood-control reservoir. *In a period of 60 days 62,600 acre-ft of water had disappeared into the ground while periodic readings were being made in several hundred observation wells* used for monitoring the effect of groundwater on crop production and managed irrigation in the valley. Unknowingly data were being gathered in a monumental field percolation test. A plan of the valley which shows the locations of observation wells and displays typical well hydrographs and river hydrographs at both ends of the area is contained in Fig. 2.23. The Los Angeles District of the U.S. Army Corps of Engineers had made the normal types of foundation exploration which included drill holes, undisturbed samples, laboratory classification tests, several well pumping tests, and a number of test pits. The massive field percolation test, which was discovered in a routine search for any recorded information that could be useful in estimating overall mass permeabilities of the formations, proved to be of great value in estimating the permeabilities of the foundations for the proposed levees. To study the spreading groundwater the area was divided into 26 subareas, and the average permeability of the gravels under each was estimated from the volume of water entering the soil. These calculations produced permeabilities ranging from 300 ft/day (0.1 cm/sec) to 3,000 ft/day (1.0 cm/sec). A good check on the overall validity of the calculations is the fact that the amount of water entering the soil, estimated from well readings and soil porosity, was 63,800 acre-ft and that a quantity of 62,600 acre-ft was obtainable from the hydrographs for river flows measured at the upper and lower ends of the project area.

2.9 SUMMARY COMMENTS

Permeability is one of the most vital properties of the earth formations on which engineering works need to be constructed as well as the materials used in constructing earthworks, including drainage materials for protecting these facilities from water damage. Minor-appearing soils and geologic details can damage or cause failure of dams, reservoirs, and other hydraulic structures. Therefore, thorough soil and permeability investigations must always be made for projects that can be harmed by inadequate control over the water in them or in their environments.

Because permeabilities can vary over such a great range (several billion times), extreme care is needed in evaluating all factors influencing permeability and seepage and drainage behavior. Designers must always insist on adequate subsurface investigations and testing programs of all in-place or constructed features of project facilities to try to obtain realistic interpretation of conditions as they will really exist in foundations and facilities that will be exposed to water.

No potential source of information about the materials influencing seepage behavior should be overlooked by those designing facilities that will be subject to water damage. An alert investigator can sometimes find recorded information on foundation and water conditions at or near a planned project that can be of great value in revealing large-scale properties of formations affecting the seepage behavior of a project (e.g., see Fig. 2.23).

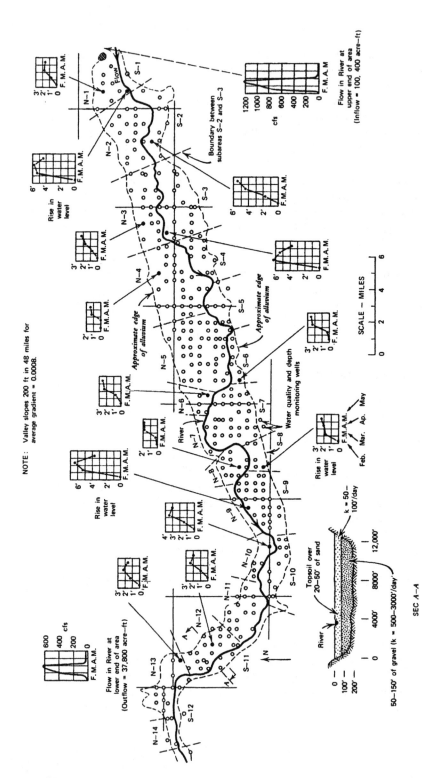

FIG. 2.23 Large-scale "field permeability" test in an irrigated valley in Arizona desert (Welton-Mohawk Irrigation District). (Printed with permission of Los Angeles District, U.S. Army Corps of Engineers). Valley slopes 200 ft in 48 miles for average gradient of 0.0008.

REFERENCES

Bittinger, Morton W., and D. E. L. Maasland (1963), "Selected References on Mathematical Developments in Transient Groundwater Hydraulics," *Proceedings,* Symposium on Transient Ground Water Hydraulics, Colorado State University, Fort Collins, Colo., July 25 to 27, 1963, Appendix F, pp. 220-3 (78 references).

Bouška, M. (1973), "Preparation of De-aerated Water and Samples for the Laboratory Permeability Test," *Geologicky Průzkum,* Prague, Vol. 15, No. 8, pp. 243-246.

Cary, A. S., Boyd H. Walter, and Howard T. Harstad (1943),"Permeability of Mud Mountain Dam Core Material," *Transactions,* A.S.C.E., Vol. 108, pp. 719-728.

Darcy, H. (1856), *Les fountaines publiques de la ville de Dijon.*

Engineering News-Record (1977), "Teton Dam Failure is Blamed on Bu Rec Design Deficiencies," January 13, 1977, pp. 8-9.

Glover, R. E. (1966), "Ground-water Movement," Engineering Monograph No. 31, U.S. Department of the Interior, Bureau of Reclamation, Denver, second printing, April 1966, pp. 4-18.

Hazen, A. (1911), "Discussion of 'Dams on Sand Foundations,' by A. C. Koenig," *Transactions,* A.S.C.E., Vol. 73, p. 199.

Hvorslev, M. J. (1951), "Time Lag and Soil Permeability in Ground-Water Observations," Bulletin No. 36, Waterways Experiment Station, Corps of Engineers, Vicksburg, Miss.

Johnson, A. I. (1963), "Selected References on Analog Models for Hydrologic Studies," *Proceedings,* Symposium of Transient Ground Water Hydraulics, Colorado State University, Fort Collins, Colo., July 25 to 27, 1963, Appendix F, pp. 212-219 (163 references).

Lambe, T. W. (1960), "Compacted Clay Structure," *Transactions,* A.S.C.E., Vol. 125, Part I, p. 682.

Londe, Pierre (1972), "The Mechanics of Rock Slopes and Foundations," Rock Mechanics Research Report No. 17, Imperial College of Science and Technology, University of London, April 1972.

Muskat, Morris (1937), *The Flow of Homogeneous Fluids through Porous Media,* McGraw-Hill, New York, p. 68.

Neely, W. J. (1974), "Field Measurements of Permeability Using Auger Holes," *Ground Engineering,* London, Vol. 7, No. 1, pp. 38-42.

Richter, Raymond C. (1963), "Permeability, Coefficients of Transmissibility, Coefficients of Storage, Annotated Bibliography Through 1961," State of California Department of Water Resources, Sacramento, Calif., pp. 1-49 (207 references).

Seed, H. Bolton, and C. K. Chan (1961), "Structure and Strength Characteristics of Compacted Clays,"*Transactions,* A.S.C.E., Vol. 126, Part I, pp. 1344-1385.

Strohm, W. E., E. H. Nettles, and C. C. Calhoun (1967), "Study of Drainage Characteristics of Base Course Materials," Highway Research *Record* No. 203, Highway Research Board (now Transportation Research Board).

Taylor, D. W. (1948), *Fundamentals of Soil Mechanics,* Wiley, New York, p. 99; 122-123.

Terzaghi, Karl (1943), *Theoretical Soil Mechanics,* Wiley, New York, p. 274.

Todd, David K. (1959), *Ground Water Hydrology,* Wiley, New York.

Uhliarik, A. (1973), "Determination of Coefficient of Permeability from Pumping Tests," *Inženýrské Stavby,* Prague, Vol. 21, No. 10, pp. 443–445.

U.S. Department of the Army, Corps of Engineers (1951), "Time Lag and Soil Permeability in Ground-Water Observations," Bulletin No. 36, Waterways Experiment Staton, Vicksburg, Miss. (written by M. Juul Hvorslev).

U.S. Department of the Army, Corps of Engineers (1954), "Waterways Experiment Station Relief Well Flow Meter." Misc. Paper No. 5–83, Vicksburg, Miss., April 1954.

U.S. Department of the Army, Corps of Engineers (1957), "Manual for Subsurface Investigations," pp. 53–57, December 1, 1957.

U.S. Department of the Army, Corps of Engineers, Office of the Chief of Engineers (1986), "Seepage Analysis and Control For Dams," Engineering Manual EM 1110-2-1901, p. 2–30, 30 September 1986.

U.S. Department of the Interior, Bureau of Reclamation (1960), "Studies of Ground Water Movement," Technical Memorandum 657, 180 pages, March 1960.

U.S. Department of the Interior, Bureau of Reclamation (1966), "Ground-water Movement," Engineering Monograph No. 31, Denver, second printing, April 1966 (written by R. E. Glover).

U.S. Department of the Interior, Bureau of Reclamation (1973), "Design of Small Dams," U.S. Government Printing Office, Washington, D.C., pp. 193–196, 2nd ed.

U.S. Department of the Interior, Bureau of Reclamation (1974), "Earth Manual," Designation E-18, pp. 573–578, and Designation E-19, pp. 578–593, 2nd ed.

U.S. Department of the Interior, Geological Survey (1942), "Methods for Determining Permeability of Water-Bearing Materials," U.S. Government Printing Office, Washington, D.C. (written by L. K. Wenzel, with a bibliography on permeability and laminar flow by V. C. Fishel, Water-Supply paper No. 887).

U.S. Department of the Navy, Naval Facilities Engineering Command (1982), "Soil Mechanics," NAVFAC Design Manual 7.1, May 1982.

Weber, Ernest M., and A. A. Hassan (1972), "Role of Models in Groundwater Management," Water Resources Bulletin, Journal of the American Water Resources Association, Minneapolis, Minn., Vol. 8, No. 1, pp. 198–206.

Wenzel, L. K. (1942), "Methods for Determining Permeability of Water-Bearing Materials with a Bibliography on Permeability and Laminar Flow by V. C. Fishel," U.S. Geological Survey Water-Supply paper 887, U.S. Government Printing Office, Washington, D.C., pp. 20–50 (more than 400 references).

CHAPTER THREE

SEEPAGE PRINCIPLES

3.1 NEED FOR KNOWLEDGE OF SEEPAGE PRINCIPLES

All engineers who work with the design of civil engineering structures must understand the meaning and importance of the fundamental equilibrium equations: $\Sigma H = 0$; $\Sigma V = 0$; and $\Sigma M = 0$, as all of the forces and overturning moments acting on a structure must add up to zero, otherwise it would not be at rest. In these equations, ΣH is the summation of all horizontal forces, ΣV is the summation of all vertical forces, and ΣM is the summation of all overturning moments.

Another equation, $Q = kiA$ (Darcy's law) can be as important as the equilibrium equations for structures subject to detrimental forces of water (see Sec. 3.2). Hence, a thorough knowledge of seepage principles is a necessity for all engineers designing structures which can be damaged by water pressures or seepage forces. This chapter explains Darcy's law and other principles governing the behavior of water as it affects the safety and economy of engineering works. The emphasis is on practical applications to everyday problems rather than on complex theory.

Every soil and rock mass, whether formed by nature or modified by man, has a specific but often complex structure with definite water-conducting capabilities. In developing solutions to seepage and drainage problems, it is necessary to work with simplified cross sections that hopefully are reasonably representative of the real-life sections. To a degree every cross section analyzed is an approximation. We must try to make the best possible determination of the real water-transmitting properties of formations and masses that will influence flow. Then, with the judicious application of seepage principles, we have the best probability of developing structures and water-control systems that will perform as expected.

Seepage principles offer a way to understand the basic nature of seepage and drainage and ways to develop the right kind of solution to a given problem. The U.S. Army et al. (1971) say, "An approximate solution to the right problem is more desirable than a precise solution to the wrong problem." In the past, when reasonable evaluations were not made of factors influencing seepage behavior, damage and even failure of structures was more the rule than the exception. Designers often did not know the *order of magnitude* of a problem. Helping to avoid serious mistakes by providing ways to develop realistic solutions is a primary objective of this book.

A good understanding of fundamental seepage and drainage principles will help engineers avoid serious mistakes in selecting the *best kinds of systems to control seepage and groundwater.* As an example, the most efficient way to drain water out of many kinds of facilities (such as described in Chap. 9) is by *vertically downward flow* into a system that can freely discharge the water to safe exits. This includes pavements of all kinds, saturated earth slopes, saturated mining waste dumps, fly-ash dumps, dredger sludges, etc. *Downward drainage* or *bottom drainage* is the most effective kind when it can be employed because it works *with* Nature, not *against* Nature. A careful study of the principles described in this book will enable engineers to understand why bottom drainage can be so effective.

Anyone who is alert and observing can find unlimited opportunities to see the differences between predictions and actual behavior, because every dam, spillway, retaining wall, below-water structure, and pavement is a real, full-scale experimental model whose performance should reveal the accuracy or inaccuracy of assumptions and methods of analysis.

3.2   DARCY'S LAW

Laminar Flow and Darcy's Law

In Chapter 2 the interdependence of Darcy's law and permeability is described. Purely for purposes of convenience, *permeability* is defined as a factor with units of a velocity, which when multiplied by the total cross-sectional area of soil, gives the seepage quantity. Not only does Darcy's law provide a means for determining permeability (Sec. 2.2), but it has a great many other practical and theoretical uses. This simple relationship may have a number of forms:

$$Q = kiA$$

$$Q = kiAt$$

$$q = kiA = v_d A$$

$$v = ki = \frac{q}{A} = v_d \qquad (3.1)$$

$$v_s = \frac{ki}{n_e}$$

In these expressions Q or q is the seepage quantity, t is time, k is Darcy's coefficient of permeability, i is the hydraulic gradient $= \Delta h/\Delta l = h/L$, A is the total cross-sectional area normal to the direction of flow, including both void spaces and solids, v_d (or v) is the discharge velocity, and v_s is the seepage velocity. The effective porosity n_e is the ratio of the actual volume of pore spaces through which water is seeping to the total volume. Usually it is somewhat less than the true porosity. Limitations in the range of validity of Darcy's law are described in Chapter 2 (Sec. 2.3). Descriptions are also given of its use in determining permeability in the laboratory (Sec. 2.4) and in the field (Sec. 2.6 and 2.7).

Newton's law of friction (Sec. 2.3), together with the classical Navier-Stokes equations of hydrodynamics, provide a basis for studying the behavior of fluids under motion in porous media. Unfortunately the resulting equations become so complex that their solutions are impracticable. Recognizing these problems, Darcy (1856) resorted to experimental study of the flow of water through sand filters and developed a quantitative representation of the behavior of fluids flowing through porous media (see also Sec. 2.3). His classic experiments led to the simple relationships (Eq. 3.1) that eliminate the errors inherent in the theoretical methods.

Although mathematical representations of the behavior of fluids in porous media are for practical purposes completely unsurmountable, the theory of dimensions may be drawn on (Muskat, 1937) to understand more fully the "law of flow" expressed by Darcy. Applying rules of this theory, Muskat obtained the following type of expression:

$$\frac{\Delta h}{\Delta l} = \text{constant } \frac{\mu^2}{\gamma d^3} F\left(\frac{dv\,\gamma}{\mu}\right)$$

The argument $dv\,\gamma/\mu = \text{Re}$ of the function F is a well-known term in hydrodynamics relating to the flow of fluids through pipes. If d represents the pipe diameter, $dv\,\gamma/\mu$ is the Reynolds number. For low velocities, low densities, or small pipe diameters the function F equals its argument; hence

$$\frac{\Delta h}{\Delta l} = \text{constant } \frac{\mu v}{d^2} \qquad (3.2)$$

Muskat concludes that this expression represents linear "viscous flow." It is seen that the average velocity v is proportional to the hydraulic gradient $\Delta h/\Delta l$. Likewise, the gradient is proportional to fluid viscosity μ.

As values of γ, d, v, or $1/\mu$ increase and the Reynolds number increases beyond a critical value, around 2000, flow in pipes suddenly changes from a smooth, orderly, streamline condition to one of "turbulence" in which fluctuating, irregular eddies develop. Muskat points out that for this kind of flow the function F becomes proportional to the square of its argument, and Eq. 3.2 has the form

$$\frac{\Delta h}{\Delta l} = \text{constant } \frac{\gamma v^2}{d} \tag{3.3}$$

Here the gradient is independent of viscosity but proportional to the square of the velocity, whereas for streamline flow it is directly proportional to velocity.

Fundamental considerations of the nature of flow in porous media have led investigators to conclude that Darcy's law of proportionality of *macroscopic* velocity and hydraulic gradient is an accurate representation of the "law of flow" as long as velocities are low. Although it is generally concluded that the range of validity cannot be definitely established, Darcy's law is widely considered to be infinitely superior to methods that adhere strictly to basic laws but because of their complexity are beyond practical application.

Muskat (1937a), Taylor (1948), Leonards (1962), and others have presented excellent discussions of permeability and Darcy's law. Taylor (1948a) points to the slow transition in soils from purely laminar flow to a slightly turbulent state and concludes that under a hydraulic gradient of 100% uniform soils with a grain size of 0.5 mm or less always have laminar flow. For a gradient of 800% the diameter is 0.25 mm. This admittedly is a conservative approximation, based on a Reynolds number of 1.0.

Jacob (1950) concludes from experiments with natural and artificial sands of nearly uniform spherical grains that the transition from laminar to turbulent flow requires at least a thousandfold increase in velocity to reach the limit of fully established turbulence. He states that, ". . . a tenfold increase above the approximate critical velocity results in about 50 percent error in the hydraulic gradient as predicted by Darcy's law."

Fishel (1935) reports that experiments with very low heads indicate that, "for the material tested (Fort Caswell sand) the rate of flow varies directly as the hydraulic gradient, down to a gradient of 2 or 3 inches to the mile, and there are indications that Darcy's law holds for indefinitely low gradients."

In the analysis of seepage in coarse sands and gravels Darcy's law is not strictly applicable. Forchheimer (1902) found the frictional resistance of pervious gravel to be

$$\frac{\Delta h}{\Delta l} = \frac{1.77}{10^3} V + \frac{3.18}{10^4} V^2 \tag{3.4}$$

In Eq. 3.4 V is the velocity in meters per day.

The general form of Eq. 3.4 is

$$\frac{\Delta h}{\Delta l} = aV + bV^2 \tag{3.5}$$

According to Eq. 3.4 and 3.5, head losses in gravels are greater than indicated by Darcy's law. If permeability tests can be made under conditions similar to those that will exist in a prototype, the errors will tend to be neutralized; however, this is not always possible. It is therefore desirable to allow liberal factors of safety in the design of drainage systems that contain coarse, clean aggregates in which semiturbulent or turbulent flow may develop (see also Sec. 3.7).

Applications of Darcy's Law

Applications of Darcy's law to permeability determinations are described in Chapter 2.

The validity of Darcy's law is an essential assumption in the following soil mechanics theories and methods.

1. The theory of consolidation of clays.
2. Quantitative theory of laminar flow of homogeneous fluids through porous media.
3. Practical solutions to Laplace equations by flow nets.

The validity of Darcy's law is an essential assumption for all laminar flow solutions presented in this text:

1. Flow nets for steady seepage through earth cross sections of one or more different permeabilities for both isotropic and anisotropic conditions (Darcy's law enters into the derivation of the basic differential equation).
2. Calculations involving the *velocities* of masses of water in porous media under steady seepage conditions. (These computations involve the *seepage velocity* $v_s = ki/n_e$.) The *seepage velocity* of moving groundwater can be used as an index of permeability (Sec. 2.7); its magnitude in any water-bearing material or drainage layer is a useful criterion of the rate of movement of water.
3. Approximate nonsteady seepage applications of the flow net to moving saturation lines. (These computations involve the additional use of the *seepage velocity* $v_s = ki/n_e$, which depends on Darcy's law.)
4. Calculations for seepage quantities through saturated soil and rock formations and other porous media. (These determinations involve the *discharge velocity* $v_d = ki$, determined from Darcy's law.)

5. Determination of the discharge capacities of porous aggregate drains, chimney drains, and sand-filled wells. (These determinations also make use of the *discharge velocity*, defined in Sec. 3.4). Methods of correcting for turbulent flow are given in Sec. 3.7.

The relationships represented by Darcy's law, though simple, represent some of the most powerful tools available to the soils engineer and the drainage engineer. Unfortunately their great benefits are not always realized. Practical examples of the application of Darcy's law in the solution of countless everyday engineering problems relating to seepage and drainage are to be found throughout this book. For instance, simple calculations given in this new edition help explain why serious groundwater contamination problems went on so long without being noticed (see Sec. 11.1).

3.3 FLOW NETS

Basic Solutions to Seepage Problems

The flow of water through soil is but one of a number of forms of streamline flow that obey similar fundamental relationships and can be represented by the Laplace equation. Muskat (1937) demonstrates that the hydrodynamics of steady-state fluid flow through porous media follows the same basic laws as the problems of steady-state heat flow, electrostatics, and current flow in continuous conductors. Other types of problem, such as certain cases in the theory of torsion of elastic rods and the flow of viscous liquids, are also governed by Laplace's equation.

To develop the Laplace equations for flow of water through porous media, let the following be assumed:

1. The soil is homogeneous.
2. The voids are completely filled with water.
3. No consolidation or expansion of the soil takes place.
4. The soil and water are incompressible.
5. Flow is laminar and Darcy's law is valid.

It can then be shown (Terzaghi, 1943) that the quantity of water entering an element must equal that leaving and the equation of continuity takes the form

$$\frac{\partial u}{\partial x} + \frac{\partial v}{\partial y} + \frac{\partial w}{\partial z} = 0 \qquad (3.6)$$

The terms u, v, and w are discharge velocity components in directions x, y, and z in cartesian coordinates.

According to Darcy's law ($v_d = ki$), the components of the discharge velocity are

$$u = -k\frac{\partial h}{\partial x}, \qquad v = -k\frac{\partial h}{\partial y}, \qquad w = -k\frac{\partial h}{\partial z}$$

Substituting in Eq. 3.6,

$$\partial\frac{-k(\partial h/\partial x)}{\partial x} + \partial\frac{-k(\partial h/\partial y)}{\partial y} + \partial\frac{-k(\partial h/\partial z)}{\partial z} = 0 \qquad (3.7)$$

and if k is a constant

$$\frac{\partial^2 h}{\partial x^2} + \frac{\partial^2 h}{\partial y^2} + \frac{\partial^2 h}{\partial z^2} = 0 \qquad (3.8)$$

This is the common form of the Laplace equation for three-dimensional flow of water through porous media. In two dimensions it has the form

$$\frac{\partial^2 h}{\partial x^2} + \frac{\partial^2 h}{\partial z^2} = 0 \qquad (3.9)$$

Equation 3.9 may be represented by two families of curves that intersect at right angles to form a pattern of "square" figures known as a *flow net*. One set of lines is called the *streamlines* or *flow lines*, the other the *equipotentials*. The flow lines represent paths along which water can flow through a cross section. The equipotential lines are lines of equal energy level or head.

Although mathematical solutions have been developed for a number of cases of flow, the solutions are cumbersome and often approximate. Leliavsky (1955) says, ". . . the analytical method, although rigorously precise, is not universally applicable, because the number of known functions on which it depends is limited. Moreover, except in a few elementary cases, the analytical method lies beyond the mathematics of practising design offices."

For those who wish to pursue the mathematical approach Harr (1962) presents an interesting treatment of seepage theory, giving both rigorous and approximate mathematical solutions to seepage under weirs and other structures. He describes conformal mapping, the velocity hodograph, and other special mapping techniques, such as the Zhukovsky functions and the Schwarz-Christoffel transformations.

Solutions to flow problems are also obtained by the use of the *electric analogy*, small-scale *viscous fluid* models, (see Fig. 3.27), sand models, and trial-and-error sketching methods. Sand models, with colored dye inserted, give a graphic picture of seepage that is convincing to the skeptic (see Fig. 3.2). The electric analogy is adaptable to two- and three-dimensional problems.

Mathematical solutions to practical multi-permeability cross sections can be extremely complex and difficult to interpret and are little more than fancy exercises unless they can be put to practical use. Computerized solutions to the Laplace equation are being developed, and some very good ones are available to those having access to the programs. Biedenham and Tracy (1987) have prepared a technical report, No. ITL-87-6, a Users Guide Finite Element Method Seepage Package for solving certain steady-state problems. Inquiries should be made to the Waterways Experiment Station, U.S. Army Engineer Division, Vicksburg, Miss., 39180-0631. Using the program requires knowledge of computerized solution methods and thorough familiarity with the details for setting up individual problems. Other programs have been developed. At the present time (1988) there is no clear-cut source of information on all programs available and under what conditions, so anyone wishing to make use of programs must do some searching on his or her own to try to determine which are available and under what terms.

In a thorough review of numerical methods and computer solutions in ground engineering, Christian (1987) points out that there are a number of computer programs available to the general user for solving problems of fluid flow. As long as non-linear properties or phreatic surfaces are not involved, general-purpose finite element programs can be used, he says. Many of these programs are capable of treating problems of thermal conductivity, which is mathematically identical to the problem of flow through porous media when expressed in terms of total head. He also points out that a large number of programs have been written specifically for flow through porous media and many are available from geological surveys or other groups working with groundwater hydrology. Under "Conclusions" he says, "The engineer contemplating employing numerical methods should be familiar with the actual physical problem to be investigated, the numerical techniques to be used, and the limitations and capabilities of the software." He emphasizes that if a person contemplating numerical solutions is not familiar with all of these aspects, "he should consult someone who is."

Christian also says that numerical and computer methods have become so complicated that they "begin to form a discipline within the larger engineering field." While problems exist in the preparation of input to and interpreting output from computer programs, and shortcomings in reliable, documented software exist, ground engineers can expect important developments (in these methods) over the next few years, according to Christian.

Weber and Hassan (1972) discuss methods of developing models to simulate the behavior of groundwater basins under changed conditions brought on by increased pumping or other management changes. One approach divides an entire groundwater basin into a mesh of subareas in which transmission and storage factors can be estimated. Based on a knowledge of past behavior patterns, these models predict changes under the different conditions that may develop in the future.

When a problem is set up for electronic computer or model analysis, exactly repeatable answers are obtained. It must be remembered, however, that the answers are no better than the input. Useful though these methods may be the graphic sketching methods for solving the Laplace equation are preferred by many, because they are so extremely versatile. Essentially all of the flow nets presented in this book were developed by sketching methods. Step-by-step procedures for sketching flow nets are given in Chapter 4.

Any flow net must meet certain requirements, of which the following are basic:

1. Flow lines and equipotential lines must intersect at right angles to form areas that are basically squares.
2. Certain entrance and exit requirements must be met.
3. A basic deflection rule must be followed in passing from a soil of one permeability to a soil of a different permeability.
4. Adjacent equipotentials have equal head losses.
5. The same quantity of seepage flows between adjacent pairs of flow lines.

The last two are fundamental requirements *indirectly* entering into the construction of flow nets. Another requirement may be stated: the quantity of seepage flowing through a section must be constant throughout the section unless additional water enters or some is removed by drains. It will be seen that in dams and other cross sections in which seepage emerges on a slope a portion of the flow net is cut off and this rule does not apply.

At first glance the sketching method may appear to present obstacles to a beginner, but once the initial obstacles are overcome one seldom resorts to other means of solving seepage problems. The frequent, thoughtful examination of well-constructed flow nets is helpful in becoming familiar with their general shapes. Numerous examples may be found in this text.

Flow Lines and Equipotential Lines

When water is forced through porous soil, particles of water flow through all of its interconnected pore spaces. Although the number of individual paths is nearly infinite, the flow lines drawn in any flow net are but a few of the many paths that water can take in flowing through a cross section. The nature of flow lines is illustrated in simple form in Figure 3.1. A small volume of sand is contained within a horizontal tube, to which reservoirs are attached at both ends. If a difference in head h of any magnitude exists, water will flow through the soil from left to right. If several small diameter tubes are inserted a small distance into the sand at the left reservoir and dye is fed into these tubes, colored streaks, marking the *flow lines*, will form. An infinite number of lines is possible. Obviously, if dye is inserted at a great many locations, the whole section will become colored and the individual paths will be lost.

In Figure 3.1 the flow lines are straight because the confining boundaries

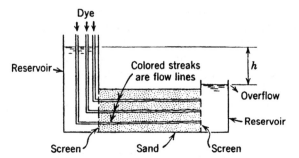

FIG. 3.1   Illustration of flow lines.

are parallel. In most flow problems, however, both flow lines and equipotential lines are curved. An example of a curved dye line obtained in a model is shown in Figure 3.2. Numerous examples of flow nets made up largely of curved lines are to be found throughout this book.

The second kind of line in a flow net, the *equipotential line,* is illustrated in Figure 3.3. As water flows through the small mass of sand from left to right the energy is gradually consumed by friction. If small piezometers are inserted at several positions along the tube of sand, water will rise to line A–A '. Thus in the vertical plane above point 1 it rises to $a,$ any position above point 2 it rises to $b,$ and at point 3 it rises to level $c.$ In the vertical plane above point 2, piezometers are installed at three elevations within the tube of sand; however, water rises to exactly the same level in all three piezometers. The rule here is that *water rises to the same level in piezometers installed anywhere along a given equipotential line.* This must be, because *the same energy level exists everywhere along a given equipotential line.*

In Figure 3.3 the equipotential lines are straight lines. As with the flow lines in Figure 3.1 an almost infinite number could be selected; to define a flow net, however, only enough lines are selected to give a readable pattern. Too many

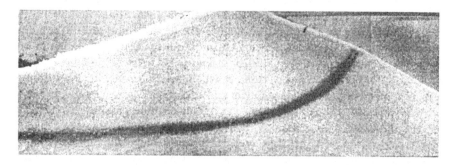

FIG. 3.2   Model test on Ottawa sand showing curved flow line.

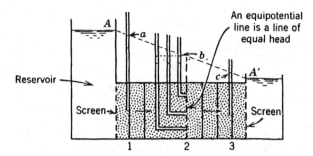

FIG. 3.3 Illustration of equipotential lines.

lines simply confuse things. On the other extreme, if too few are used, detail may be lost; additional detail can always be secured in important areas by subdividing squares into smaller squares (Fig. 3.5).

If the flow lines of Figure 3.1 are combined with the equipotential lines of Figure 3.3 a simple flow net is obtained (Fig. 3.4). Though this flow net is extremely simple, it possesses the properties of a true flow net. The total net head h is divided equally among the nine equipotential drops, and the shape factor (Sec. 3.4) is equal to $n_f/n_d = 4/9 = 0.44$. The flow lines intersect the equipotential lines at right angles and the net is composed of squares.

Figures 3.1 and 3.3 illustrate the basic meaning of flow lines and equipotential lines and Figure 3.4 shows a simple flow net. Most flow nets are composed of curves and the "square" figures are not true squares but *curvilinear* squares. *The requirement that a net must be composed of square figures is met if the average width of an area is equal to its average length.* Thus, in the portion of a flow net shown in Figure 3.5 the distance between points 9 and 10 must be equal to the distance between points 11 and 12. As the sizes of larger "squares"—such as 1-2-3-4--are subdivided into smaller and smaller figures, they look more and more like true squares.

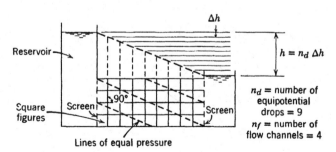

FIG. 3.4 Simple flow net.

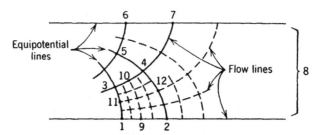

FIG. 3.5 Subdivision of peculiar "squares" provides check on accuracy of original lines. (After A. Casagrande, *Seepage Through Dams,* 1937.)

Boundary Conditions

Casagrande (1937) describes important conditions that must be met by flow nets at points of entry, discharge, and transfer across boundaries between dissimilar soils. If flow is beneath a sheet pile wall or through the foundation of a dam or other structure impounding water, all boundary conditions are fixed, but if the seepage is through an earth dam, levee, or other embankment the *upper boundary* or upper *line of seepage* is not known in advance of the flow net construction. At the uppermost line of seepage equipotential lines must intersect the free water surface at equal vertical intervals (Fig. 3.6). This requirement permits determination of the free surface or uppermost line of seepage *simultaneously while a flow net is being constructed.* The step-by-step procedure is shown in Chapter 4.

Important entrance, discharge, and boundary conditions that must be met by flow nets are given in Figure 3.7. An important condition existing at the boundaries between soils of different permeabilities is discussed in detail in the following section.

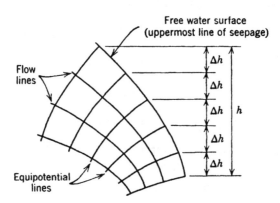

FIG. 3.6 General condition for line of seepage. (After A. Casagrande, *Seepage Through Dams,* 1937.)

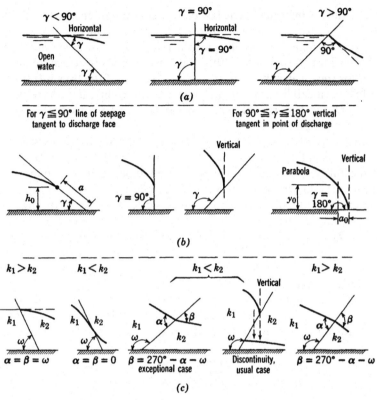

FIG. 3.7 Entrance, discharge, and transfer conditions of line of seepage. (*a*) Conditions for point of entrance of line of seepage. (*b*) Conditions for point of discharge of line of seepage. (*c*) Deflection of line of seepage at boundary between soils of different permeability. (After A. Casagrande, *Seepage Through Dams,* 1937.)

Flow through Sections of More Than One Permeability

Many practical seepage and drainage problems can be studied by constructing flow nets for sections with a single permeability. In some instances, however, a great deal can be learned by studying seepage patterns in cross sections with soils of more than one permeability. Surprisingly, such studies have revealed important shortcomings in some commonly accepted beliefs about seepage and drainage. *The study of composite cross sections is one of the most worthwhile and rewarding applications of the flow net.* At first glance flow nets for sections with more than one permeability may appear rather formidable, but attention to the transfer conditions at boundaries between soils of different permeabilities (Eqs. 3.10 and 3.11) greatly facilitates their construction.

When water flows across a boundary between dissimilar soils, the flow lines bend much in the way that light rays are deflected in passing from air into water or from air into glass. The law of conservation of energy forces all natural phenomena to take the line of least resistance. Thus, when water flows from a soil of high permeability into a material of lower permeability, the pattern develops in such a way that the flow remains in the more permeable material for the greatest possible distance. Likewise, if the flow is from a material of low permeability into one of higher permeability, it deflects as soon as possible into the material of higher permeability. To conserve energy water seeks the easiest paths to travel.

Another way of looking at the behavior of seepage in sections with more than one permeability is the concept that, other factors being equal, the higher the permeability, the smaller the area required to pass a given volume of water. Conversely, the lower the permeability, the greater the area required.

In relation to the amount of energy needed to force water through porous media, the higher the permeability, the lower the energy needed, and vice versa. In seepage the rate of loss of energy is measured by the steepness of the hydraulic gradient; steep hydraulic gradients should be expected in the zones of low permeability, and flat gradients in zones of high permeability. This, in fact, does take place.

The way flow lines deflect when they cross boundaries between soils of different permeabilities is shown in Figure 3.8. The flow lines bend to conform to the following relationship:

$$\frac{\tan \beta}{\tan \alpha} = \frac{k_1}{k_2} \tag{3.10}$$

Simultaneously, the areas formed by the intersecting lines either elongate or shorten, depending on the ratio of the two permeabilities, according to the following relationship:

$$\frac{c}{d} = \frac{k_2}{k_1} \tag{3.11}$$

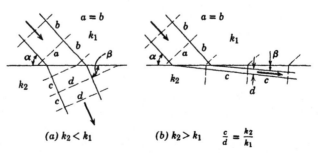

(a) $k_2 < k_1$ (b) $k_2 > k_1$ $\frac{c}{d} = \frac{k_2}{k_1}$

FIG. 3.8 Transfer conditions at boundaries between soils of different permeabilities. (After A. Casagrande, *Seepage Through Dams*, 1937.)

This relationship can easily be remembered by thinking in terms of the energy required to force water through soil. *When water flows from a soil of low permeability into a soil of higher permeability, the rectangles must elongate because less area is required to accommodate the same quantity of water and lower gradients are needed.* If the flow is from high permeability into low permeability, the figures must shorten because steeper gradients are required and relatively more area is needed to accommodate the flow.

In Figure 3.8*a* the second permeability is lower than the first; hence *shortened* rectangles are formed in the second. In Figure 3.8*b* the second permeability is greater than the first; so *elongated* rectangles form in the second material. The deflection relationship expressed by Eq. 3.10 should be kept in mind; however, the change in the shapes of areas as expressed by Eq. 3.11 is extremely useful, as it provides an exact check of the accuracy of flow nets for sections with more than one permeability. *When constructing flow nets for sections with more than one permeability, it is desirable to measure lengths and widths of figures regularly with an engineer's scale* to be sure that this fundamental requirement is being satisfied.

Study of a number of the flow nets in this book will increase understanding of the meaning of the relationships just described.

Figure 3.9 is a simplified illustration of seepage through soils of different permeabilities. A horizontal tube is filled with two kinds of sand. From A to B and from C to D sand number 1 is used, but in the central part, from B to C, a finer sand is placed. Flow takes place from left to right. The permeability of sand number 2 is 1/20 or 0.05 times that of sand number 1; hence, in crossing the boundary, equipotentials must squeeze together in the lower permeability sand to 1/20 of the spacing of equipotentials in the more permeable sand on either end. If each pair of equipotential lines in the outer zones produces one equipotential drop, there will be five drops from A to B and five from C

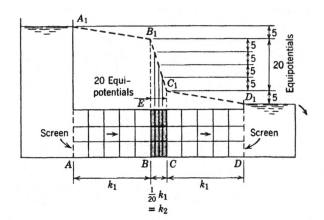

FIG. 3.9 Simple two-permeability flow net.

to D, adding up to 10 in the more pervious sand. The single column of squares between B and C in the less pervious sand must use up 20 equipotential drops, making the overall total 30. The amount of head used up between any two adjacent equipotentials is 1/30, the total differential head h. If piezometers were installed in the tube, the head would reduce in accordance with line A_1-B_1-C_1-D_1. Two-thirds, or 20/30 of the total head would be used up in the narrow middle zone of low permeability sand.

If steady flow is maintained in soil and rock formations when distinct differences in permeabilities exist, little of the energy can be consumed in the high permeability materials. This fact accounts for the manner in which minor geological details, such as fissures and open cracks, can cause serious uplift conditions unless plugged by grout or the seepage is controlled by drainage.

Under flow conditions similar to those in Figure 3.9 no deflection of flow lines can take place, for the approaching flow lines are perpendicular to the boundary (line B-E). Furthermore, all flow lines are forced by the boundary conditions to be straight lines parallel to the sides of the tube.

A similar condition sometimes occurs in portions of cross sections being studied by flow nets. The observation that natural boundaries restrict the pattern of flow often aids in the construction of flow nets.

An important rule for flow-net construction, which gives an added check on basic correctness, is this: *In any flow net the number of flow channels must remain the same throughout the net.* When this rule is followed, a flow net must also satisfy the basic requirement that it be *composed entirely of squares.* In the construction of flow nets for composite sections Eq. 3.11 furnishes an additional important check on the correctness of the work. Section 4.6 describes in more detail the ways that the accuracy of the construction of multi-permeability flow nets may be checked.

Anisotropic Soil Conditions

The Laplace equation (Eq. 3.9) for two-dimensional flow is based on the assumption that permeabilities are equal in the x (horizontal) and z (vertical) directions. Most compacted embankments and many natural soil deposits are more or less stratified, often with horizontal beddings that make horizontal permeabilities much greater than the vertical. By recalling that, according to Darcy's law, discharge velocities are proportional to permeabilities Eq. 3.9 for two-dimensional flow can be written

$$k_x \frac{\partial^2 h}{\partial x^2} + k_z \frac{\partial^2 h}{\partial z^2} = 0$$

To reduce this expression to a Laplace equation the coordinate system must be transformed to that of $\bar{x}, \bar{y}, \bar{z}$, defined as $\bar{x} = x/\sqrt{k_x}$; $\bar{y} = y/\sqrt{k_y}$; $\bar{z} = z/\sqrt{k_z}$; it then follows that

$$\frac{\partial^2 h}{\partial \bar{x}^2} + \frac{\partial^2 h}{\partial \bar{z}^2} = 0 \tag{3.12}$$

Equation 3.12 expresses the anisotropic seepage condition as a Laplace equation. To construct flow nets for this condition *it is necessary only to shrink the dimensions of a cross section in the direction of greater permeability.* If the average vertical permeability is expressed as k_v and the horizontal as k_h, the natural horizontal distances are multiplied by $\sqrt{k_v/k_h}$, as shown in Figure 3.10b.

If the horizontal permeability is greater than the vertical, the transformed section will always be shrunk to a narrower horizontal dimension. If the reverse were true, it would be lengthened horizontally.

Terzaghi (1943a) proves that if a foundation is composed of several layers having individual permeabilities $k_1, k_2 \ldots k_n$, the effective permeability parallel to the bedding planes k_I is

$$k_I = \frac{1}{H}(k_1 H_1 + k_2 H_2 + \cdots + k_n H_n) \tag{3.13}$$

And for flow perpendicular to the bedding planes, the effective permeability is

$$k_{II} = \frac{H}{\dfrac{H_1}{k_1} + \dfrac{H_2}{k_2} + \cdots + \dfrac{H_n}{k_n}} \tag{3.14}$$

To obtain a flow net for anisotropic soil conditions, the cross section is redrawn to a reduced horizontal scale, as indicated in Figure 3.10b. On this section, which is the *transformed* section, the flow net is constructed in the ordinary manner. Having obtained the flow net, it is then reconstructed on the cross section drawn to the natural scale. Typical flow nets on transformed and natural sections are given in Figures 3.11, 3.22, and 6.18.

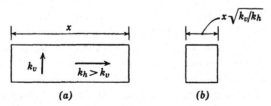

FIG. 3.10 Adjustment in scale to compensate for anisotropic soil conditions. (*a*) Natural section (redraw flow net here). (*b*) Transformed section (draw flow net here).

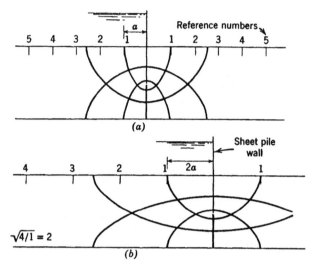

FIG. 3.11 Flow nets constructed on transformed section and redrawn on natural sections (special case). (*a*) Natural section for $k_h = k_v$, also is transformed section for anisotropic conditions, which is true only for cross sections of infinite width. (*b*) Natural section for flow net in (*a*) for $k_h = 4k_v$.

Seepage quantities can be calculated by counting the number of flow channels and equipotentials in the usual manner and using $\bar{k} = \sqrt{k_h k_v}$ as the effective permeability. The shape factor is the same on both natural and transformed cross sections.

Hydraulic gradients for anisotropic conditions must be determined in flow nets redrawn on sections at the natural horizontal scale because the true distance over which a given amount of head is used can be measured only on a true section.

This discussion emphasizes a point that must not be overlooked in the construction of flow nets for anisotropic soil conditions: *The section must be transformed before the flow net is constructed.* An easy way to check a finished flow net is to take a look at the net and see if it appears reasonable, remembering that stratification tends to pull or stretch the flow net in the direction of greater permeability. *On the natural cross section the flow net will not be composed of squares but of rectangles elongated in the direction of greater permeability.* This check should always be made to determine that a mistake has not been made. It is easy to become confused.

A cross section can also be transformed by holding the horizontal dimensions to the natural scale and multiplying the vertical distances by the factor $\sqrt{k_h k_v}$. This procedure usually produces a cross section of impractically large size; however, it will be geometrically similar to one obtained by the first procedure.

Horizontal stratification tends to stretch seepage out horizontally. In homogeneous types of dam this is highly undesirable, as it raises the level of saturation in the downstream area. Its influence on saturation in earth dams and levees is shown in Figures 6.17 and 6.18.

3.4 SEEPAGE QUANTITIES

In hydraulic flow in pipes and open channels the quantity of flow is

$$q = Av$$

In this expression A is the cross-sectional area and v, the *mean* velocity.

In the flow of water through porous media the quantity of seepage can also be expressed as

$$q = Av \tag{3.15}$$

Here, A is the cross-sectional area and v, the *discharge* velocity. The discharge velocity is

$$v = ki \tag{3.16}$$

By combining Eqs. 3.15 and 3.16 the quantity of seepage in the soil is

$$q = kiA \tag{3.1}$$

which is a familiar form of Darcy's law (Sec. 3.2). With Darcy's law seepage can be estimated if at a common location k, i, and A are known or can be estimated with reasonable accuracy.

Numerous practical examples of the use of Darcy's law in the study of seepage in soils and both aggregate drains and geotextile drains are given in this book.

Figure 3.12 illustrates an agricultural application of Darcy's law to a seepage quantity calculation. Assume that each of a system of sprinklers (Fig. 3.12a) irrigates 400 sq ft of ground. If the vertical permeability of the soil is 1 ft/day, what maximum rate of sprinkler discharge q can be continuously absorbed by the soil? Assume that the surface is kept continuously wet and consider only gravitational forces.

As shown in section A-A (Fig. 3.12b), the depth of penetration of saturation l just equals the hydraulic head h that is forcing water downward into the soil. If capillarity is neglected, the hydraulic gradient i is equal to h/l, or unity, and the capacity of 400 sq ft of soil to absorb water is $q = kiA = 1$ ft/day(1.0) (400 sq ft) $= 400$ cu ft/day $= 2.1$ gpm. Hence, for the assumed conditions, sprinklers discharging more than 2 gpm on 400 sq ft of area would waste water.

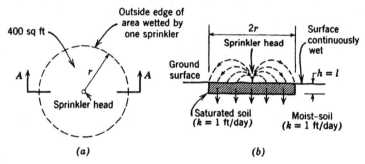

(a)   (b)

FIG. 3.12 Study of seepage quantity by Darcy's law. (a) Plan of area wetted by a sprinkler. (b) Section A-A.

With Darcy's law a simple equation can be derived that permits seepage quantities to be calculated from flow nets. Referring to Figure 3.13, which gives a flow net for seepage under a weir, let the number of flow channels equal n_f, the number of equipotential drops equal n_d, and the quantity flowing between any two adjacent flow lines equal Δq. In the flow net the total quantity $q = \Sigma \, \Delta q = n_f \, \Delta q$.

Anywhere in the flow net $i = \Delta h/\Delta l$; hence in any square between points a and b in Figure 3.13

$$\Delta q = k \frac{\Delta h}{\Delta l} a$$

Because a flow net must be composed of squares, the width a of a square must be equal to Δl, its length; hence

$$\Delta q = k \frac{\Delta h}{a} a = k \, \Delta h \tag{3.17}$$

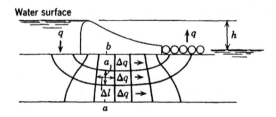

FIG. 3.13 Flow net for seepage under a weir (used in deriving Eq. 3.18).

However, $\Delta h = h/n_d$, so

$$\Delta q = k \frac{h}{n_d}$$

but $q = \Delta q n_f$, hence

$$q = kh \frac{n_f}{n_d} \qquad (3.18)$$

Equation 3.18 is the basic equation for computation of seepage quantities from flow nets. The ratio n_f/n_d is called the *shape factor*. From a flow net the seepage quantity thus is the product of the permeability, the net head, and the *shape factor*. Equation 3.18 applies for isotropic soil conditions. (Horizontal and vertical permeabilities are equal.) If the soil is stratified and the horizontal permeability k_h is greater than vertical permeability k_v, the seepage quantity is estimated by using the effective permeability $\overline{k} = \sqrt{k_h k_v}$. For anisotropic conditions the seepage quantity therefore is

$$q = \overline{k}h \frac{n_f}{n_d}$$

or

$$q = \sqrt{k_h k_v} \, h \frac{n_f}{n_d} \qquad (3.19)$$

If a flow net is for a cross section with more than one permeability, care must be taken to use consistent shape factors and permeabilities. The permeability used for seepage quantity calculations must be in the material in which true squares, not elongated or foreshortened rectangles, are constructed. Thus, if the quantity flowing through the tube in Figure 3.9 were to be computed with k_1, the number of equipotentials would be $5 + 20 + 5$, or 30, and the seepage quantity would be $k_1 h \, n_f/n_d = k_1 h \, 3/30 = 1/10 \, k_1 h$. If the quantity were computed with k_2, the number of equipotential drops would be $5/20 + 1 + 5/20 = 1.5$. Now the seepage quantity would be $k_2 h \, n_f/n_d = k_2 h \, 3/1.5 = 2k_2 h$. These two determinations produce the following quantities:

1. Using k_1: $q = 1/10 k_1 h$
2. Using k_2: $q = 2k_2 h$

but $k_2 = 1/20 k_1$

Substituting in (2), $q = 2k_2 h = 2(1/20 k_1)h = 1/10 k_1 h$, which is identical to the answer obtained by using k_1.

3.5 SEEPAGE FORCES AND UPLIFT PRESSURES

General

Seepage theory and drainage practice, as developed in this book, deal almost entirely with "free water." This is the water that is neither in capillary tension nor under excess pore pressures in partly consolidated soils.

Free water trapped behind retaining walls, under pavements, or under numerous other "impervious" types of structure may cause serious damage due to hydrostatic pressures.

Engineering works can also be damaged by the forces of seeping water that are transferred to the soil grains by friction.

The Energy of Water under Pressure

Water is a relatively heavy fluid, nearly 100% incompressible, and has almost no shearing strength. These properties combine to make water a troublesome fluid since it can penetrate readily into the tiniest crack and exert tremendous forces.

According to the basic equation of hydrostatics,

$$h = \frac{p}{\gamma_w} + z \tag{3.20}$$

In this expression h, the *energy head,* is equal to the *position head z,* plus the *pressure head* or *piezometric head p/γ_w.* The term γ_w is the density of water.

According to hydraulic principles, the pressure at any point in a liquid under gravity depends on the depth of the point below the surface, the density of the liquid, and the pressure on its surface. When atmospheric pressure exists equally everywhere on a given system, it can be ignored. If the velocity head is small (as when water flows in porous media), it also can be neglected. The pressure at any point, therefore, is equal to $h\gamma_w$ and the pressure head is p/γ_w.

The piezometric head is the height to which water eventually rises in small diameter wells or *piezometers.* The piezometric head is a measure of the energy level or energy head, as defined above, excluding the position head.

If a piezometer is inserted into a body of earth, the level to which free water rises is a measure of the energy head at the piezometer tip. It is a measure of the potential energy at that point. If several piezometers are inserted in the soil beneath the bottom of a lake (Fig. 3.14) water rises to the same level in all of the piezometers—to the level of the lake. Because the energy level is the same at all points, the water in the soil is motionless. There is no seepage because seepage in soil occurs only if a difference in energy level exists between points. It always flows from a point of higher energy to a point of lower energy.

It is important to distinguish between potential energy represented by the

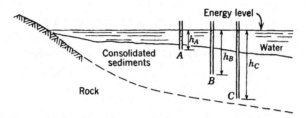

FIG. 3.14   Hydrostatic conditions in sediments beneath a lake.

piezometric head just described and pressure. Differences in pressure can exist between points in a soil without differences in total energy. Referring again to Figure 3.14, we can see that the three piezometers extend to different depths and the pressures at points A, B, and C are all different. The pressure at A is $\gamma_w h_A$; at point B it is $\gamma_w h_B$, and so on. Although the pressure increases in direct proportion to depth below the free-water surface, the energy level is exactly the same at A, B, and C.

Methods For Determining Seepage Forces

Seeping water imparts energy to individual soil grains by friction. The force acting on a given volume of soil is equal to the volume in cubic feet multiplied by the unit weight of water in pounds per cubic foot and the hydraulic gradient; thus

$$F = 62.5iV \tag{3.21}$$

If the hydraulic gradient is 1.0, the seepage force in pounds per cubic foot is equal to the unit weight of water.

To aid in visualizing the manner in which the energy of seeping water is transferred to soil, Figure 3.15 shows a small volume of sand confined in a

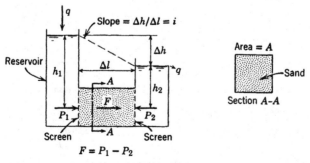

FIG. 3.15   Seepage forces in soil.

tube fitted with a reservoir on the left and another on the right. Head h_1, presses on the sand at the left and head h_2, on the right. According to hydraulics the water pressure in pounds on the left side of the sand is

$$P_1 = 62.5h_1 A$$

in which A is the cross-sectional area of the sand normal to the direction of flow. Likewise, the pressure on the right side is

$$P_2 = 62.5h_2 A$$

The *resultant* force $P_1 - P_2 = 62.5A(h_1 - h_2)$. This force must be consumed in friction; therefore the seepage force F must be equal to $P_1 - P_2$, and

$$F = P_1 - P_2 \tag{3.22}$$

For a given difference in head and a given area the energy loss is a fixed amount $(P_1 - P_2)$ regardless of the distance over which it is consumed. Figure 3.16 shows sand prisms of three lengths acted on by equal differences in head. The energy losses are exactly the same, although the lengths vary considerably. Equation 3.21 proves this to be true because $F = 62.5iV = 62.5(\Delta h/\Delta l)A\,\Delta l$, and $F = 62.5\Delta hA$. Thus the seepage force is proportional to the differential head and the cross-sectional area of the soil.

Seepage forces in earth masses are easily determined if frictional forces are expressed in relation to *hydraulic gradient i* in Eq. 3.21. This expression can be derived as follows.

In Figure 3.15 the the hydraulic gradient in the soil is $i = (h_1 - h_2)/\Delta l$, which can be expressed as $\Delta h/\Delta l$. The force exerted on the soil grains by fric-

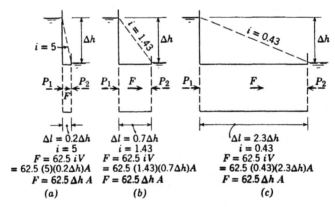

FIG. 3.16 Illustration of validity of basic seepage force equation. $F = 62.5iV$. (*a*) Steep gradient. (*b*) Medium gradient. (*c*) Flat gradient.

tion must be equal to the difference in energy on the front and back faces; hence $F = P_1 - P_2 = h_1\gamma_w A$. Therefore

$$F = (h_1 - h_2)\gamma_w A = \Delta h \gamma_w A \qquad (3.23)$$

but the volume of soil $V = A\,\Delta l$ and $A = V/\Delta l$. Substituting $V/\Delta l$ *for A* in Eq 3.23,

$$F = \Delta h \gamma_w A = \frac{\Delta h \gamma_w V}{\Delta l} = \gamma_w \left(\frac{\Delta h}{\Delta l}\right) V$$

Hence

$$F = \gamma_w i V$$

and

$$F = 62.5 i V$$

The differential head Δh is always a measure of the loss in energy between two points. As previously noted, *pressure differences* can exist without seepage.

The magnitude of the seepage force acting on a volume of soil must always be equal to that indicated in Eq. 3.21, $F = 62.5iV$. In practice, it may be determined by either of two methods which require a flow net and give exactly the same result if the work is carried out accurately: (1) Calculate directly from Eq. 3.21, using hydraulic gradients measured in flow nets. (2) Determine graphically, using the boundary pressure method.

Gradient Method. To determine seepage forces directly from hydraulic gradients it is necessary to take several steps.

1. Determine the average hydraulic gradient in a soil element.
2. Calculate the magnitude of the seepage force acting in the element.
3. Determine the *direction* of the seepage force.
4. Determine the *line of action* of the seepage force.

A flow net is needed for determining the hydraulic gradient and for positioning the force. The use of some judgment is required in all of these determinations. It is illustrated by reference to Figure 3.17 which shows the downstream portion of an earth dam in which steady seepage has developed under full reservoir conditions. This drawing has been simplified to facilitate the presentation of techniques for determining seepage forces, and does not represent a recommended design. A potential curved failure plane *AB* intersects the lower portion of the slope at *B*. The seepage force *F* acting on each of a num-

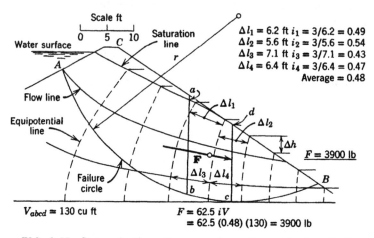

FIG. 3.17 Determination of seepage forces by *gradient method.*

ber of soil elements, such as *abcd*, is required for calculation of the factor of safety of the circular soil mass *ABC* by the *Swedish method,* which is described in most soil mechanics texts, later in this section, and in Sec. 8.1.

In the flow net in Figure 3.17 Δh is the vertical distance between points of intersection of equipotential lines with slope *BC*. The hydraulic gradient at any point in a flow net is equal to $\Delta h/\Delta l$. Because Δh is constant in any given flow net, the hydraulic gradients are proportional to $1/\Delta l$; hence it is possible by the inspection of any flow net to know the degree of uniformity or nonuniformity of hydraulic gradients. Points of concentration can be readily observed. Furthermore, *the magnitude of the hydraulic gradient at any point in a flow net* may be determined by measuring the distance Δl and dividing Δh by Δl to obtain the ratio $\Delta h/\Delta l$.

In the present problem (Fig. 3.17) it is seen that slight variations in hydraulic gradients exist because distances between equipotential lines vary somewhat. To determine the *average* hydraulic gradient in the saturated portion of element *abcd* the distance Δl is measured at several places and the corresponding hydraulic gradients are calculated. By inspection or by an appropriate averaging method the mean gradient within the element is determined. In this problem the average or mean gradient is determined to be approximately 0.48. A simple, yet reasonably accurate, method is to place an engineer's scale on the drawing and measure Δl at a number of locations and by judgment select a mean Δl distance for the element under consideration. A single computation is then made for the hydraulic gradient. Other procedures, no doubt, could readily be devised. By the use of reasonable judgment it should be possible to determine average gradients with little error.

When the mean hydraulic gradient of an element has been determined, the volume of the saturated soil is then calculated. The volume is taken as its area times 1 ft of thickness. Having the gradient and the volume in cubic feet, the seepage force in pounds for the element is calculated from Eq. 3.21, $F = 62.5iV$. For element *abcd* in Figure 3.17 the force $F = 62.5(0.48)$ (130 cu ft) = 3900 lb, as determined by the *hydraulic gradient* method.

The *direction* of the resultant seepage force F can usually be determined reasonably accurately by inspection. It should be drawn parallel to the mean direction of flow in an element. Its *position* should be established by first locating the center of gravity of the portion of the element in which water is seeping. If the gradients are uniform throughout an element, the force F will pass through the center of gravity of the portion subjected to seepage. If the gradients are nonuniform, the force F should be shifted slightly toward the higher gradients. This procedure is not exact, but if reasonable judgment is used the errors need not be large.

For anisotropic soil conditions the flow net must be drawn on a *transformed section* (Sec. 3.3) and then reconstructed on the natural section before magnitude and direction of seepage forces can be determined.

Boundary Pressure Method. The second procedure for determining seepage forces is illustrated in Figure 3.18. Soil element *abcd* of Fig 3.17 is reproduced in Figure 3.18*a*. Seepage force F acting on element *abcd* must be the resultant of all boundary water pressures plus a force W_o equal to the *unit weight of water times the total soil volume in which seepage is taking place.* The boundary water pressures are obtained by determining from the flow net the head at each corner of an element (and at an intermediate point if the flow net converges greatly). The head at a corner is determined by drawing an intermediate equipotential through the point to the free water surface. For point *b* line *bb'* is such a line drawn consistent with the shapes of adjacent equipotentials in the flow net. The head at *b* is the vertical distance between points *b* and *b'*. A horizontal line from *b'* to *e* establishes distance *be,* which is the head at point *b*. Arc *ee'e"* is drawn with a compass, and distances *be, be',* and *be"* are radii of this arc. Line *be',* drawn horizontally, is the maximum horizontal head acting on side *ab* of the soil element. Triangle *abe'* is the pressure diagram on the left side of the element. Pressure P_1 is equal to the area of triangle *abe'* times the unit weight of water. It is equal to (17 ft) (16 ft) (62.5 lb/cu/ft)/2 = 8500 lb. This is for a soil element 1 ft thick.

Similarly, forces P_2 and P_3 are determined by making lines *be"* and *cf"* perpendicular to line *bc* and line *cf'* perpendicular to line *cd*. Force W_o is equal to the volume of the saturated soil times the unit weight of water. This volume is the same as that used in the *gradient method* (Fig. 3.17), 130 cu ft.

Having determined forces P_1, P_2, P_3, and W_o as described, a force polygon is constructed (Fig. 3.18*b*). The closing distance is the resultant of all of the water forces acting on the element and is the *seepage force F.* As determined in Figure 3.18*b*, F is 4200 lb, compared with 3900 lb arrived by the gradient method. The difference, 300 lb, represents about 7%.

The amount of work required for determining seepage forces by the boundary pressure method is several times that required by the *gradient method.* For

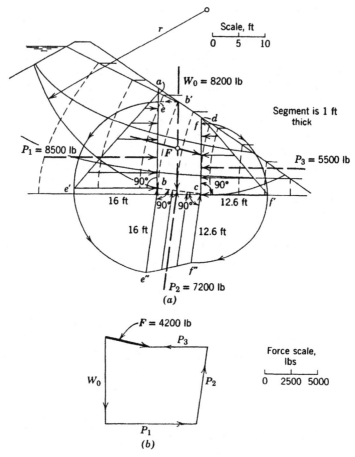

FIG. 3.18 Determination of seepage force by *boundary pressure method*. (*a*) Flow net, segment, and water forces. (*b*) Force diagram giving resultant seepage force *F*.

most applications the gradient method is sufficiently accurate; however, in cases in which the gradient method can be applied only with difficulty the boundary pressure method can always be used with confidence as the primary or referee method.

Resultant Body Force with Seepage

Calculations for the design of structures acted on by several forces can be simplified by determining first the *resultant* of the group of forces acting. In some cases the position of the resultant is a measure of the degree of stability of a structure; for example, if the resultant of the forces acting on a concrete gravity dam lies within the middle third of its base, it is considered amply safe against overturning. Frequently resultants are separated into certain *components* that are used in stress or stability calculations as in earth structures.

Problems in the stability of earth structures and foundations often make use of all forces acting on masses of earth, called *body forces.* These resultants are separated into components that are *normal* and *tangential* to trial failure surfaces. Factors of safety against rupture of the soil along assumed circular failure surfaces may be calculated from the equation (Sec. 8.1)

$$G_s = \frac{\Sigma N \tan \phi + cl}{\Sigma T} \qquad (3.24)$$

In Eq. 3.24 ΣN is the sum of all components *normal* or perpendicular to an assumed failure circle, ΣT is the sum of all *tangential* components, ϕ is the angle of friction of the soil, c is the unit cohesive strength of the soil, and l, the length of arc over which there is cohesive strength.

In the computation of factors of safety by this method, called the *Swedish* or *circular arc method,* the mass of earth above a trial failure surface is subdivided into a number of vertical slices or segments. A *body force* and its N and T components are determined for each slice. These values are added up for all slices, and the total ΣN and ΣT values are used in Eq. 3.24 for determining the factor of safety, G_s. Terzaghi (1943*b*), Taylor (1948b), Bridges and Cedergren (1962), Chowdhury (1987), and others describe procedures for using the Swedish method and other methods for determining the stability of earth slopes and foundations. Equation 3.24 applies to the simplified method that ignores the forces acting on the sides of individual slices. For some conditions these forces can have an appreciable influence on the calculated factor of safety. Methods that take side forces into consideration are described by Taylor (1948c), Bishop (1955), Bishop and Morgenstern (1960), and others. Janbu (1973) presents an excellent review of the theoretical and practical aspects of the generalized procedure of slices which includes forces between slices. Solutions can be obtained by slide rule or computerized calculations.

Soil in which water is flowing is made up of a framework of mineral particles interspersed with water. Taylor (1948d) points out that the resultant *body force* of the soil and water can be determined in several ways.

1. Using the submerged soil weight and the seepage force.
2. Using the total weight of soil and water and a resultant boundary neutral force.
3. Using the true weight of soil grains, the true weight of water, and a resultant boundary neutral force.

According to standard procedures in soil mechanics, the true weight of the soil grains in saturated soil can be expressed by G, the specific gravity of the grains, and the true weight of the water contained in the soil voids can be expressed by e, the void ratio. In the metric system the total unit weight of saturated soil is $G + e$ in a volume equal to $1 + e$. The unit saturated weight is therefore $(G + e)/(1 + e)$, the unit dry weight is $G/(1 + e)$, and the unit submerged weight is $(G - 1)/(1 + e)$. Unit weights in pounds per cubic foot are obtained by multiplying by 62.5 lb/cu ft, the unit weight of water.

Consider the forces acting on a small element of soil subjected to seepage (Figure 3.19a). In Figure 3.19b-1 the downward forces due to the weight of the soil and water are made up of the true weight of the soil grains (distance AC) and the true weight of water (distance CD); or in Figure 3.19b-2 the submerged soil weight (distance AB), the uplift on the soil grains (distance BC), and the weight of the water (distance CD). In Figure 3.19b-3 the total downward force is represented by the submerged soil weight (distance AB) and the total buoyancy (distance BD).

In Figure 3.19c the resultant body force (distance AE) may be obtained by combining the forces in several different ways, the two most practical combinations of which are the following:

1. The submerged soil weight and the seepage force.
2. The total weight of soil and water and the resultant boundary neutral force.

The way these combinations can be made is shown in Figs. 3.19d and 3.19e. In (d) the submerged soil weight is combined with the seepage force and in (e) the saturated soil weight is combined with the resultant boundary neutral force. The resultant R is exactly the same in both cases.

Detailed applications of seepage forces to stability problems are beyond the scope of this work; however, the manner in which this can be done is illustrated in the following sections.

Influence of Seepage on Soil Stability

Seepage forces can combine with soil weights to improve stability or worsen it, depending on the direction in which the forces act in relation to the geometric cross section; for example, consider the influence of seepage forces on soil elements a and b (Fig. 3.20a) at the ground surface on the water side and the land side of a sheet pile wall, respectively. Acting on element a (Fig. 3.20b) is a downward force W_o equal to its submerged weight and a *downward seepage force F* equal to 62.5 iV. These two forces are acting together to produce resultant *body force R* measured by distance AC. Now, consider element b at the land side of the sheet pile wall, where seepage is emerging upward. The submerged weight of element b produces downward force W_o, equal to distance AB in Figure 3.20c, and seepage produces *upward seepage force F* equal to distance BC. Under this condition the two forces are acting against each other to produce resultant *body force R'*, equal to distance AC in Figure 3.20c. If force F equals force W_o in element b, *flotation* takes place and the wall will

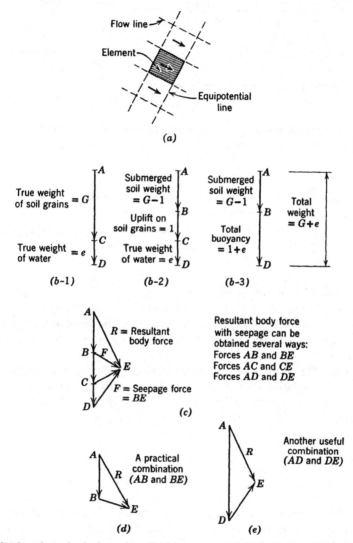

FIG. 3.19 Resultant body force in soil with seepage. (*a*) Soil element. (*b*) Three possible combinations of soil and water. (*c*) Combining forces. (*d*) Submerged soil weight and seepage force. (*e*) Total weight of soil and water with resultant boundary neutral force. (See *Fundamentals of Soil Mechanics*, by D. W. Taylor, Wiley, New York, 1948, Fig. 9-20, p. 201.)

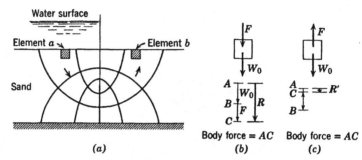

FIG. 3.20 Seepage forces can improve or worsen stability. (*a*)Sheet pile wall in fine sand. (*b*) Element *a*. (*c*) Element *b*.

be undermined. Failure could be prevented by placing a *weighted filter* over the ground surface at the right side of the wall.

Referring again to Figure 3.20, we see that the effective stresses on element *a* are increased by the seepage force and the shearing strength is therefore increased. At element *b* the seepage forces oppose the forces of gravity, neutralizing part or all of the weight of the soil and thereby lowering effective stresses and reducing shearing strength.

A common method for calculating factors of safety for earth slopes has already been described briefly. It is noted that the resultant *body forces* of vertical segments of earth are resolved into *normal* and *tangential* components that are used to calculate factors of safety from Eq. 3.24.

The ratio N/T is an index of *relative stability* (Chapter 8, Slopes). By examination of the relative values of N/T in slopes with and without seepage, it is readily seen that seepage can have a major destabilizing influence on the stability of slopes.

Seepage Forces in Earth Dams

In preceding pages methods for determining seepage forces and for combining them with soil weights to determine *body forces* have been described. The forces of seeping water often enter into calculations for the design of many types of civil engineering works. If the forces acting on large earth masses are not in stable equilibrium, massive failures can take place. If the forces acting on individual soil particles are large and these particles are not firmly held in place, internal erosion and piping failures can take place. Dams and other important works influenced by seepage should be designed with suitable filters and drains (Chapter 5).

It is generally assumed that if filters are provided there need be no concern for the magnitude of hydraulic gradients because properly designed filters can provide 100% protection for gradients of any magnitude in these structures. Assuming that every part of an impervious zone is protected with correctly designed filters, this conclusion is true, but it must be admitted that the average workman who builds dams and other works is not aware of the difference between a properly constructed filter and one that can lead to failure.

The importance of protecting *every square foot* of erodible zones cannot be overemphasized. With proper control of construction there is no harm in designing a dam with high internal gradients, but a design having high seepage gradients should offer substantial benefits to warrant consideration. All other factors being equal, *designs that hold internal gradients to low levels are safer than those that develop exceptionally high gradients.*

The influence of size and shape of impervious cores on the escape gradients at the downstream sides of cores is shown in Figure 3.21. These gradients vary

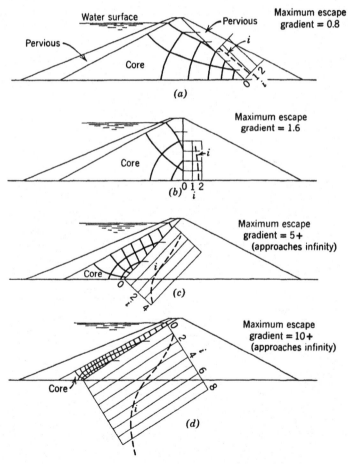

FIG. 3.21 Influence of size of impervious core on escape gradients in zoned dams. Filters and drains are not shown in these simplified cross sections.

from less than unity for wide cores (Fig. 3.21*a*) to 10 plus for dams with narrow inclined cores (Fig. 3.21*d*). The relative magnitudes of the gradients are apparent from a glance at the flow nets in Figure 3.21.

Flow nets are also useful for studying uplift gradients in the foundations of earth dams and levees. (See also Chapter 6.) Figure 3.22 shows a study of a large earth dam on a semipervious rock foundation underlying sand and gravel. A trench into the foundation cuts off seepage through the sand and gravel. The rock is somewhat stratified; hence a flow net was constructed on a *transformed section* (Sec. 3.3), assuming that $k_h = 4k_v$, (Fig. 3.22*a*), and redrawn on the natural section (Fig. 3.22*b*). Stratification causes the equipotentials to spread out horizontally and raises uplift pressures under the toe to higher values than would exist for homogeneous conditions.

Observation wells were installed into the rock at two depths, as shown in Figure 3.22*b*. Readings that were made at several partial pool levels were plot-

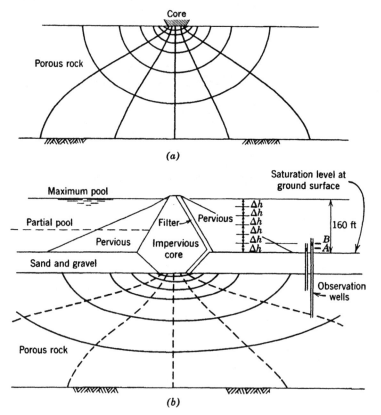

FIG. 3.22 Study of uplift pressures in rock foundation of an earth dam. (*a*) Transformed section. $k_h = 4k_v$. (*b*) Natural section. Flow net is drawn consistent with well readings at partial pool level.

ted in a simple diagram and extended to maximum pool level. The extrapolated levels at maximum pool are shown in Figure 3.22b. They indicate that the level in the shallow well can be expected to rise to point A and in the deep well, to point B. These values will be verified by readings taken at higher reservoir stages. If the predicted uplift pressures at maximum reservoir were considered unsafe, supplemental measures would be taken to protect the dam.

To be safe against uplift the downward forces should exceed the upward forces. Methods for determining both have been described in this chapter. To be reasonably safe calculated factors of safety should be at least 2.5 or 3.0.

Uplift Pressures

Considerable damage to civil engineering structures occurs when hydrostatic pressures behind floor slabs, retaining walls, and the like, are greater than these structures can withstand. If no seepage is taking place, the uplift pressure head is simply the height to the free-water table. If seepage occurs, the magnitude of the uplift pressures can be estimated with flow nets. In Chapter 10 methods of controlling groundwater and seepage under structures are described.

The general procedure for determining uplift pressures with a flow net is illustrated in Figure 3.23. Just as the flow net permits determination of the pressure on any plane within a mass of saturated soil (Fig. 3.18a), it also offers a means of calculating the pressure against impermeable objects in contact with saturated soil. Figure 3.23 gives the flow net for seepage through a permeable foundation under a concrete weir. The total differential head h acting on the weir is divided into 10 equal parts by the flow net, for the net has 10 equipotential lines. It necessarily follows that an amount of head $\Delta h = h/10$ is consumed by friction between each adjacent equipotential. To determine the distribution of uplift pressure at the bottom of the structure parallel horizontal lines are drawn on the cross section dividing the vertical distance between the reservoir and water surfaces at the toe into 10 equal parts. Vertical lines such

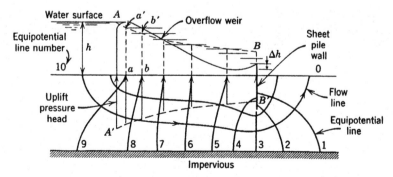

FIG. 3.23 Determining uplift pressures with a flow net.

as *aa'* are then drawn from the points of intersection of equipotentials with the bottom of the structure, to the correct horizontal line. Thus distance *aa'* is equal to the pressure head at point *a*; *bb'* is equal to the head at point *b*, and so on. Line *AB* gives the hydrostatic pressure distribution at the bottom of the weir. Its mirror image, line *A'B'*, shows the uplift pressure head in correct relation to the structure.

Structures must always be designed to resist uplift pressures safely. The flow net is a handy tool for making uplift studies for proposed structures and for studying conditions under completed structures.

3.6 MOVING SATURATION LINES

General Considerations

Some of the commonest uses of the flow net are for the solution of problems of steady seepage; it can also be used in developing approximate solutions to certain nonsteady problems relating to moving saturation lines.

Under steady seepage conditions seepage velocity v_s can be estimated at any point in a flow net from the relationship

$$v_s = \frac{\text{discharge velocity}}{\text{effective porosity}}$$

According to Darcy's law, the superficial *discharge velocity* of seepage through porous media is

$$v_d = ki$$

The discharge velocity multiplied by the cross-sectional area gives the seepage quantity

$$Q = v_d A = kiA$$

But seepage occurs only in the pores or voids; hence the discharge velocity multiplied by the total volume must equal the seepage velocity multiplied by the volume of voids. If the volume of voids is e and the total volume $1 + e$,

$$v_d(1 + e) = v_s(e)$$

and

$$v_s = v_d \frac{1 + e}{e}$$

but soil porosity n is equal to $e/(1 + e)$; hence

$$v_s = \frac{v_d}{n} = \frac{ki}{n} \tag{3.25}$$

Porosity is always less than unity; therefore the seepage velocity must always be greater than the discharge velocity.

When water is flowing into or out of soil, the change in moisture content is less than the amount required to fill the voids with water. The rate of progress of a saturation line into or out of soil depends on the actual volume of voids, filled or emptied. The volume establishes the "effective" porosity. Equation 3.25 may therefore be written as

$$v_{sl} = \frac{ki}{n_e} \tag{3.26}$$

According to Eq. 3.26, $v_{sl} = ki/n_e = (k/n_e)i$. If it is assumed that the ratio k/n_e is constant, the velocity v_{sl} is proportional to i. Curves for v_{sl} at various levels of saturation will therefore be identical in shape to the curves for i.

Terzaghi (1943c) expresses the relationships for the velocity of falling saturation lines in terms of G_a, the *air space ratio* of sand after drainage:

$$v = -\frac{dz}{dt} nG_a \tag{3.27}$$

In this expression v is the discharge velocity, $-dz/dt$, the rate of fall of a saturation line, n, the porosity, and G_a, the air-space ratio of the sand after drainage. The term G_a is defined as the ratio of the volume of air-filled voids e_a to the total voids e. It is also equal to $1 - (S\%/100)$, in which $S\%$ is the degree of *saturation*.

Transposing terms in Eq. 3.27,

$$-\frac{dz}{dt} = \frac{v}{nG_a} \tag{3.28}$$

Equation 3.28 is identical in meaning to Eq. 3.26.

If water is draining out of saturated soil, menisci form at the boundary between the drained and undrained parts, producing tension throughout the pore water. The maximum negative pressure is at this boundary. It is equal to the capillary head h_c multiplied by the unit weight of water. The negative head acts in opposition to the pull of gravity, slowing down the flow of water out of the soil.

If saturation is moving *into* unsaturated soil, the capillary head is working *with* the forces of gravity to increase the hydraulic gradient. This condition is illustrated by Figure 3.24, which shows a horizontal tube filled with sand. A reservoir at the left produces head h to push water into the sand. At some arbitrary time t the saturation line will have moved into the sand a distance L, as shown in Fig. 3.24a. The capillary head h_c is now *pulling* on the water while the head h is *pushing* (Fig. 3.24b), thus producing a total head $h + h_c$ to force water into the sand. For this condition the hydraulic gradient is

$$i = \frac{h + h_c}{L} \tag{3.29}$$

When seepage fills the reservoir at the right, the menisci and capillary head will disappear.

Transient Flow Net Method

From elementary physics the time required for matter to travel a *distance* is equal to the *velocity* multiplied by the *time*. It follows that the time Δt for particles of water to move a distance Δl may be calculated from the equation

$$\Delta t = \frac{\Delta l}{v_{sl}} \tag{3.30}$$

Equation 3.30 provides the basis for the transient flow net method as developed in this section.

Whenever a zone of saturation is moving into or out of soil, a fringe exists

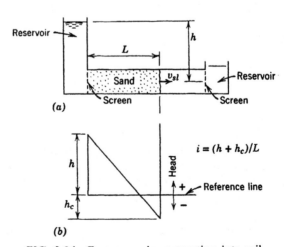

(a)

(b)

FIG. 3.24   Forces moving saturation *into* soil.

along which the soil is changing from a moist or dry state to a saturated state or from a saturated state to a partially saturated state. Within the zone of saturation the equation of continuity (Eq. 3.6) is satisfied; hence a flow net can be constructed. This flow net differs from the steady seepage flow net because flow lines do not necessarily parallel the phreatic line (upper level of saturation) but may in fact intersect it. In a lowering zone of saturation water is being supplied by drainage from the soil above; therefore flow lines must originate at the phreatic line (Fig. 3.28). If saturation is penetrating into soil, as in the saturation of a dam or a rising water table, flow lines terminate in the forward portions of the phreatic line (Fig. 3.25b, c, and d).

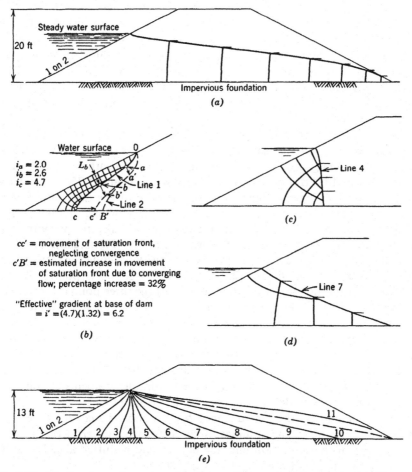

FIG. 3.25 Study of moving saturation lines with transient flow nets. (a) Flow net after complete development of saturation. (b) Method illustrated. (c) Flow net during saturation. (d) Flow net during saturation. (e) Progress of saturation through embankment.

In Sec. 3.5 it is shown that in any flow net hydraulic gradients are inversely proportional to the spacing of equipotential lines ($i \sim 1/\Delta l$). The *relative* seepage velocities at various points in a flow net are directly proportional to gradients, and the *time* Δt required for a given distance of movement is equal to $\Delta l/v_{sl}$ (Eq. 3.30). It is evident therefore that changes in the shape of moving saturation lines with time can be studied with transient flow nets.

By means of transient flow nets hydraulic gradients are determined at a number of points on a moving saturation line. Each successive position of the saturation line is determined by moving several points by distances that are proportional to the hydraulic gradients at each point. Adjustments can be made for nonparallel flow. With a moderate amount of practice reasonable solutions can be obtained. The intermediate flow nets and the developed shapes of the saturation lines at various positions give a visual picture that aids in understanding the manner in which saturation can develop in earth masses. Actual times required for the saturation of dams or other specific movements of saturation lines can be estimated if permeability and porosity data are available for the soils. Cedergren (1941, 1948) developed the method described in this section. It applies to cases in which capillarity is small enough to be ignored.

Figure 3.25 illustrates the basic procedure for studying changes in the position and shape of moving saturation lines with transient flow nets. The cross section of a homogeneous dam is given in Figure 3.25a with the steady seepage flow net. The first transient flow net is shown in Figure 3.25b. This net was obtained by dividing the total width of the base of the dam into about 10 parts and drawing an initial saturation line, but recognizing that both *seepage velocities* and *seepage distances* in a given time are directly proportional to hydraulic gradients. A transient flow net was drawn (Fig. 3.25b), and the hydraulic gradients were determined at several positions along the seepage front. Thus at point a the gradient is 2.0, at point b, 2.6, and at point c, 4.7. The distance L from the face of the dam to each of the three points is then scaled. If these distances are not *approximately* proportional to the hydraulic gradients and the L/i ratios are not approximately equal, the initial saturation line should be redrawn, a new flow net constructed, and new L/i ratios determined. Having established a compatible initial saturation line (line Oc, Fig. 3.25b), aa', bb', and cc' are drawn parallel with the directions of flow at each point. Their lengths are proportional to the hydraulic gradients at points a, b, and c. Examining the flow net in Fig. 3.25b it is seen that flow lines converge in the lower part of the net; distance $b'c'$ is shorter than distance bc. To maintain a balance in the volumes of soil filled with water at the lower and upper parts of the saturation line, the bottom end is shifted ahead to a point B'. In this example distance $c'B'$ is 0.3 distance cc'.

In the study of the time of saturation of this dam the varying hydraulic gradients along its base are used. In areas in which flow is converging, as in Figure 3.25b, the gradients at the base are multiplied by the ratios of distance

cB' to distances cc'. This adjustment makes the time of saturation consistent with the transient saturation lines such as $Oa'B'$ in Figure 3.25b.

If flow were diverging, a negative correction would be needed. Some engineers may wish to ignore the effects of converging or diverging flow; however, attempts to make reasonable allowances will reduce errors, especially when extreme convergence or divergence is taking place. In the present example (Fig. 3.25) these corrections are of minor magnitude.

The transient flow net method requires the construction of a flow net for each of a reasonable number of intermediate positions of a moving saturation line. Additional typical flow nets for this example are shown in Figures 3.25c and 3.25d, and the changing shape of the saturation line developed by 10 transient flow nets is shown in Figure 3.25e. The shape of a moving saturation line in a homogeneous section is independent of the soil permeability, provided the permeability remains constant along the moving saturation line. Therefore, if capillarity is small enough to be ignored, the moving saturation line will take the same shapes regardless of the absolute permeability.

In summary, the shape of a moving saturation line can be studied with transient flow nets by the following steps.

1. Divide the total distance the saturation line will move into a reasonable number of parts, generally 5 to 10 (the approximate number of parts will vary with the shape being studied); for example, in Figure 3.25e 10 increments were used, whereas in Fig. 3.28 five were sufficient.

2. Draw the first trial saturation line and construct a flow net for the boundary conditions existing momentarily.

3. Determine the hydraulic gradients at several points along the saturation line: i_a, i_b, i_c, etc.

4. Measure the approximate distances water has traveled to reach points a, b, c, etc. Note these distances as L_a, L_b, L_c, and so on.

5. Determine the ratios L_a/i_a, L_b/i_b, L_c/i_c, etc. If these ratios are not roughly equal, change the shape of the first transient saturation line by increasing L at points at which the ratio is low or decreasing it where the ratio is high. Draw a new flow net. Repeat if the new L/i ratios are not approximately equal.

6. Select the next position for the saturation line, moving it ahead from the first position by amounts that are proportional to the *relative* hydraulic gradients at the selected points on the first line. Adjust for nonparallel flow if necessary, as described in this section.

7. Repeat for the number of increments needed.

To verify the validity of transient flow nets for studying moving saturation lines as described here Cedergren (1941) constructed a small scale viscous fluid model of the dam shown in Fig. 3.25a. Several intermediate positions of the

moving saturation line observed in the model and several positions obtained
with the transient flow nets (theoretical solution) are shown in Figure 3.26*b*.
The steady seepage saturation line observed in the model is shown in Figure
3.26*a* which also contains the steady seepage theoretical flow net.

It is apparent that the model and transient flow net solutions agree quite
well. Shannon (1948), Casagrande and Shannon (1952), and others have obtained close agreement between viscous fluid models and theoretical seepage
solutions. Todd (1959) reviews the theory of viscous flow between closely
spaced parallel plates and points out that this type of seepage model was first
used by Hele-Shaw in England in 1897.

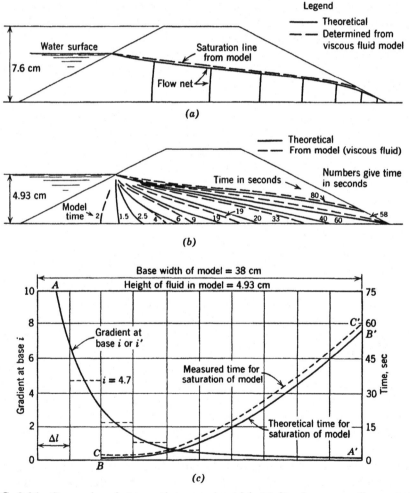

FIG. 3.26 Comparison between theory and model. (*a*) Steady seepage flow net. (*b*)
Progress of saturation. (*c*) Hydraulic gradients and time for saturation.

The study of the *time* required for movement of a saturation line is illustrated in Figure 3.26c which is for the model having the same shape as the dam shown in Figure 3.25. First a plot is made of the changing hydraulic gradients (line AA', Fig. 3.26c), and the total width of the model is divided into a convenient number of *equal* increments Δl as shown. The average hydraulic gradient in each increment is then determined; for example, in the second increment in Fig. 3.26c the average hydraulic gradient is 4.7. The time required for the saturation line to move each incremental distance Δl is determined from Eq. 3.30, $\Delta t = \Delta l/v_{sl}$. Appropriate values of n_e and k are used for calculating v_{sl} from Eq. 3.26 ($v_{sl} = ki/n_e$). By summation the total time is

$$T = \sum \Delta t = \sum \frac{\Delta l}{v_{sl}} = \frac{n_e}{k} \sum \frac{\Delta l}{i} \qquad (3.31)$$

Line BB' shows the theoretical time for saturation to progress through the model by using a coefficient of permeability obtained experimentally for the model. Line CC' gives the time actually recorded for the model. The agreement is good.

Another example of drainage of embankments was studied with transient flow nets and a viscous fluid model by Cedergren (1948). A simple plywood model with strips of linoleum to hold a sheet of glass a suitable distance from the plywood was used for the experiments (Fig. 3.27), and a viscous fluid (glycerine) was allowed to flow in the narrow annular space between the glass and

FIG. 3.27 Simple viscous fluid model used in drawdown tests. Sheet of glass (removed for photo) clamped to front of model provides thin annular space for flow of glycerine.

the plywood. An enlarged reservoir with a quickly removable bottom permitted *rapid drawdown* conditions to be simulated. The base width of the model was 10¼ in. Typical transient flow nets are shown in Figures 3.28*a*, *b*, and *c*. The shape of the saturation line at various intervals, as determined theoretically and by observation of the model, are compared in Figure 3.28*d*. Typical paths of flow of fluid out of the cross section, both theoretically and by observation of the model, are given in Figure 3.28*e*. It is seen that the theoretical solutions agree substantially with the experimental results. Other cases that have been studied have shown comparable agreement between transient flow net solutions and model performance.

Browzin (1961) investigated the nonsteady state flow after rapid drawdown in homogeneous dams on impervious foundations by model tests (Hele-Shaw flume tests) and confirmed the basic validity of the method of continuous succession of steady states. From theoretical considerations he developed

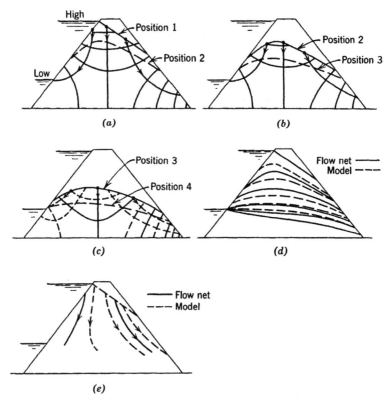

FIG. 3.28 Study of drainage from embankment after drawdown by flow nets and viscous fluid model. (*a*) First flow net. (*b*) Second flow net. (*c*) Third flow net. (*d*) Shape of saturation line during drainage. (*e*) Paths of flow during drainage.

equations that predicted with a high degree of accuracy the shapes taken by the observed phreatic lines at various intervals of time.

The method described in this chapter for the study of moving saturation lines can produce approximate solutions to a variety of nonsteady seepage problems. If used with an awareness of its limitations, it has considerable practical value.

Another application of these principles, described in Sec. 2.8, "Field Methods that Depend on Observation of Spreading or Receding Groundwater Mounds," utilizes transient flow nets to estimate coefficient of permeability of formations in which the amount of movement of a saturation line in a known amount of time is known. In that section Eq. 3.31 is expressed in a form that permits the coefficient of permeability k to be calculated (see Eq. 2.25).

3.7 SEMITURBULENT AND TURBULENT FLOW

In this book a wide variety of practical solutions to seepage and drainage problems is developed with flow nets and Darcy's law. The Laplace equations (and flow nets) depend on the validity of Darcy's law and the assumption that flow is laminar, not turbulent. In coarse natural formations and coarse open-graded aggregates in drainage systems it is likely that flow will sometimes be semiturbulent to turbulent. This section provides a method for adapting Darcy's law to this kind of flow.

When flow is laminar, the velocity of flow increases in direct proportion to the hydraulic gradient. If the gradient is doubled, the velocity is doubled and so on. When flow is semiturbulent to turbulent, the seepage velocity increases at a smaller rate than the gradient. For fully turbulent flow, as in large pipes or open channels, the velocity increases in proportion to the *square root* of the slope or hydraulic gradient, as expressed by the well-known Chezy formula:

$$v = C\sqrt{Rs} \tag{3.32}$$

A fourfold increase in hydraulic gradient only doubles the velocity.

When water is flowing in highly permeable materials (such as clean rock zones or open-graded drainage materials), the condition can vary from laminar or nearly laminar flow at very small hydraulic gradients to semiturbulent or nearly turbulent as the gradient and the size of the aggregate and pore spaces become larger.

Various investigators have looked into flow conditions in coarse gravel and rock and have developed numerous formulas to represent the average velocity in the pore spaces and other factors. Leps (1973) gives a good summary of the work done and presents a number of formulas for estimating the flow velocities in clean gravel or rock, such as that used in some rock fill dams. He is quoted as saying that all investigators (1971) appeared to agree that the basic equation for flow through rockfill is a formula for turbulent flow, which can be expressed as follows:

$$V_v = Wm^{0.5}i^{0.54} \tag{3.33}$$

In Eq. 3.33 V_v is the average velocity of water in the voids of the rockfill, W is an empirical constant for a given material depending primarily on shape and roughness of the rock particles and the viscosity of the water, m is the hydraulic mean radius, and i is the hydraulic gradient.

Another approach to the analysis of flow in open-graded materials is to make permeability tests at the actual gradient expected in a prototype. When this can be done, errors from turbulence are largely eliminated because a measured coefficient—*although not a true Darcy coefficient*--should have the correct magnitude for estimating seepage quantities and velocities at the test gradient. For such a procedure to be used the hydraulic gradients in a prototype should be relatively constant. If wide variations in gradient occur, a cross section of perfectly uniform (highly permeable) material becomes in effect a *variable-permeability* cross section.

Still another procedure is to make permeability tests for a range of hydraulic gradients, including a very low gradient that produces nearly laminar flow, and larger gradients that produce various degrees of turbulence. By this procedure plots of "effective" permeability versus hydraulic gradient can be developed. When flow is semiturbulent to turbulent, Darcy's law does not apply; however, a quasi-Darcy relationship can be written as follows:

$$q = k'iA = (kC)iA \tag{3.34}$$

and

$$v_s = \frac{k'i}{n_e} = \frac{(kC)i}{n_e} \tag{3.35}$$

In these expressions k is the laminar (or almost laminar) coefficient of permeability determined by tests at a very small hydraulic gradient, i is the actual hydraulic gradient, C is an experimental factor that varies with hydraulic gradient (as determined from tests), and k' is the "effective" permeability, equal to kC, for semiturbulent to turbulent flow.

This experimental approach (which is similar to Darcy's approach) was used by the author with open-graded American River (California) crushed river gravels. In these experiments three different gradations of clean crushed gravel with no fines were tested in a small flume that could be tilted to allow measurements to be made of seepage quantities and velocities at several different hydraulic gradients to a maximum of about 0.3. After steady flow had developed in each test, colored dye was inserted at the upper end and the travel time was recorded. In addition, seepage quantities were measured for each test. In this way independent calculations were made from seepage velocities and seepage quantities.

From Darcy's law

$$\frac{q}{i} = kA \tag{3.36}$$

For laminar flow q/i is a constant regardless of hydraulic gradient, but for semiturbulent to turbulent flow q/i diminishes with increasing hydraulic gradient and can be expressed as

$$\frac{q}{i} = k'A = (kC)A \tag{3.37}$$

Because a constant cross section was used in the tests, A was constant and the "effective" permeability $k' = kC$ is proportional to the measured values for q/i. *Relative* values for q/i (which are proportional to C), determined by the tests on the American River crushed gravels (with q/i at $i = 0.01$ taken as unity), are summarized in Table 3.1. Examination of its data shows that q/i and C diminish with increased hydraulic gradient and aggregate size, which should be expected because turbulence increases with both factors. For the American River crushed gravel C varies from 0.96 for 3/8 in. gravel at hydraulic gradient of 0.02 to 0.40 for 3/4 in. gravel at a hydraulic gradient of 0.30.

A detailed study of the test data for the American River gravel showed that "effective" coefficients of permeability for semiturbulent to turbulent flow are related to the laminar (or nearly laminar) Darcy coefficients according to the following expression:

$$\frac{k_x}{k_0} = \left(\frac{i_0}{i_x}\right)^y \tag{3.38}$$

and

$$k_x = k_0\left(\frac{i_0}{i_x}\right)^y \tag{3.39}$$

TABLE 3.1 Relative Values of $q/i = C$ Obtained in Flow Tests with Crushed American River Gravels with No Fines

Nominal size of aggregate:	Hydraulic gradient							
	0.01	0.02	0.05	0.08	0.10	0.15	0.20	0.30
3/8 in.	1.0[a]	0.96		0.71		0.62	0.57	0.53
1/2 in.	1.0[a]	0.92	0.77	0.67		0.55	0.50	0.45
3/4 in.	1.0[a]	0.88	0.66		0.53		0.45	0.40

[a]Assumes laminar flow at $i = 0.01$.

In these expressions k_0 is the true Darcy coefficient (or a nearly true coefficient at a small hydraulic gradient), k_x is the "effective" permeability coefficient at a larger gradient i_x, and y is an exponent that varies with the hydraulic gradient.

Figure 3.29a gives "relative" permeability $= C$ values obtained in the tests

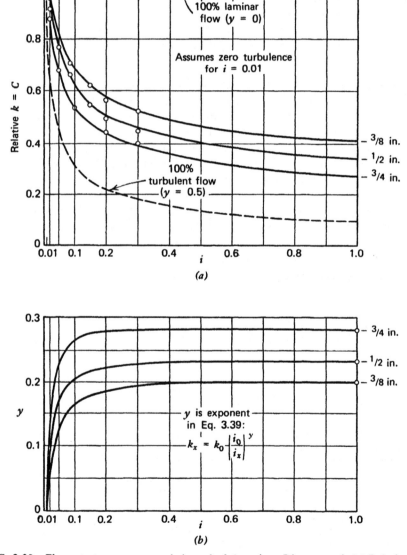

FIG. 3.29 Flume tests on open-graded crushed American River gravel. (a) Relative k versus hydraulic gradient. (b) Exponent y versus hydraulic gradient.

on the American River gravel, and Figure 3.29*b* is a plot of corresponding values of *y* determined from these tests. It is seen that values for the exponent *y* start from zero at a small hydraulic gradient (representing laminar flow) and quickly rise toward a limiting value (0.2 for the 3/8 in. aggregate, 0.235 for the 1/2 in. aggregate, and 0.285 for the 3/4. aggregate). Somewhat different values could be expected for other aggregates of different particle shape and degrees of compaction.

The validity of Eq. 3.39 can be demonstrated in part by considering the limiting conditions of flow. If flow is laminar, *q/i* does not change; hence there should be no reduction in "effective" permeability. This is demonstrated with Eq. 3.39:

$$k_x = k_0 \left(\frac{i_0}{i_x}\right)^y = k_0 \left(\frac{i_0}{i_x}\right)^0 = k_0\,(1.0) = k_0$$

(The proof is that *k* does not vary with hydraulic gradient.) The upper dashed line in Figure 3.29*a* represents a constant coefficient of permeability for the laminar condition ($y = 0$).

Now, if flow is 100% turbulent, Eq. 3.39 shows that

$$k_x = k_0 \left(\frac{i_0}{i_x}\right)^y = k_0 \left(\frac{i_0}{i_x}\right)^{0.5}$$

For 100% turbulent flow the "effective" permeability varies with the square root of the reciprocal of the hydraulic gradients, and the larger the hydraulic gradient, the smaller the "effective" *k*.

The bottom dashed line in Figure 3.29*a* represents the lower limiting envelope for "relative" *k* values (for 100% turbulent flow). It is seen that increasing the hydraulic gradient 100 times (from 0.01 to 1.0) reduces the effective permeability to one-tenth of the laminar value.

Figure 3.30 was prepared to assist in the application of these test data to a wider range of conditions than had been tested. Equation 3.39 aided in extrapolating beyond the test data. Tests on graded Ottawa sand gave values for a material finer than those tested and also aided in the extrapolations. The limiting envelope (100% turbulence) gives the extreme limit for materials coarser than those tested. Figure 3.30 is presented as a rough guide for estimating values of *C* to be used in calculating "effective" permeability $k' = kC$ to be used in seepage quantity and velocity computations for semiturbulent to turbulent flow. Somewhat different values no doubt would be obtained for other aggregates and for other degrees of compaction.

Equations 3.34, 3.35, and 3.39 and Figures 3.29 and 3.30 have been presented to illustrate a practical method for making allowance for semiturbulent to turbulent flow. Examination of Figures 3.29*a* and 3.30 shows that turbulence is of relatively little importance in studying seepage quantities and veloc-

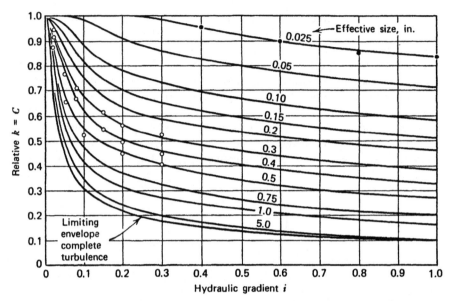

FIG. 3.30   Rough guide for estimating reductions in *k* of narrow size-ranged aggregate caused by turbulence. A preliminary guide only—based primarily on flume tests (unconfined flow) on crushed American River gravel. ○ Flume tests on crushed American River gravel. • Constant head *k* tests on graded Ottawa sand.

ities in flat-lying layers (such as open-graded bases under pavements or flat drainage blankets or line drains in dams) when hydraulic gradients are in the order of 0.02 or less, and errors in such computations (because of turbulence) will be moderate even if no correction is made for turbulent flow. However, when hydraulic gradients are much greater than about 0.02, substantial reductions in "effective" permeability and thus in seepage quantities and velocities can be expected.

When water flows through coarse rock fill under semiturbulent to turbulent conditions, head losses concentrate in regions of increased hydraulic gradient. Under such conditions (as previously noted), a uniform cross section becomes in fact a *variable-permeability* cross section, with effective permeability diminishing in areas of high gradients. Because *seepage forces* are proportional to hydraulic gradient, turbulence tends to concentrate them at critical exit regions and increases tendencies for slipouts to occur in these areas if the water is not collected in properly located and designed drains built into these structures.

REFERENCES

Biedenham, Debra G. and Fred T. Tracy (1987), "Finite Element Method Package for Solving Certain Steady-State Seepage Problems", Technical Report ITL-87-6, Final

Report, Department of the Army, Waterways Experiment Station, Corps of Engineers, P. O. Box 631, Vicksburg, Miss., 39180-0631, May, 1987.

Bishop, A. W. (1955), "The Use of the Slip Circle in the Stability Analysis of Slopes," *Geotechnique,* Vol. V, No. 1, March, 1955, pp. 7-17.

Bishop, A. W., and Norbert Morgenstern (1960), "Stability Coefficients For Earth Slopes," *Geotechnique,* Vol. X, No. 4, December 1960, pp. 129-150.

Bridges, R. A., and Harry R. Cedergren (1962), "Soil Engineering, Street and Highway Applications," University of California, Institute of Transportation and Traffic Engineering, pp. 40-41.

Browzin, Boris F. (1961), "Nonsteady-State Flow in Homogeneous Earth Dams after Rapid Drawdown," *Proceedings,* 5th International Conference on Soil Mechanics and Foundation Engineering, Paris, July 17-22, 1961, Vol. II, pp. 551-554.

Casagrande, A. (1937), "Seepage Through Dams," Harvard University Pub. 209, reprinted from *Journal of the New England Water Works Association,* June 1937.

Casagrande, A., and W. L. Shannon (1952), "Base Course Drainage for Airport Pavements," *Transactions,* A.S.C.E., Vol. 117, p. 807.

Cedergren, Harry R. (1941), "Discussion of 'Saturating an Earth Dam' by K. P. Karpoff," *Civil Engineering,* August 1941, p. 499.

Cedergren, Harry R. (1948), "Discussion of 'Investigation of Drainage Rates Affecting Stability of Earth Dams' by F. H. Kellogg," *Transactions,* A.S.C.E., Vol. 113, pp. 1285-1293.

Chowdhury, R. N. (1987), "Stability of Soil Slopes," *Ground Engineers' Reference Book,* Edited by F. G. Bell, Butterworths, London, Chap. 11, pp. 11/1 to 11/16.

Christian, J. T. (1987), "Numerical Methods and Computing in Ground Engineering," *Ground Engineer's Reference Book,* Edited by F. G. Bell, Butterworths, London, 1987, Chapter 57, pp. 57/13 to 57/17.

Darcy, H. (1856), "Les Fontaines publiques de la ville de Dijon."

Fishel, V. C. (1935), "Further Tests of Permeability with Low Hydraulic Gradients," *Transactions,* American Geophysics Union, p. 503.

Forchheimer, Philip (1902), "Discussion of "The Bohio Dam," by George S. Morison," *Transactions,* A.S.C.E., Vol. 48, p. 302.

Harr, Milton E. (1962), *Groundwater and Seepage,* McGraw-Hill, New York.

Jacob, C. E. (1950), "Flow of Groundwater," in *Engineering Hydraulics,* Hunter Rouse (Ed.), Wiley, New York, Chapter 5, pp. 322-323.

Janbu, Nilmar (1973), "Slope Stability Computations," in *Embankment-Dam Engineering, Casagrande Volume,* Wiley, New York, pp. 47-86.

Leliavsky, Serge (1955), *Irrigation and Hydraulic Design,* Chapman and Hall, London, p. 139.

Leonards, G. A. (1962), *Foundation Engineering,* McGraw-Hill, New York, pp. 107-139.

Leps, Thomas M. (1973), "Flow Through Rock Fill," in *Embankment-Dam Engineering-Casagrande Volume,* Wiley, New York, 1973, p. 90.

Muskat, M. (1937), *Flow of Homogeneous Fluids Through Porous Media,* McGraw-Hill, New York (lithographed by Edwards Bros., Ann Arbor, Mich., 1946), pp. 55-58; (1937a) pp. 55-120.

Shannon, W. L. (1948), "Discussion of 'Investigation of Drainage Rates Affecting Stability of Earth Dams,' by F. H. Kellogg," *Transactions,* A.S.C.E., Vol. 113, pp. 1302-1307.

Taylor, D. W. (1948), *Fundamentals of Soil Mechanics,* Wiley, New York, pp. 97-123; (1948a) pp. 122-123; (1948b) pp. 431-479; (1948c) pp. 433-438; (1948d) pp. 200-204.

Terzaghi, K. (1943), *Theoretical Soil Mechanics,* John Wiley & Sons, Inc., New York, pp. 239-240; (1943a) pp. 243-244; (1943b) pp. 155-175; (1943c) pp. 317-318.

Todd, David K. (1959), *Ground Water Hydrology,* Wiley, New York, pp. 314-316.

U.S. Army, Navy, and Air Force (1971), "Dewatering and Groundwater Control for Deep Excavations," Army TM 5-818-5; Navy NAVFAC P-148; Air Force AFM 88-5, Chapter 6, April 1971, p. 37.

Weber, Ernest M., and A. A. Hassan (1972), "Role of Models in Groundwater Management," *Water Resources Bulletin, Journal,* American Water Resources Association, Urbana, Ill., Vol. 8, No. 1, pp. 198-206.

CHAPTER FOUR

FLOW-NET CONSTRUCTION

4.1 INTRODUCTION

Flow nets are a very practical and useful tool for solving many kinds of seepage and drainage problems, and those who possess skill in their construction have at their disposal an extremely powerful technique for designing safe and economical structures exposed to the damaging effects of water. This chapter contains helpful suggestions and step-by-step examples to enable the reader to draw suitable flow nets for a wide variety of conditions he may encounter. Also Section 3.3 contains detailed information about the properties and characteristics of flow nets and their many practical uses.

The thoughtful study of well-constructed flow nets can be of much benefit to the beginner because it can help him develop a general appreciation for the nature of seepage through porous media. In this text flow nets are presented for a wide variety of cross sections (nearly 100 examples are given). Also, a recent publication contains many examples of well constructed flow nets (U.S. Department of the Army, Corps of Engineers, 1986). If a flow net can be found for a cross section approximately similar to one being studied, much time can be saved because the general pattern is already known, and, as additional flow nets are constructed for approximately similar conditions, less and less time is required.

Study of the examples and the suggestions that are given in this book, together with a reasonable amount of practice, should enable engineers to learn to construct flow nets that are sufficiently accurate for many practical purposes. If the important checks described in this chapter are correctly made, flow nets will be fundamentally accurate; however, the beginner will be less likely to develop bad habits if his work is checked by someone who can distinguish errors.

The analysis of seepage by flow nets contributes to the proper design and construction of many kinds of engineering structure. It is generally recognized that the development of important works involving seepage should be executed in the following important steps:

1. Thorough field investigations establishing soil and geologic conditions at project sites.
2. Thorough experienced evaluation of field conditions.
3. Adequate studies and analyses.
4. Economical and sound designs.
5. Adequate specifications and construction.
6. Adequate instrumentation and observation of finished works.

Flow-net studies fit primarily into step 3; however, they are also useful in the study of seepage in completed projects (step 6).

All the flow net examples in this chapter are related to steady-state seepage. Use of flow nets to study moving saturation lines is described in Sec. 3.6. Several of the examples in this chapter are for cross sections simplified to facilitate the presentation of flow net construction techniques, and do not represent recommended designs (e.g., Figs. 4.4, 4.5, 4.6, 4.7, and 4.8). All dams should be designed with positive internal drainage systems as shown in Chapter 6 (Figs. 6.7, 6.12, 6.14, 6.21, and 6.23).

4.2 GENERAL SUGGESTIONS

This section contains a number of practical time-saving suggestions about the construction of flow nets. Much time and effort can be saved in obtaining accurate flow nets if the following procedures are observed:

1. Draw the cross section to be studied on good tracing paper; turn the sheet over and construct the flow net on the reverse side. After the flow net has been completed by freehand sketching trace it on the front side or on fresh paper. Use French curves if a smooth finished product is desired for illustrative purposes. *Freehand lines do not detract from the accuracy or usefulness of a flow net if the basic rules have been properly followed.*
2. Be practical in the number of lines drawn. Do not clutter up the drawing with too many lines, but do not use so few that essential features are lost. Remember that parts of a flow net can be subdivided to any degree that is required to emphasize detail (Fig. 3.5 and Fig. 4.2d).

3. Be practical in selecting a scale for the drawing. A scale that is too large wastes time and erasers. An 8½-by-11-in. sheet is a good size for many flow nets, although a sheet several feet wide may sometimes be needed for the analysis of a levee or dam with a wide berm or a wide upstream blanket and for groundwater studies of wide cross sections.

4. Before starting to sketch a flow net look for important boundary conditions and for prefixed flow net lines (Figs. 4.2a and 4.3a). In every flow net some flow lines and some equipotentials are established by the boundary conditions before the flow net is started. *Noting the prefixed lines helps one to get the general shape somewhat correct in the beginning.*

5. In drawing flow nets for composite sections (those having more than one permeability) look for dominating parts of a cross section. Highly pervious or highly impervious parts sometimes have a major influence on flow patterns (Figs. 4.7 and 4.8).

6. Do not overlook the overall shape while working on details and never refine a small portion of a flow net before other parts have been fairly well developed. Corrections made in one part have some influence on every other part.

7. Study the basic rules that must be carried out and follow them (Sec. 3.3).

When the basic rules are *not* followed serious errors can be made. Figure 4.1 shows two of the common errors made by beginners. Figure 4.1a shows extraneous equipotentials, both upstream and downstream from the structure, and Figure 4.1b shows flow lines disappearing into the boundary below the permeable formation. By studying a few correctly constructed flow nets such errors should be eliminated (i.e., Figs. 4.2d and 4.3e).

4.3 TYPES OF FLOW NET

Flow nets can be of several types, depending on the configuration and the number of zones of soil or rock through which seepage is taking place. A primary subdivision can be made that depends on the following conditions:

1. *Flow is confined* within known saturation boundaries and the phreatic line is therefore known.

2. *Flow is unconfined* and the upper level of saturation (the phreatic line) is *not* known.

A second subdivision can be made that depends on whether *(a)* the cross section can be drawn as one zone or unit of a single permeability or *(b)* the cross section contains two or more zones or units of different permeabilities. The latter is described in this book as a *composite section.*

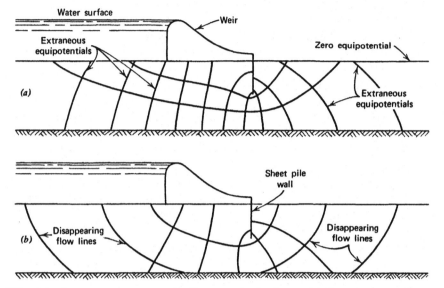

FIG. 4.1 Some common errors in flow nets by beginners.(*a*) Extraneous equipotentials. (*b*) Disappearing flow lines.

These criteria give four possible combinations of flow conditions:

1a. *Confined flow in single permeability sections.*
1b. *Confined flow in composite sections (those having two or more permeabilities).*
2a. *Unconfined flow in single permeability sections.*
2b. *Unconfined flow in composite sections.*

Any of these types can be constructed for *isotropic* soil (horizontal and vertical permeabilities equal) or for *anisotropic* soil (horizontal and vertical permeabilities *not* equal). The rule for correcting for anisotropic conditions in which a single transformation factor is applied to an entire cross section is given in Chapter 3 (Sec. 3.3).

Step-by-step procedures for constructing flow nets are discussed in the remaining part of this chapter, starting with the simplest type and progressing to the more complex.

4.4 CONFINED FLOW SYSTEMS (Phreatic Line Known)

Example 1 Seepage Under Sheet Pile Cutoff Wall (single permeability section). Figure 4.2 shows a vertical sheet pile wall in a sandy foundation. This example represents type 1*a* flow because all seepage boundaries are defined in

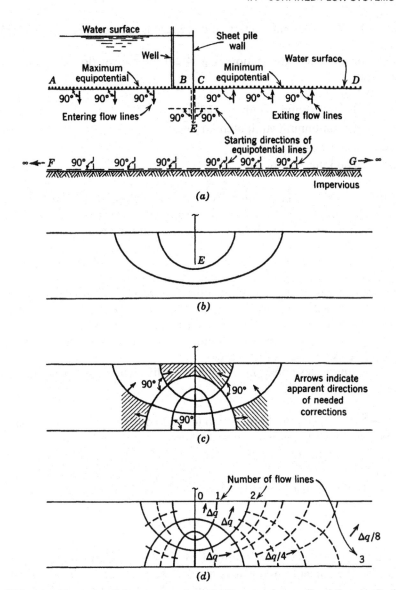

FIG. 4.2 Example of the construction of a flow net for confined flow in single permeability section (Example 1). (*a*) Identify prefixed conditions, noting starting directions of lines. (*b*) Draw trial family of flow lines (or equipotentials) consistent with prefixed conditions. (*c*) Keeping the lines drawn in (*b*), sketch first trial flow net. Make all lines intersect other set of lines at 90°. (*d*) Erase and redraw lines until all figures are square. Subdivided as desired for detail and accuracy.

advance of the flow net construction and seepage is restricted to the uniform sand layer.

By inspection of the cross section (Fig. 4.2*a*) the known flow net lines are the following:

Line *AB:*   The maximum equipotential line.
Line *CD:*   The minimum equipotential line.
Line *BEC:*   The shortest flow line.
Line *FG:*   The longest flow line.

With the knowledge that equipotential lines and flow lines must intersect one another at right angles, the directions of lines in the flow net are established at certain points. Thus all flow lines must enter the foundation at right angles to line *AB* and exit perpendicular to the line *CD;* also, equipotential lines must intersect lines *BE, EC,* and *FG* at right angles. The directions of these intersecting lines are shown at several random points in Figure 4.2*a*. Though the positions of the lines shown do not necessarily coincide with any that will be developed in the flow net, their directions give useful clues regarding the shape of the completed flow net. With the known restrictions in mind and knowing that all the lines will begin to curve soon after leaving fixed boundaries, the sketching can be started.

Flow nets can be developed in a number of different ways, including the following:

1. Immediately sketching both flow lines and equipotential lines, working from one part of a flow net into another.
2. Sketching a plausible family of flow lines before starting to sketch equipotentials.
3. Sketching a plausible family of equipotential lines before starting to sketch flow lines.
4. Combinations of the above or any other suitable procedures.

In Figure 4.2*b* the flow net is started by drawing two intermediate flow lines, making the entrance and exit angles 90° and keeping in the mind that seepage tends to concentrate at a focal point such as the bottom of the wall at point *E*. The next step (Fig. 4.2*c*) is to draw a family of equipotentials, *making all intersections with flow lines right angles and trying to make some of the figures squares*. This procedure, applied to this example, gives the first trial flow net (Fig. 4.2*c*). On cursory examination this flow net might appear to be fairly satisfactory, but a detailed examination shows that a number of the figures are not squares. If all intersections are 90° angles, the figures will be rectangular, as desired; however, if diamonds or other off-shaped figures are formed, lines should be deliberately shifted until all intersections are 90°

angles and all figures are basically rectangular in shape. The figures that are not squares and the apparent directions in which lines need to be shifted to convert the figures into squares should now be noted. The word "apparent" is used because we cannot always be sure at this stage just which of the initial adjustments will prove to be correct, since the shifting of any line alters the amount of the subsequent corrections required of intersecting lines. In Figure 4.2c some of the figures that definitely are not squares are crosshatched. The apparent directions of needed corrections of this flow net are indicated by the arrows.

By repeatedly erasing and redrawing portions of flow lines and equipotential lines and studying the overall pattern and individual parts a finished flow net can be developed (Fig. 4.2d). Portions of this flow net have been subdivided one or more times to give greater detail. (See also Fig. 3.5.)

Sometimes, when drawing flow nets, it is desirable to erase all lines, leaving only a trace of the most recent for redrawing and improving the entire flow net.

If a flow net has a whole number of flow channels (2.0, 3.0, 4.0, etc.), the corresponding number of equipotential spaces depends on the shape of the cross section and will not necessarily be a whole number. The finished flow net in Figure 4.2d has a whole number of flow channels and equipotential spaces; but if the shape had been slightly different, the number of equipotential spaces might have been 5.6, 6.5, or some other fractional number.

Likewise, if a whole number of equipotential spaces was selected for a flow net, the number of flow channels could be a fractional number. In this example (Fig 4.2) a slightly different shape could have produced 2.7, 3.5, or some other fractional number of flow channels. The flow net in Figure 4.4c with 1.2 flow channels is an example.

There is no objection to the construction of flow nets with fractional numbers of flow channels and equipotential spaces, but usually one or the other should be made a whole number.

Example 2   Flow Beneath Concrete Weir (single permeability section).   Figure 4.3 is another example of confined flow in a section of one permeability (type 1a). Here, as in Example 1, all seepage boundaries are known before the flow net is started. The known flow net lines are the following (Fig. 4.3a).

Line *AB:*	The maximum equipotential line.
Line *CD:*	The minimum equipotential line.
Line *BEFGHC:*	The upmost flow line.
Line *IJ:*	The longest flow line.

Known directions of lines in the flow net that intersect these prefixed lines at right angles are shown at arbitrary points in Figure 4.3b.

In Figure 4.3c the flow net is started by drawing two intermediate flow lines,

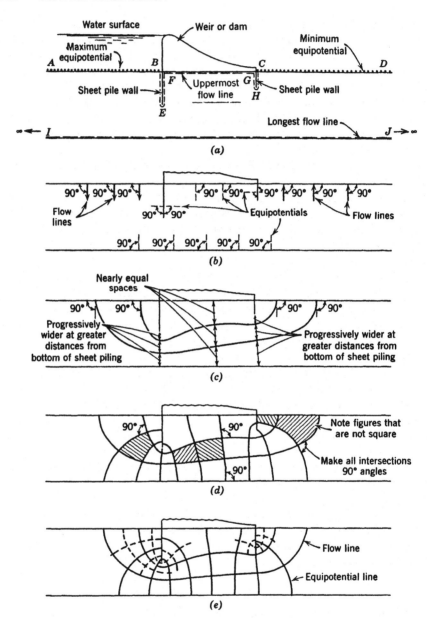

FIG. 4.3 Another example of a method of constructing flow nets with all boundaries known (Example 2). (*a*) Identify prefixed flow lines and equipotential lines. (*b*) Look for prefixed starting directions of lines.(*c*) Draw trial family of flow lines (or equipotentials) consistent with prefixed conditions. (*d*) Keeping the lines drawn in (*c*), sketch first trial flow net. Make all lines intersect other set of lines at 90° angles. (*e*) Erase and redraw lines until all figures are squares. Subdivide as desired for detail and accuracy.

noting that flow radiates around the bottoms of the cutoff walls and that the lines tend to be somewhat equally spaced under the centers of wide structures.

In Figure 4.3*d* a trial family of equipotentials has been added to obtain a trial flow net, and several nonsquare figures have been crosshatched. The flow net obtained by progressive correction of errors is given in Figure 4.3*e*.

4.5 UNCONFINED FLOW SYSTEMS (Phreatic Line Unknown)

In the examples given in Sec. 4.4 the line of saturation, or *phreatic* line, is known in advance, which makes the procedure for obtaining the flow nets simpler than it is for cross sections in which the upper saturation line is not known. Cross sections with an unknown phreatic line are considered the most challenging because the phreatic line must be located simultaneously with the drawing of the flow net. The procedures described in this section permit flow nets of this kind to be constructed with minimum time and effort.

Example 3 Seepage Through Homogeneous Earth Dam.   The cross section and known conditions for a type 2*a* flow net are given in Figure 4.4*a*. Line *AB,* the face of the dam, is the maximum equipotential line; line *AC,* the base of the dam, is a flow line. The exact position of the phreatic line is unknown, but it can reasonably be expected to lie somewhere within the shaded zone *BDE.* The general condition at the free surface is known (Fig. 3.6) and it is known that the net must be composed of squares.

Before starting to construct a flow net with an unknown phreatic line, the total head h should be divided into a convenient number of equal parts Δh, and light guidelines should be drawn across the region in which the phreatic line is expected to lie. In Figure 4.4*a* four intermediate guidelines (for convenience, called *head* lines in this section) are drawn at a vertical spacing $\Delta h = h/5$.

The conditions that establish the position of the phreatic line in Fig. 4.4 are the following:

1. Equal amounts of head must be consumed between adjacent pairs of equipotential lines.
2. Equipotential lines must intersect the phreatic line at the correct elevation.
3. To satisfy requirements 1 and 2, each *equipotential* line must intersect the phreatic line at the appropriate *head* line. this key requirement must be satisfied by all flow nets having an unknown phreatic line.

After drawing the horizontal guidelines or *head* lines across the region in which the phreatic line is expected to lie a *trial saturation line* (line *ab,* Fig. 4.4*b*) should be drawn; make a reasonable guess about its probable location

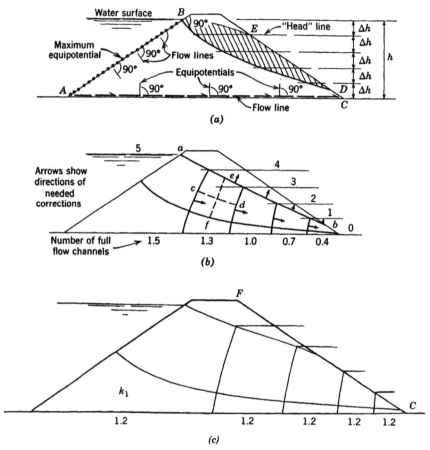

FIG. 4.4 Flow net for homogeneous dam on impervious foundation (Example 3). (*a*) Known conditions. (*b*) Trial saturation line and flow net. (*c*) Final flow net.

and shape. Next a family of equipotential *lines should be drawn, making all intersections with flow lines right angles.* Then draw one or more intermediate flow lines to establish a trial flow net as shown in Figure 4.4*b*. If all intersections are right angles, the first trial net will be composed of rectangular figures. *Some* of the figures may happen to be squares, but most will probably be elongated rectangles, as in this example.

Usually it is possible by inspection alone to note the kinds of correction that must be made and to develop accurate flow nets with an indeterminate phreatic line (e.g., Fig. 4.4*c*). At this point it is well to review a systematic checking method that takes the guesswork out of flow nets for unconfined seepage. This check is based on the rule: *In any flow net the number of flow channels must remain the same throughout the net.* This rule simply states that all the water entering a cross section must flow through the section and emerge at the low potential side. It applies to all sections that have no secondary sources feeding water into them or drains removing water from them.

If a flow net has been correctly constructed and *is composed entirely of squares,* the rule given here is observed automatically. An example is the flow net in Figure 4.3*e,* which has two intermediate flow lines and eight intermediate equipotential lines. To satisfy this rule each adjacent pair of equipotential lines must encompass the same number of "square" figures, a square being a figure that meets the requirement described in Sec. 3.3. A "square" is not usually a true square, for it is enclosed in curved lines, but *it satisfies the fundamental requirement if its average width equals its average length.* In Figure 4.3*e* each pair of equipotential lines encompasses three squares; hence the basic rule is met. A flow net that does *not* fulfill this rule is the trial flow net in Fig. 4.4*b;* another that *does* fulfill it is the finished flow net in Figure 4.4*c.* Because the last two flow nets are somewhat similar in appearance, how is it possible to detect inaccuracies and make the necessary corrections? It can be done in two ways.

1. By inspection note the figures in a trial flow net that are *not* squares and shift lines in directions that appear reasonable; for example, the arrows in Figure 4.4*b* show initial adjustments that are judged from visual inspection to be necessary. If this procedure is correctly applied, subsequent trial flow nets will become progressively more accurate until *the flow net is composed of squares and has the same number of flow channels throughout* (Fig. 4.4*c*). If you are not confident enough to improve flow nets by inspection alone, then use the second method.

2. Use an engineer's scale to measure the widths and lengths of figures in a trial flow net and calculate the number of flow channels between pairs of equipotentials. A figure that has a width-to-length ratio of 1.0 contains 1.0 flow channel; one that has a width-to-length ratio of 0.5 contains 0.5 flow channel, and so on. Thus in Figure 4.4*b* lines *cd* and *ef* are equal and the ratio of *ef* to *cd* is 1.0; hence one full flow channel is contained in this figure. the fractional space below this square has a width-to-length ratio of about 0.3; therefore, the number of flow channels between equipotentials 3 and 4 is 1.0 + 0.3, or 1.3. This number is written below the base of the dam as shown. A similar procedure of measuring figures and recording the total number of flow channels is carried out for the balance of the flow net, as indicated in Figure 4.4*b.* If the numbers written down are not all nearly equal, the flow net is out of balance and must be redrawn until these numbers are the same. In this trial flow net (Fig 4.4*b*) these numbers are not at all equal. The need for the adjustments indicated by the arrows is now apparent. This adjustment will raise the phreatic line above the trial position in Figure 4.4*b.* Generally, if the "squares" above a given point are elongated horizontally, the phreatic line is *too low* above that point; if the figures are elongated vertically, it is *too high* above the point.

To correct a trial flow net that has an indeterminate phreatic line the line should be raised if the first guess is too low or lowered if the first guess is too high. This correction automatically forces the equipotential lines to be moved in the correct direction. After the phreatic line is raised or lowered, a new family of equipotentials should be drawn and the new flow net, examined. If the figures are not all squares and the calculated number of flow channels is not the same throughout, the procedure should be repeated until a satisfactory flow net is obtained (Fig. 4.4c). With experience accurate flow nets can be developed with the help of visual inspection alone. If any doubt exists, the lengths and widths of the figures should be measured with an engineer's scale and the number of flow channels calculated at several places in the section, as described in the preceding paragraph. If this check is properly made, accuracy is guaranteed. As a flow net is improved and all figures become true squares, the phreatic line will be forced into its correct position. In developing this kind of flow net we are solving simultaneously for the flow net and the position of the phreatic line.

When seepage emerges along a sloping surface, such as the downstream slope of a dam (Fig. 4.4c), some of the water flows down the slope and a portion of the flow net is thereby cut off. When counting the number of flow channels in the extremities of flow nets, cut-off portions should be included in the totals. Thus in Figure 4.4c the finished flow net has 1.2 flow channels throughout, even though a portion of the right end of the flow net is cut off. If line CF in Figure 4.4c were a boundary between zones of two different permeabilities k_1 to k_2, the shape of the flow net would depend on the ratio of k_2 to k_1. If the second zone is considerably more permeable than the first $(k_2/k_1 = 1000$ or more), the influence of the second zone on the flow net can usually be ignored. If the ratio k_2/k_1 is in a moderate range, several hundred or less, the method described in Sec. 4.6 for *composite sections* should be used in developing the flow net.

4.6 COMPOSITE SECTIONS WITH PHREATIC LINE UNKNOWN

In this section a method of checking the accuracy of flow nets for composite sections with unknown phreatic lines is described and examples of shortcut methods are given.

Example 4 Zoned Earth Dam. Figure 4.5 illustrates a basic procedure for drawing composite flow nets for sections with unknown phreatic lines. The transfer conditions at boundaries between soils of different permeability materials were described in Sec. 3.3. When water flows across a boundary into a soil of different permeability, the figures in the second soil must elongate or shorten to make the length-to-width ratios of the figures satisfy Eq. 3.11:

$$\frac{c}{d} = \frac{k_2}{k_1} \tag{3.11}$$

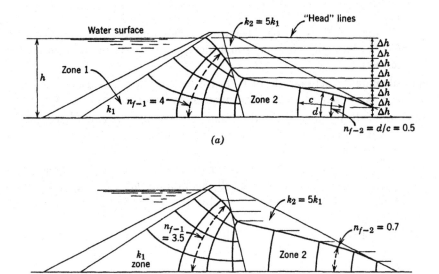

FIG. 4.5 Method for constructing flow nets for composite sections (Example 4). (*a*) First trial flow net (not correct). (*b*) Completed flow net.

In Eq. 3.11 c is the length and d, the width of the figures in the second soil; the permeability of the first soil is k_1, that of the second k_2.

Figure 4.5 shows a common type of composite cross section with an indeterminate phreatic line. It is for a zoned earth dam with a soil of slightly lower permeability in the upstream part than in the downstream part. This section is a type 2b (Sec. 4.3) because it is for unconfined flow in a composite section. In this example the downstream zone is assumed to be five times as permeable as the upstream zone.

To develop a flow net for this type of section these steps should be followed.

1. Locate the *reservoir level* and the *tail water level*, noting the difference in head as h and dividing h into a convenient number of equal parts or increments Δh. Draw a series of light horizontal guide lines (*head* lines) at intervals of Δh across the downstream part of the section (Fig. 4.5a).

2. *Guess* a trial position for the phreatic line in both zones and draw a preliminary flow net as shown in Figure 4.5a, making squares in zone 1 and rectangles in zone 2. Make the length-to-width ratios of all of the rectangles in zone 2 approximately equal by adjusting the shape of the saturation line, using an engineer's scale to measure the lengths and widths of the figures as in Example 3, Figure 4.4. When this step is completed, the trial flow net should be reasonably well drawn. It should satisfy the basic shape requirements of a flow net, but the length-to-width ratio of the shapes in zone 2 probably will not satisfy Eq. 3.11.

Although the flow net has been drawn for a composite section, the ratio of k_2/k_1 probably does not equal the k_2/k_1 ratio originally assumed for the section.

3. Calculate the actual ratio of k_2/k_1 for the trial flow net just constructed. To make this important check proceed as follows:

(a) Count the number of full flow channels between any two adjacent equipotential lines in zone 1 and call this number $n_{f\text{-}1}$. In the trial flow net in Figure 4.5a $n_{f-1} = 4.0$.

(b) Count the number of full flow channels between any two adjacent equipotential lines in zone 2 and call this number $n_{f\text{-}2}$. In Figure 4.5a n_{f-2} is equal to the width-to-length ratio of the figures in zone 2, d/c and equal to 0.5.

(c) The actual value of k_2/k_1 for the trial flow net in Figure 4.5a can now be determined from the equation

$$k_2 = k_1 \frac{n_{f-1}}{n_{f-2}}$$

or $\hspace{10cm}$ (4.1)

$$\frac{k_2}{k_1} = \frac{n_{f-1}}{n_{f-2}}$$

(d) If the calculated k_2/k_1 ratio is *too high,* the saturation line in zone 2 is too low and must be raised. If the calculated k_2/k_1 ratio is *too low,* the saturation line in zone 2 is too high and must be lowered. *Raise* or *lower* the general level of the saturation line in zone 2 as indicated and construct another trial flow net.

(e) Repeat steps (a) through (d) until a flow net of the desired accuracy is obtained. (Usually a few trials will be sufficient.)

By applying Eq. 4.1 to the first trial flow net in this example (Fig. 4.5a) $k_2/k_1 = 4.0/0.5 = 8.0$. Because the ratio of k_2/k_1 for this example was assumed to be 5, the k_2/k_1 ratio of the trial flow net is *too high;* hence the general level of the saturation line in zone 2 is too low and must be raised. For the second trial flow net (Fig. 4.5b) $n_{f-1} = 3.5$ and $n_{f-2} = 0.7$. The *calculated ratio of $k_2/k_1 = 3.5/0.7 = 5.0$,* the value originally assumed.

Equation 4.1 may be derived by recalling that the quantity of seepage in zones 1 and 2 (Fig 4.5) must be equal. Using Eq. 3.18,

$$q = kh \frac{n_f}{n_d}$$ $\hspace{5cm}$ (3.18)

in zone 1, $q = k_1 h(n_{f-1}/n_d)$, and in zone 2, $q = k_2 h(n_{f-2}/n_d)$. For a given head h, $q \sim k_1(n_{f-1}/n_d) \sim k_2(n_{f-2}/n_d)$ and $q/n_d \sim {}_1 k_1 n_{f-1} \sim k_2 n_{f-2}$. Therefore $k_1 n_{f-1} = k_2 n_{f-2}$ and $k_2 = k_1(n_{f-1}/n_{f-2})$. This expression (Eq. 4.1) can be used for determining the permeability ratios k_2/k_1, k_3/k_2, and so on, for any composite flow net being examined for accuracy. It is an essential criterion to be used in constructing accurate flow nets for composite sections.

The fundamental relationships represented in Eqs. 3.11 and 4.1 should be kept in mind when flow nets for composite sections are being studied or constructed.

More than one system of lines can be selected for a flow net. This point is illustrated by Figure 4.6 which gives three sets of lines for the flow net developed in Figure 4.5. Although the three nets appear to be different, they are actually identical. The total head h has been divided into eight parts equal to Δh. As drawn in Figure 4.6a, the portion in zone 1 has been constructed with solid lines that form squares, the portion in zone 2 with elongated rectangles

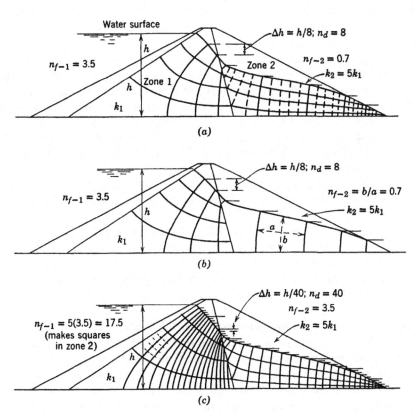

FIG. 4.6   Three forms of one flow net.

(solid lines) with length-to-width ratios of 5. The dashed lines in zone 2 divide the elongated rectangles into squares. For this flow net $n_{f-1} = 3.5$, $n_{f-2} = 0.7$, and $n_d = 8$. The seepage quantity q can be computed by using k_1 or k_2. Using k_1, $q = k_1 h(n_{f-1}/n_d) = k_1 h(3.5/8) = 0.44 k_1 h$. Using k_2, $q = k_2 h(n_{f-2}/n_d) = k_2 h(0.7/8) = 0.088 k_2 h$, but $k_2 = 5 k_1$, and $q = 0.088(5 k_1)h = 0.44 k_1 h$.

The line system in Figure 4.6a is typical of that frequently used for this type of flow net. The dashed lines may be omitted unless they are needed in stability studies or to bring out detail.

Sometimes all intermediate lines in one or more zones may be omitted, as in zone 2, Figure 4.6b, In this flow net, as in Fig. 4.6a the head h is divided into eight parts equal to Δh and the number of equipotential drops $n_d = 8$. In zone 1 the number of flow channels $n_{f-1} = 3.5$; in zone 2 the number of flow channels $n_{f-2} = 0.7$ (as in Fig. 4.6a). Seepage quantities determined by using k_1 and k_2 are identical to those made from the flow net in Figure 4.6a.

A third system of lines for this flow net shown in Figure 4.6c was obtained by dividing the total head h into 40 parts, each one-fifth as large as Δh in Figures 4.6a and 4.6b. The flow net in Figure 4.6c has been constructed with squares in zone 2 and shortened rectangles in zone 1. Because $k_2 = 5 k_1$, the figures in zone 1 must foreshorten to lengths equal to one-fifth of their widths. This relationship can be verified readily by Eq. 3.11. With the flow net drawn as shown in Figure 4.6c, $n_{f-2} = 3.5$ and $n_{f-1} = 3.5(5) = 17.5$, for it is determined by the number of *squares* between equipotentials. The number of equipotential drops $n_d = 40$. Computing the seepage quantity in zone 1, using k_1, $q = k_1 h(n_{f-1}/n_d) = k_1 h(17.5/40) = 0.44 k_1 h$, which was obtained from the flow net drawn in Fig. 4.6a. Computing the seepage quantity in zone 2, using k_2, $q = k_2 h(n_{f-2}/n_d) = k_2 h(3.5/40) = 0.088 k_2 h$, which was also obtained for the flow net drawn in Figure 4.6a. Recalling that $k_2 = 5 k_1$, $q = 0.088(5 k_1)h = 0.44 k_1 h$, as obtained before.

By inspection of the three forms of the flow net in Figure 4.6 we see that the flow patterns and hydraulic gradients are identical in all three. This should be true because this is but one flow net.

If a flow net is being constructed for a section in which k_2/k_1, is large, it may not be practical to subdivide the elongated rectangles in the manner described above. In such cases (Sec. 3.3) the measurement of the lengths and widths of flow-net figures with an engineer's scale can establish the accuracy of basic shapes.

Example 5   Earth Dam on Low Permeability Foundation.   Figure 4.7 shows an earth dam (or levee) on a foundation that is relatively low in permeability in relation to the embankment. In this problem it is assumed that the permeability of the foundation k_f is one-tenth of the permeability of the embankment k_e. When seepage occurs through two dissimilar but more or less parallel zones, as in Figure 4.7, flow through the more permeable zone dominates the combined seepage pattern. Time can often be saved by constructing a flow net for the more permeable part, the dam in this case, assuming temporarily that

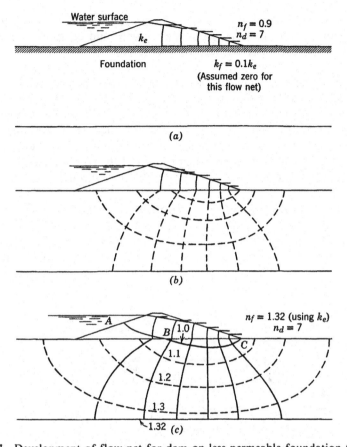

FIG. 4.7   Development of flow net for dam on less permeable foundation (Example 5). (*a*) Construct flow net for dam, assuming foundation is completely impervious. (*b*) Extend equipotential lines into foundation without adjusting lines of net in dam. (*c*) Adjust equipotentials and flow lines until a balanced flow net is obtained.

the other part is completely impervious (Fig.4.7*a*). The equipotentials are then extended down into the less permeable part (Fig. 4.7*b*), and by trial and error a balanced flow net is developed (Fig. 4.7*c*). When developing and using flow nets of this kind, fundamental relationships previously discussed should be kept in mind. In the completed flow net in Figure 4.7*c* one flow channel in the foundation conducts only one-tenth as much seepage as one flow channel in the more permeable embankment. For this reason the flow lines in the foundation are shown as dashed lines because they do not enclose full flow channels. The embankment in Figure 4.7 accommodates nearly one full flow channel; however, the first full flow channel (solid line) dips slightly into the foundation. Seepage beneath this flow line adds only about 0.32 of a full flow channel, thus raising the total value of n_f to 1.32. The number of equipotential drops in this flow net is 7 and the shape factor is 1.32/7 or 0.19. Through the embankment alone (Fig. 4.7a) the shape factor would be approximately 0.9/ 7, or 0.13; hence the relatively deep foundation increases the water losses only about 50%.

In constructing a flow net for a section of the type represented by this example, it is helpful to think in terms of the water-conducting capabilities of soils of different permeabilities. Thus, if a foundation has lower permeability than an embankment, as in this example, a greater thickness of the lower permeability soil is needed to conduct the same amount of water. Because most of the flow is through the embankment, the position of the equipotential lines in the embankment is influenced only slightly by the flow through the foundation.

Example 6   Earth Dam on a Highly Permeable Foundation.   Figure 4.8 illustrates the same physical cross section used in Figure 4.7, but here the foundation is 10 times more permeable than the embankment. In this example the general shape of the net is controlled more by the foundation than by the dam.

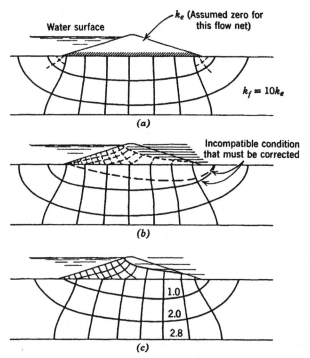

FIG. 4.8   Development of flow net for dam on more permeable foundation (Example 6). (a) Construct flow net assuming dam is completely impervious. (b) Extend equipotentials up into dam locating initial position of saturation line in dam. (c) Adjust equipotentials and flow lines until a balanced flow net is obtained.

A good starting procedure is to draw a flow net for the foundation, assuming temporarily that the embankment is completely impervious. Such a net (Fig. 4.8a) is for a *confined flow system* in which the upper saturation line (the phreatic line) is known. This trial net is the type illustrated in Figure 4.3. After this trial net has been drawn (Fig. 4.8a) the equipotential lines are extended up into the embankment (Fig. 4.8b). The lines in both dam and foundation are then adjusted until a flow net (Fig. 4.8c) compatible with the assumed differences between the permeabilities of the two soil units is obtained. Before extending the flow net into the embankment, the total head h should be divided into the correct number of parts (eight in this example) to conform to the number of equipotential drops in the trial flow net in the foundation. As in Figure 4.4a, horizontal *head* lines are drawn across the downstream part of the dam and each equipotential line must intersect the correct head line at the phreatic line. (See also Figs. 3.6 and 4.4c.)

In developing flow nets of the type illustrated in Figures 4.7 and 4.8, the position of the phreatic line must be adjusted simultaneously with the refinement of the flow net. This type of flow net is the most difficult to construct. At the start of the construction, time taken to appraise the broad nature of the flow pattern can save much effort. Frequently shortcuts of the kind illustrated in Figures 4.7 and 4.8 help to obtain correct solutions with minimum work. Initial approximations of the type made in these two examples should never be permitted to overshadow the basic rules governing flow nets.

4.7 EXAMPLES OF COMPLEX FLOW NETS

When the principles illustrated in the preceding examples are carefully followed, interesting and useful flow nets can be developed for a wide variety of cross sections. Two examples of fairly complex flow nets are given here in Figures 4.9 and 4.10. When large differences in permeabilities exist in various zones as in Figure 4.9 considerable detail may be needed in one material but only a few lines may be needed in another when the basic checks are carried out properly. The flow net in Figure 4.9 is for an earth dam and foundation assumed to have horizontal permeabilities nine times vertical; hence the flow net is constructed on a *transformed section* (Fig. 4.9b) in which the horizontal dimensions have been reduced by the square root of the ratio of vertical-to-horizontal permeabilities, or $\sqrt{1/9} = 1/3$ (see sec. 3.3). the flow net is then redrawn on the natural section, as in Figure 4.9a.

Figure 4.10 gives two flow nets (constructed on *transformed sections* for $k_h = 25\ k_v$) in a study of seepage beneath a dam built in a natural saddle separating a reservoir on the left from a lower valley on the right. Beneath a relatively thin layer of low permeability soil is a stratum of highly permeable sand and gravel to a depth of about 50 ft. Beneath the pervious sand and gravel layer is a 50- to 60- ft thick silty formation, which in turn is underlain by fairly permeable gravelly materials. Far to the right is a fairly steep slope (about 1

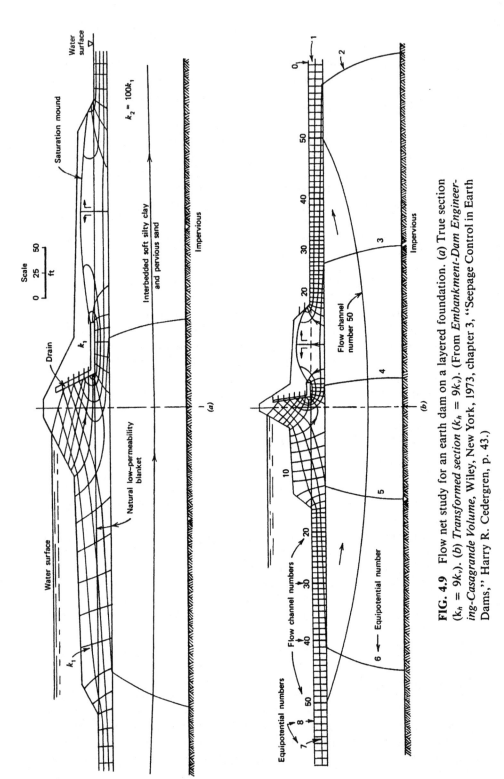

FIG. 4.9 Flow net study for an earth dam on a layered foundation. (*a*) True section ($k_h = 9k_v$). (*b*) *Transformed section* ($k_h = 9k_v$). (From *Embankment-Dam Engineering-Casagrande Volume*, Wiley, New York, 1973, chapter 3, "Seepage Control in Earth Dams," Harry R. Cedergren, p. 43.)

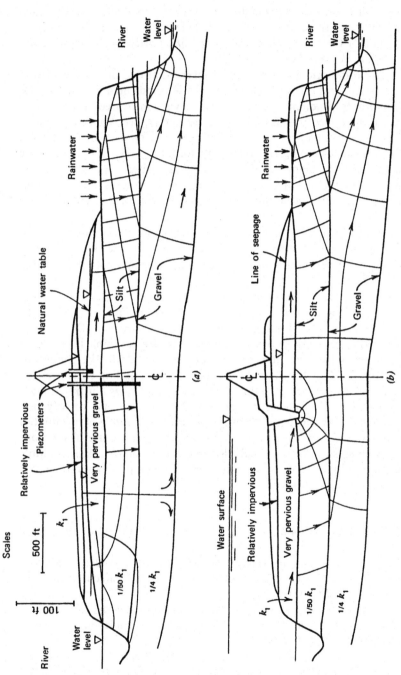

FIG. 4.10 Flow net study for a saddle dam on a layered foundation, transformed sections, $k_h = 25\, k_v$. (a) Preexisting conditions, (b) Full reservoir, steady state. (From *Embankment Dam-Engineering-Casagrande Volume*, Wiley, New York, 1973, Chapter 3, "Seepage Control in Earth Dams," Harry R. Cedergren, p. 33.

on 1) down to a river. On the transformed section this slope appears much steeper than it really is because horizontal dimensions have been shrunk to 1/5 of natural (for the assumption that $k_h = 25k_v$, horizontal dimensions are multiplied by $\sqrt{k_h/k_v} = \sqrt{1/25} = 1/5$).

Preexisting piezometric levels had been observed at two or more depths at a number of locations at this site. These observations indicated the existence of downward hydraulic gradients, as smaller heads were measured in the bottom gravels than in the upper gravels at a given piezometer. Seepage evidently was escaping freely from the lower gravels where they emerged on the river bank at the right. Springs emerging from the silty formation relatively high above the river evidently were being fed by infiltering rainwater in the general area being studied. Without such a supplemental source of water, the line of seepage could not have emerged at such high levels unless the soil cross section were very different from that shown.

The flow net in Figure 4.10a, which is for preexisting conditions, was constructed so that the equipotential lines were consistent with the observed piezometer readings. Flow nets constructed for a variety of design options indicated that the safest procedure would be to excavate a cutoff trench through the upper permeable gravels, into the silt layer, and tie the cutoff backfill into the impervious core of the dam, as shown by the flow net in Figure 4.10b, which indicates very safe conditions at the downstream toe of the dam. Details such as locations of filters are not shown in this drawing. This example is presented here as it illustrates the way flow nets should be adapted to the conditions existing at proposed sites for dams or other water-impounding structures.

In the construction of complex flow nets, such as the ones illustrated here, the most important single check is that provided by Eq. 3.11, namely that figures be elongated or shortened in proportion to the *relative* permeabilities of the various zones in a cross section. Many other extremely useful solutions to everyday problems can be obtained with some of the simplest kinds of flow net; hence one need not develop skill with the more complex types to be able to make good use of the flow net.

SUPPLEMENTAL READING

Casagrande, A., "Seepage Through Dams," *Journal,* New England Water Works Association, June 1937, pp. 131–170; also publication No. 209, Graduate School of Engineering, Harvard University.

Cedergren, Harry, R., "Use of Flow Net in Earth Dam and Levee Design," *Proceedings,* 2nd International Conference on Soil Mechanics and Foundation Engineering, Rotterdam (1948), Vol. 5, pp. 293–298.

U.S. Department of the Army, Corps of Engineers (1986), "Seepage Analysis and Control for Dams," Engineering Manual EM 1110-2-1901, Office of the Chief of Engineers, Washington, D.C., 30 September 1986.

REFERENCES

U.S. Department of the Army, Corps of Engineers (1986), "Seepage Analysis and Control for Dams," Engineering Manual EM 1110-2-1901, Office of the Chief of Engineers, Washington, D.C., 30 September 1986.

PART II

APPLICATIONS

CHAPTER FIVE

FILTER AND DRAIN DESIGN

5.1 BASIC REQUIREMENTS OF FILTERS AND DRAINS

Properly designed filters and drains are essential for the safety and economy of essentially all civil engineering facilities exposed to the damaging actions of water in their foundations or other supporting soil or rock formations. But if filters and drains are *not* properly designed (to ensure both good filter protection for erodible materials, and adequate discharge capacity), both the safety and economy of the facilities can be endangered. This chapter shows how to achieve both requirements.

The process by which percolating water or groundwater is removed from soils and rocks by natural or artificial means is called *drainage*. The fundamental principles that apply to seepage also apply to drainage. When an analysis is being made of the best means of controlling water in engineering works, it is important to try to identify the source or sources of the water. In some cases it may be possible to reduce or entirely cut off the inflows by means of seepage-reducing methods such as blankets, linings, cutoffs, and grout curtains. In most cases, however, the safest, most economical and satisfactory solution is achieved by drainage systems (often in combination with seepage-reducing methods). Frequently drainage is the primary if not *only* means of control. Of great importance in drainage design is the need for developing systems capable of removing all the water that reaches them without excessive head build-up (Sec. 5.4) and without clogging or piping (Sec. 5.2). *Furthermore, designers should analyze every component of a drainage system (filters, conducting layers, collectors, outlets, and so on) to ensure that the entire system will have the necessary capacity and will function as intended.*

Sound rocks can usually be drained by allowing water to escape freely at exposed surfaces in drain wells, tunnels, etc., because these materials have sufficient cohesion to resist erosion. But soft, weathered rocks and most soils present more difficult drainage problems, as unprotected surfaces of these materials can be eroded by the forces of escaping water. If the erosion process is permitted to begin, it can lead to clogging of filters and drains and, in extreme cases, piping failures (E.N.-R., 1976) (Civil Engineering Magazine, 1981). Consequently, drainage surfaces of erodible soils and rocks must be covered with special protective filter layers that allow the water to escape freely but hold the particles firmly in place.

Good quality aggregates are virtually indestructible, relatively incompressible, readily available in many areas, and relatively inexpensive. When used correctly, porous drainage aggregates can have a vital part in the permanent performance of a great many kinds of civil engineering works. They are frequently used in drainage systems in conjunction with slotted, jointed, or porous pipes, which assist in the collection and removal of seepage.

Since about 1965 synthetic filter fabrics have taken on increased importance in drainage systems. Although they have not been fully time-tested, there is considerable evidence that they are viable construction materials for many kinds of civil engineering works. Seemel (1976) says that filter fabrics have three basic functions: (1) *separation* of two widely different natural granular materials to prevent mixing under the influence of water, (2) *filtering*, and (3) *reinforcement*. In drainage systems filter fabrics are often used as a substitute for a fine aggregate filter in regions in which good quality aggregates are scarce or costly or in situations in which a filter fabric may be easier to install than a fine aggregate filter. They are becoming widely used for enveloping strong, highly permeable plastic cores in "fin" drains and "wick" drains (see Sec. 5.7). There are many types of filter fabric with a wide range in filaments and textures. Designers should make sure that any they specify have the required properties to function as intended for the life of an individual project.

If filters and drains are to serve their intended purpose, the materials used in their construction must have the correct gradation (Sec. 5.2) and they must be handled and placed with care to avoid contamination and segregation (Sec. 5.8). Close control is required in the production, handling, and placement of the materials, for even a single improperly constructed portion of a filter can lead to problems.

When small quantities of seepage are to be removed, a single layer of well-graded, moderately permeable material meeting both requirements may sometimes serve the dual roles of filter and drain, but usually one or more filter layers are needed for the prevention of piping and a coarse layer, for the removal of water. Such systems are called *graded filters* or multiple-layer drains. Filters covered with surcharges to prevent uplifting by seepage forces are called *loaded* filters or *weighted* filters.

Criteria that will ensure proper functioning of drains while meeting the basic *piping criterion* (No. 1) are presented in Sec. 5.2 and illustrated in Sec. 5.3.

Criteria and analysis methods ensuring that drains will have sufficient discharge capacity to meet the *permeability requirement* (No. 2) are presented in Sec. 5.4. Frequently the *permeability requirement* can be met only with coarse, open-graded aggregates. When such materials are used for drainage of erodible soils or rock formations, fine filters must be used to prevent piping failures and clogging problems. Fine filters for drains serve two different functions: (a) as true *filters* to allow free flow of water into the coarse layer of a drain while holding back the erodible material or (b) as *separators* or barriers to keep fine erodible materials out of coarse drainage layers when there is no significant flow of water across the filter.

Figure 5.1 illustrates these two functions of filters. The coarse rock buttress fill at the left and the trench drain both need true filter protection, whereas the open-graded base drainage layer under the pavement needs only a separator to prevent mixing with the subgrade. A conventional subbase can usually serve this purpose. Groundwater or sidehill seepage can be expected into the buttress and trench drain, but no significant upward flow of water into the roadbed is anticipated under the conditions shown here.

Drains and filters that are designed and built in accordance with good practices ought to function for the life of an installation with only moderate upkeep and maintenance. Nevertheless, operators of projects should be made aware of the need to avoid any action that could harm a drainage system and to be alert for any changed conditions that could reduce the efficiency of these systems.

Cases have been reported of drainage layers and pipes becoming partly clogged by iron oxide buildup, calcareous material, and other material deposited by seeping water. If seepage quantities and uplift pressures are thoroughly monitored, as recommended elsewhere in this book, these readings should forewarn operators of any major decreases in the efficiency of drainage systems.

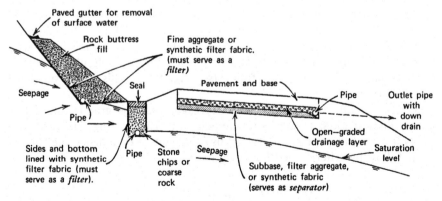

FIG. 5.1 Illustration of *filter* and *separator* functions of protective "filters" for drains.

Grass and MacKenzie (1972) discuss procedures for restoring subsurface drain performance for two types of blockage: (1) silt and roots and (2) mineral deposits. They say that large roots in drain lines can be removed with long, flexible, rotating rods equipped with cutting blades. High-pressure water jets have been successfully used for removing silt from drains. These investigators also say that mineral deposits have been removed by injecting a 2% mixture by weight of SO_2 gas and water after flushing with water to remove loose materials.

5.2 PREVENTION OF PIPING

General

To prevent piping, water-bearing erodible soils and rocks must never be in direct contact with passageways large enough to allow appreciable loss of the erodible material.

Grading of Drainage Aggregates to Control Piping

To prevent the movement of erodible soils and rocks into or through filters, the pore spaces between the filter particles should be small enough to hold some of the larger particles of the protected materials in place. Taylor (1948) shows that if three perfect spheres have diameters greater than six and one-half times the diameter of a smaller sphere (Fig. 5.2a) the smaller sphere can move through the larger. Soils and aggregates are always composed of *ranges* of particle sizes, and if the pore spaces in filters are small enough to hold the 85% size (D_{85}) of adjacent soils in place the finer soil particles will also be held in place (Fig. 5.2b). Exceptions are gap-graded soil and soil-rock mixtures. When filter fabrics are used, the protective filter is only the thickness of the fabric, which may be as little as 1/20 in. (about 1 mm). It is therefore extremely important that no holes, tears, or gaps be allowed to form in the fabric. In addition, the openings between the filaments of a fabric should not be so large that significant loss of soil can occur. If the D_{85} of a soil is larger than the near maximum opening size (P_{85} of fabric), little soil should be able to move through the mesh of the fabric (see Fig. 5.2c). This is my suggested tentative criterion.

Bertram (1940), with the advice of Terzaghi and Casagrande, made laboratory investigations at the Graduate School of Engineering, Harvard University, to test filter criteria suggested by Terzaghi; he established the validity of the following criteria for filter design:

$$\frac{D_{15}(\text{of filter})}{D_{85}(\text{of soil})} < 4 \text{ to } 5 < \frac{D_{15}(\text{of filter})}{D_{15}(\text{of soil})} \tag{5.1}$$

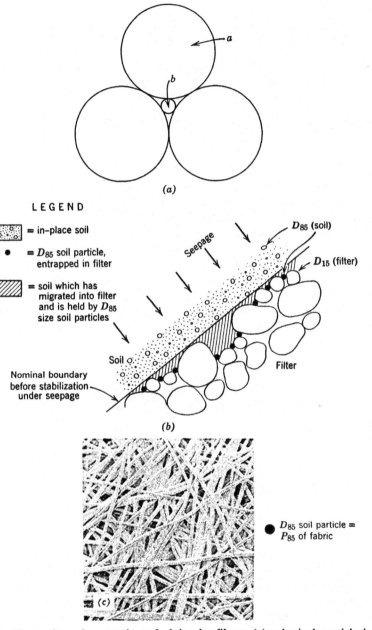

FIG. 5.2 Illustration of prevention of piping by filters. (*a*) spherical particle *b* will just pass through pore space between three spheres six and one-half times the diameter of *b* (Taylor, 1948). (*b*) Conditions at a boundary between a soil and a protective filter. (*c*) Magnified microphotograph (20 ×) of a nonwoven filter fabric, Mirafi-140 (Mirafi® is a trademark of Fiber Industries, Inc., a subsidiary of Celanese Corporation), also shows minimum D_{85} soil particle safely held in place by the fabric (assuming P_{85} of fabric should be no larger than D_{85} of protected soil; criterion suggested by author). (Photo, courtesy of Celanese Corp.)

The left half of Eq. 5.1 may be stated as follows:

Criterion 1. The 15% size (D_{15}) of a filter material must be not more than four or five times the 85% size (D_{85}) of a protected soil. The ratio of D_{15} of a filter to D_{85} of a soil is called the *piping ratio*.

The right half of Eq. 5.1 may be stated as follows:

Criterion 2. The 15% size (D_{15}) of a filter material should be at least four or five times the 15% size (D_{15}) of a protected soil.

The intent of criterion 2 is to guarantee sufficient permeability to prevent the buildup of large seepage forces and hydrostatic pressures in filters and drains. This criterion is discussed in detail in Sec. 5.4.

The work of Bertram was expanded in experiments by the U.S. Army Corps of Engineers (1941) and the U.S. Bureau of Reclamation (Karpoff, 1955). Frequently some requirements in addition to criteria 1 and 2 are placed on the grading of filter aggregates; for example, the U.S. Bureau of Reclamation limits the maximum size of filter aggregates to 3 in. to minimize segregation and bridging of large particles during placement. To prevent the movement of soil particles into or through filters the U.S. Army Corps of Engineers (1955) and the U.S. Army et al. (1971) require that the following conditions be satisfied:

$$\frac{15\% \text{ size of filter material}}{85\% \text{ size of protected soil}} \leq 5 \tag{5.2}$$

and

$$\frac{50\% \text{ size of filter material}}{50\% \text{ size of protected soil}} \leq 25 \tag{5.3}$$

It can be seen that Eq. 5.2 is another expression of the relationship given by the left half of Eq. 5.1.

The U.S. Department of the Army, Corps of Engineers (1986), says that if the 15% size of a filter is at least five times the 15% size of the protected soil, the filter will be approximately 25 times more permeable than the protected soil. In relation to filter protection, they say that the filter criteria (Eqs. 5.2 and 5.3) are applicable for all soils with gradation curves approximately parallel to those for the chosen filter material except for medium to highly plastic (CL to CH) clay soils. Where the gradation curves are not approximately parallel, the filter design is based on filtration tests. For CL and CH soils without sand or silt partings, the 15 percent size of the filter (in Eq. 5.2) may be as great as 0.4 mm, and Equation 5.3 may be disregarded. However, if the drained material should contain partings or strata of uniform nonplastic fine sand and silt sizes, the filter must be designed to meet the stability and permeability criteria.

In any instances where filter materials meeting the criteria given by Equations 5.1 to 5.3 in relation to a material being drained are too fine to prevent filter material from entering into circular or slotted holes in drain pipes (see Eqs. 5.4 and 5.5), the Army Corps of Engineers requires that multi-layered or "graded" filters be used. They say that graded filter systems may also be needed when transmitting from fine to coarse materials in a zoned embankment or where coarse material is required for improving the water-carrying capacity of a system. When filters are needed to protect gap-graded soils (in which the coarse material merely floats in the matrix of fines), the Army Corps of Engineers requires that the filter be designed to protect the fine matrix rather than the total range in particle sizes. They also use this principle in some instances where the material being drained has too wide a range of particle sizes (e.g., from coarse gravels to significant percentages of silt or clay). For major structures, such a design is checked with filtration tests.

Sherard et al. (1963) make the following additional rule for the design of filters.

> When the protected soil contains a large percentage of gravels, the filter should be designed on the basis of the gradation curve of the portion of the material which is finer than the 1-inch sieve.

Many other experimenters, in addition to Bertram, the U.S. Army Corps of Engineers, and the U.S. Bureau of Reclamation, have satisfied themselves that criterion 1 will prevent piping. In 1940 I conducted a series of experiments in which soils were mixed with water and slurries were poured over filter materials meeting criterion 1. Under these extremely severe conditions a small amount of clay and colloids washed through but nearly all of the material stayed on top of the filters.

In the construction of a military air base in the Pacific Northwest in 1942 a sudden storm washed topsoil into partly completed trench drains along the edges of the runways. Fortunately a filter layer meeting criterion 1 had been placed in the trenches. Although muddy water entered the drain pipes, topsoil penetrated the filter layer only a fraction of an inch.

In 1963 I supervised experiments in which a 2-in. layer of screenings was placed over a layer of silt. With the screenings filled with water the surface was compressed many times with a kneading compactor. These tests, which were intended to simulate the action of concentrated highway traffic on saturated subgrades, indicated that when criterion 1 was satisfied negligible intrusion occurred at the boundary between the soil and the screenings, but when the piping ratio was much above 4 or 5 substantial intrusion took place under the kneading action.

Experience indicates that if the basic filter criteria described in preceding paragraphs are satisfied in every part of a filter piping cannot occur even under extremely severe conditions. Bertram's original investigations indicated that the grain sizes of uniform filter materials may be up to 10 times those of uniform soils before appreciable amounts of soil will move through filters and that Eq. 5.1 usually is conservative. If a protected soil is a plastic clay, the *piping ratio* often can be much higher than 5 or 10, as indicated by U.S. Army Corps of Engineers practice previously noted. If cohesionless silts, fine sands, or similar soils are in direct contact with filter materials that have piping ratios much above 5 or 10, erosion is likely to occur. In 1940 I witnessed earth dam construction with loess soil being compacted adjacent to a drain composed of boulders 6 in. in diameter and having a piping ratio relative to the loess of around 2000! During the first filling of the reservoir the drain caused serious internal erosion. Eventually the drain was pumped full of cement grout to save the dam.

When coarse rock, gravel, or other materials are used in drains, erodible materials should be separated from them by two or more intervening filters, as required, each adjacent pair designed to prevent piping (for example, see Fig. 5.4). The mechanical analysis plots shown in Figures 5.4 to 5.6 offer a good visual description of the grain size distributions of individual soils and filter materials and are useful in developing filter designs.

Some designers have expressed the idea that the adequacy of the Terzaghi piping prevention criterion needs to be challenged. But Ripley (1983) says, "This is simply not so. The historic and widely accepted design practice for dams with clay cores in the Americas has been to use clean cohesionless sand-rich materials for the filter zone immediately downstream of the clay core." And, "The capability of clean cohesionless sand-rich materials to block the migration of fine particles of clay and silt sizes under seepage forces has been well established in engineering practice, initially in the early 1800's for the filtration of domestic water supplies (Babbitt and Doland, 1939), and subsequently for seepage control in dams . . . and other civil engineering works." Experience with Brazilian dams, says Ripley, "have not only confirmed the capability of clean cohesionless sand to block migration of fines by normal seepage and seepage through transverse cracks, but they have confirmed the capability of the sand to resist cracking within itself in the filter zone even when the upstream core zone has become cracked." He also says that where serious erosion and malperformance of filters of central core rockfill dams have been reported, "the filter materials have had low sand content and have been so widely graded that significant segregation during normal handling procedures was inevitable." He concludes with "the writer has not found a single case of piping or internal erosion of core fines where the core was protected with a filter zone of clean cohesionless sand-rich material for which care was taken to prevent segregation during placement, and where necessary, the filter zone itself was adequately protected by appropriate downstream zones."

Mr. Ripley's reference to the use of sand filters for the filtration of domestic water supplies is of particular interest as it says that the sand is able to block the movement of fine clay materials even when they are in suspension in the water being filtered. My own experiments with the pouring of soil slurries over filter materials, as already noted, demonstrated that only minute amounts of fine colloidal material moved through the filter, and other portions penetrated only about an inch into the filter.

Sherard et al. (1972a) point out that piping failures are likely to occur in certain types of clay which erode by a process called "dispersion" or "deflocculation." When the clay mass is in contact with water, individual clay particles are detached from the surface progressively and go into suspension. If water is flowing, the dispersed particles are carried away and erosion channels or "pipes" can form quite rapidly. Frequently the initial flow of water is along one or more cracks caused by drying shrinkage, unequal foundation settlement, and so on, or by *hydraulic fracturing* (Sherard et al., 1972b) (Sherard, 1986). A number of small homogeneous dams in Australia, the United States, and other countries have failed by these processes. In a comprehensive review of problems with dispersive clays Halliburton et al. (1975) say, "It currently is not possible to identify dispersive soil erosional problems before the fact using the results of engineering soil tests." They suggest that in areas in which there is reason to believe such problems may exist a new physical erosion test be used to try to detect in advanced of construction any tendencies toward erosion. They also say,

> The best alternative to use of the physical erosion test appears to be adequate field reconnaissance to observe the rather distinctive signs of existing dispersive soil erosion. However, this reconnaissance should be carried out no matter what identification method is used.

Various chemical tests have been suggested by other investigators to try to identify dispersive clay soils. One is "soluble salts in pore water," a standard test of agricultural soil scientists (Sherard et al., 1976). These authors say "Dispersive clays cannot be recognized by the identification tests currently employed in civil engineering practice." But, they say, "The amount of dissolved sodium relative to other salts in the pore water is the main factor determining whether clay is dispersive or not." They discuss a "pinhole test" for detecting erosional tendencies and present specific criteria that aid in identifying dispersive clays.

Detailed procedures for conducting the "pinhole erosion test" for identification of dispersive clays are given in a recently updated soil testing manual (U.S. Army Engineer Waterways Experiment Station, 1980).

The chances of piping failure in dams built on or with dispersive clay soils can be greatly reduced by providing sandy gravel filters for vertical and horizontal drains designed to collect the seepage while holding the erodible soil in place. The filter adjacent to the soil must be fine enough to hold the dispersed soil particles in place; hence two or more progressively coarser layers will be needed in such drains. Tests should be made to established the safe piping ratio (D_{15} of filter/D_{85} of soil) for all projects requiring the use of dispersive clays.

While recognizing that the prevention of erosion failures by deflocculation of dispersive clays and hydraulic fracturing may require special precautions, the designers of water-retaining facilities should be just as careful when dealing with soils that are considered relatively unsusceptible to hydraulic fracturing or dispersion failures, such as cohesionless silts and silty clays. These soils are highly erodible if not protected with well designed filters and drains, constructed without segregation.

Those who have extensive experience with dispersive clay action, seem to agree that the deflocculation process does not begin unless there is a significant flow of water, such as can occur through poorly compacted or cracked layers in an impervious core; through cracks extending in the upstream-downstream direction in a core; or along inadequately bonded contacts with rock foundations, abutment, or outlet conduits extending across the impervious core.

To safeguard *all* earth dams and other water-impounding facilities from piping failures, designers should require high standards for all facets of design and construction, use relatively wide impervious cores and other features that hold seepage quantities and hydraulic gradients to the lowest practical levels, and provide well designed and constructed filters and drains wherever needed.

Although filter criteria are almost foolproof, experience and judgment will reduce the danger of mistakes being made in their application. Several examples of "normal" designs of filters to prevent piping are given in Sec. 5.3. Precautions that must be taken in designing filters to protect gap-graded soil-rock mixtures or stratified formations, and the dangers of severe segregation in filters are described in the last part of Sec. 5.3. In these examples the primary control is assumed to be criterion 1, the ratio of the largest D_{15} size of the filter to the D_{85} size of the finest protected soil.

Pipe Joints, Holes, and Slots

When pipes are embedded in filters and drains, no unplugged ends should be allowed and the filter materials in contact with pipes must be coarse enough not to enter joints, holes, or slots. The U.S. Army Corps of Engineers (1955a) and the U.S. Army et al. (1971) use the following criteria for gradation of filter materials in relation to slots and holes:

For slots

$$\frac{85\% \text{ size of filter material}}{\text{slot width}} > 1.2 \qquad (5.4)$$

For circular holes

$$\frac{85\% \text{ size of filter material}}{\text{hole diameter}} > 1.0 \qquad (5.5)$$

The U.S. Bureau of Reclamation (1973) uses the following criterion for grain size of filter materials in relation to openings in pipes:

$$\frac{D_{85} \text{ of the filter nearest the pipe}}{\text{maximum opening of pipe drain}} = 2 \text{ or more} \qquad (5.6)$$

Equations 5.4, 5.5, and 5.6 represent a reasonable range over which satisfactory performance can be expected.

An important development in the manufacture of drainage pipes is the slotted PVC pipe (Cedergren, 1987) which has slots machined to specified widths from a minimum of 0.010 in. (0.25 mm) up to 0.10 in. (2.54 mm) or larger. Figure 5.3 shows a PVC pipe 6 in. (15.2 cm) in diameter with sawed slots of uniform size. Close control over the width of the slots ensures free flow of water into the pipe without danger of clogging with soil when the slotwidths have been correctly established with Eq. 5.4.

5.3 EXAMPLES OF FILTER DESIGNS TO PREVENT PIPING

Historical

Before the development of rational and experimental filter design criteria drain design was considered more of an art than a science. Designers depended on judgment, instinct, or precedent. In many instances coarse stone or gravel was placed in direct contact with fine-grained soils with the result that drains often became clogged or soil piped through them, thus causing structural failures.

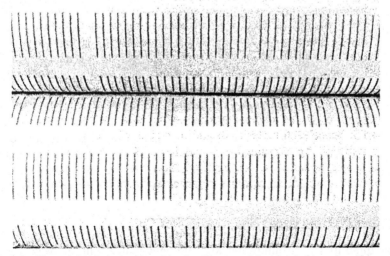

FIG. 5.3 Slotted PVC drain pipe 6-in. in diameter. Machined slots of uniform width provide good filter protection.

Such was the case with *French drains* and *macadam* rock bases used in highway construction after about 1800. Some of the early road builders, however, wisely placed fine gravel or screenings between fine soils and coarse stone bases and drains and some of the early dam builders used several layers of stone, grading from finer material in contact with the soil to coarser rock or gravel at the centers of drains. Creager et al. (1950) describe the Tabeaud Dam in California, constructed in 1902, which had a rock drain with two progressively coarser filter zones between the soil and the drain rock.

With the development of rational and experimental filter criteria the design of filters and drains has become more of a science. Several examples of the application of filter criteria to the design of filters to prevent piping are given in the following paragraphs.

Rock Slope Protection

Frequently coarse rock is placed on the banks of levees, on the upstream faces of earth dams, and in other situations in which erodible soils must be protected from fast currents and wave action. If coarse rock is placed directly on fine soil, currents and waves may wash the soil out from under the rock and lead to undermining and failure of expensive protective works or even to failure of the works being protected.

Soil erosion under rock slop protection can often be prevented by the placement of a filter layer of intermediate-sized material between the soil and the rock. Sometimes erosion can be prevented by the use of well-graded rock containing suitable fines which work to the bottom during placement. If a single layer of well-graded material or spalls between the soil and the rock is depended on for erosion prevention, the work must be carried out carefully to make sure that an unsegregated filter layer is provided; otherwise it is possible that undermining could occur under severe wave action or fast currents. A typical rock slope protection with two intervening layers is shown in Figure 5.4. The slope protection rock (curve no. 4) has a particle size range of 6 to 24 in. and a 15% size of about 7 in. A layer of rock spalls (curve 3) and a fine filter layer (curve 2) under the spalls hold the fine-graded erodible soil (curve 1) in place. If this type of construction is carried out with reasonable care, a high level of protection from undermining should be assured. In Figure 5.4 the piping ratio (D_{15} of a given layer/D_{85} of an adjacent finer layer) is always less than 5; hence each material should be stable with respect to the adjacent coarser material. An alternate design would be to eliminate the fine aggregate filter layer and substitute a suitable filter fabric on the soil while retaining the layer of rock spalls or an equivalent gravel layer under the rock (curve 3).

Drainage Blanket under Highway in Wet Cut

Large quantities of groundwater are often encountered in highway construction in hilly terrain. When highways must be built in wet cuts, blankets of permeable filter aggregate should be placed beneath the structural section to

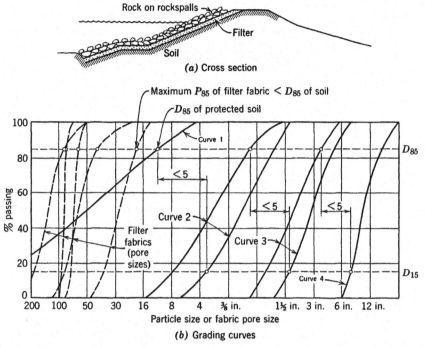

(a) Cross section

(b) Grading curves

FIG. 5.4 Rock slope protection designed to prevent undermining. Intermediate filter (curve 2) prevents erodible soil (curve 1) from washing through rock spalls (curve 3) and through coarse rock (curve 4). Care must be taken to prevent segregation of the various courses. *Alternate design* replaces fine filter (curve 2) with a filter fabric.

prevent internal flooding, pumping, and deterioration of pavements (Chapter 9). When coarse aggregates are needed for water removal, as they usually are, properly graded filter layers must be placed over the soil to prevent clogging of the coarse aggregate. Figure 5.5 shows a satisfactory combination of fine and coarse aggregates that will not become clogged when properly constructed. The coarse upper drainage layer has high permeability to ensure rapid removal of groundwater and seepage. To ensure permanent functioning of this drainage system without clogging, the fine filter layer on the subgrade (curve 2, Fig. 5.5) has a D_{15} size not more than about five times the D_{85} size of the finest soil on which the road is being built (curve 1), and the coarse filter layer (curve 3) has a D_{15} size not more than about five times the D_{85} size of the fine filter layer. The grading limits for the two filter layers shown in Fig. 5.5 satisfy criterion 1. Methods of designing drainage layers for discharge capacity are described in Sec. 5.4 and illustrated by examples in Secs. 5.5 and 5.6.

Some conditions detrimental to filter performance

The preceding examples represent normal, orthodox filter designs. If soils are poorly graded, filters are permitted to become segregated during placement,

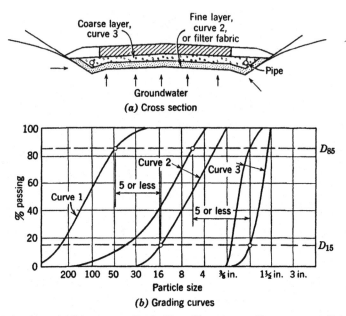

FIG. 5.5 Design of highway roadbed. Fine filter (curve 2) prevents soil (curve 1) from pumping into open-graded drainage aggregate (curve 3). There is considerable groundwater.

or other inconsistencies are permitted to occur, the theoretical piping ratio (criterion 1) may appear to be satisfied, yet a filter may be entirely unsatisfactory. Complete dependence on filter criteria without regard for localized conditions that may exist in soils or filters can lead to failures.

Difficulties can develop when filters are used for the drainage of soils and bases that are gap-graded, stratified or otherwise variable in a construction area. When filters or drains are placed against variable materials, the *finest* material must be held in place. Mechanical analysis tests can be misleading when the samples have been obtained by methods that mix the soils within some arbitrary number of feet of a hole or of the sides of a test pit. Methods of sampling that detect detailed variations in gradation and give the gradation of individual strata should be used.

Not only is it possible for protected soils to vary substantially from point to point, as just noted, but filters can also vary if wide ranges of sizes are permitted and segregation is allowed during handling or placement. The harmful effect of segregation in filters is illustrated in Figure 5.6, which shows a fine-grained soil (curve 1) that is to be protected by a filter (curve 2). The proposed filter (curve 2) has an overall grading that meets criterion 1, for its 15% size is not more than five times the 85% size of the soil. In handling and placing extensive accumulations of the coarser sizes are assumed to occur, thus allowing large pockets of coarse segregated material (curve 3) with a D_{15} size

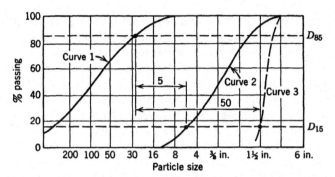

FIG. 5.6 Segregated filter materials also cause problems. D_{15} of unsegregated filter aggregate (curve 2) is five times D_{85} of soil (curve 1), but D_{15} of segregated pockets of coarse filter material (curve 3) is 50 times D_{85} of soil.

50 times the D_{85} size of the soil. Segregation in filters can lead to serious erosion of protected soils. If filters and drains are to serve their intended purposes, specifications for these facilities (Sec. 5.8) must be carefully written and strictly enforced.

5.4 PERMEABILITY REQUIREMENTS OF FILTERS AND DRAINS

The Basic Problem

The first requirement of filters and drains (Secs. 5.2 and 5.3) is that they must be safe with respect to erosion and clogging. The second requirement, which can be equally important, is that they must have sufficient discharge capacities to remove seepage quickly, without inducing high seepage forces or hydrostatic pressures.

The right half of Eq. 5.1 (Sec 5.2) was stated as criterion 2: "The 15% size (D_{15}) of a filter material should be at least four or five times the 15% size (D_{15}) of the protected soil." As noted, its intent is to ensure sufficient permeability in filters and drains to prevent the buildup of large seepage forces and hydrostatic pressures.

Criterion 2 may be checked with mechanical analysis data or curves by noting the D_{15} size of the filter and the D_{15} size of the soil and calculating the ratio: D_{15} (of filter)/D_{15}(of soil). If this ratio is less than 4 or 5, a coarser material is needed to satisfy this criterion.

Engineers often have assumed that if criterion 2 is satisfied seepage forces and hydrostatic pressures in filters and drains will always be negligible. In this section it is shown that this is not always true and that drains frequently must be analyzed hydraulically to establish their capabilities for meeting discharge needs.

In general, criterion 2 ensures that filters will be about 20 to 25 times more permeable than protected soils and will discharge seepage adequately when flow is perpendicular to the plane of a relatively thin filter layer into a highly permeable layer that discharges the water to safe exists. The filter layer shown in Figure 5.5a is such a case, as a hydraulic gradient as large as 100% across the filter would require a head only equal to the thickness of the filter. And such a moderate amount of head would not cause enough uplift to be of any consequence. But, the coarse drainage layer under the pavement must remove all of the groundwater that enters *plus* all other sources of water that can enter through upper surfaces, under a hydraulic gradient only slightly greater than the cross slope of the pavement. While criterion 2 should ensure sufficient permeability in the filter layer, it does *not* ensure adequate permeability in the drainage layer (see also Cedergren, 1962, 1987a).

When a project's design features limit the hydraulic gradients in drains to small amounts and only small areas are available for the discharge of seepage (as in the coarse layer under the roadbed shown in Fig. 5.5a), their water-removing capacities should always be evaluated by application of seepage principles. Estimates should be made of the probable rates of inflow, and liberal factors of safety should be used to allow for the uncertainties inherent in such estimates. Frequently drains can be designed with large discharge capacities for little or no more cost than low capacity drains (Sec. 9.4). When structures are to be built on formations through which seepage quantities are difficult to estimate (such as highly jointed or cracked rock or erratic alluvial or glacial deposits), designers should try to provide high reserve discharge capacities to remove water from unknown leaks, such as those that apparently destroyed the Teton Dam in Idaho in June 1976 *(E.N.-R., 1976)*, while maintaining good filter protection.

Methods for Designing for Discharge Capacity

General. Filters and drains can be designed for discharge capacity with Darcy's law or with the flow net or by combinations of these two seepage analysis methods.

1. Use Darcy's law both for approximating the rate of infiltration from the soil and designing the drain with the most reasonable values that can be assigned to the following:
   (a) The *average or effective* permeability of the soil formations to be drained. This is determined from field and laboratory tests or estimated from soil conditions by highly experienced soils engineers. It is the most important and difficult part of the work.
   (b) The average hydraulic gradients in the soil and drain.
   (c) The average areas of soil and drain material through which water is flowing (normal to the direction of seepage).

2. Use Darcy's law to design the drain after conventional flow nets have been used for estimating infiltration rates (Sec. 5.5).
3. Use *composite flow nets* (Fig. 5.9) to develop hydraulically balanced solutions for seepage in the soil and drain (Sec. 5.6).

Designing with Darcy's Law. After infiltration rates have been estimated by appropriate methods filters and drains can be designed with Darcy's law by either of the following two methods.

1. Establish a trial thickness of the water-removing part of the drain and calculate its required permeability with Darcy's law arranged as

$$k = \frac{Q}{iA} \tag{5.7}$$

Try several thicknesses (A) if desired and calculate the required permeability for each.
2. Select one or more permeabilities that represent commercially available local aggregates with acceptable gradings (Sec. 5.2) and calculate their *required thicknesses* from Darcy's law arranged as

$$A = \frac{Q}{ki} \tag{5.8}$$

In these determinations the maximum allowable hydraulic gradients in drains depend on the largest head h that can safely develop in drains without causing harmful hydrostatic pressures or spreading saturation into soil layers that must be kept free of saturation. The maximum allowable hydraulic gradient in a drain is equal to h/L, in which L is the length of the discharge path in the drain. If a longitudinal strip 1 ft wide is being considered in a seepage analysis, the cross-sectional area per running foot of drain is equal to its thickness times 1 ft and its cross-sectional area in square feet is numerically equal to its thickness in feet [under (1) above]. Care must always be taken to use compatible inflow and outflow areas.

Darcy's law can also be written.

$$\frac{Q}{i} = kA \tag{5.9}$$

In Eq. 5.9, the product of drain layer thickness ($\times$ 1 ft of length of drain $=$ A) and its permeability k, is the *transmissibility* needed in a drain to remove quantity Q under hydraulic gradient i (in the drain). With Eq. 5.9. a chart (see

Fig. 5.7) can be developed to facilitate the examination of a wide range of practical layer thicknesses and permeabilities that will provide any required *transmissibility*. Thus assume that 100 cu ft/day must be removed by a drain working with a hydraulic gradient of 0.02. Its required transmissibility is 100/0.02 = 5000 cu ft/day per linear foot of drain. By using the sloping line in Fig.5.7 for kA = 5000 cu ft/day a range of combinations can be quickly determined (e.g., 1.0 ft @ 5000 ft/day, 0.5 ft @ 10,000 ft/day, 0.25 ft @ 20,000 ft/day, and so on).

Modifying Calculations to Allow for Turbulence. When it is expected that aggregate sizes and hydraulic gradients will be so large that flow in drains will be semiturbulent to turbulent (see Sec. 3.7), two practical methods can be used to allow for the reduced efficiency caused by turbulence: (1) make permeability

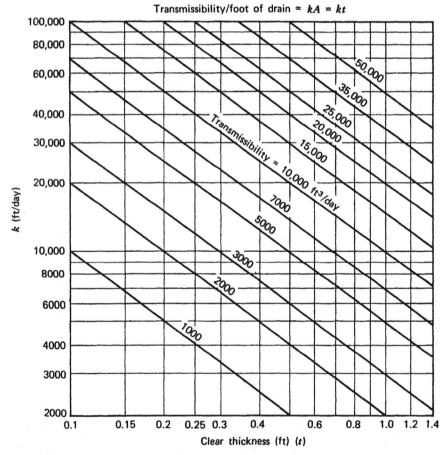

FIG. 5.7 Transmissibility chart for drainage layers (cu ft/day/ft). (From *Drainage of Highway and Airfield Pavements,* Wiley, New York, 1974, p. 149). (Updated printing by Robert E. Krieger Pub. Co., Malabar, Florida 1987).

tests at the hydraulic gradient expected and use these values in the calculations for discharge capacity; (2) establish the true Darcy coefficient by performimg tests under small hydraulic gradients that ensure laminar or near-laminar flow and apply a factor C to allow for reduced efficiency caused by turbulence at greater hydraulic gradients than used in the tests.

If permeability tests are made at the average hydraulic gradient expected in a drain (assuming relatively uniform gradients in the drain), the coefficient obtained may not be a true Darcy coefficient; however, its numerical value should be of the right order of magnitude for estimating discharge quantities under similar flow conditions in a prototype. When this is done, the quantity estimates should be reasonably correct.

If the second procedure is used, discharge capabilities of drains under semi-turbulent-to-turbulent flow can be calculated from

$$q = k'iA = (kC)\, iA \qquad (3.34)$$

in which q is the quantity per linear foot, k is the Darcy coefficient of permeability at a low hydraulic gradient, k' is the semiturbulent-to-turbulent "effective" permeability (not a true Darcy coefficient), i is the expected hydraulic gradient, A is the cross-sectional area normal to the direction of flow, and C is the reduction factor allowing for turbulence under hydraulic gradient i.

Some illustrative values for C that I estimated from tests on open-graded crushed American River gravels are given in Table 5.1 (see also Sec. 3.7). It is shown that if flow is in relatively coarse open-graded aggregates under hydraulic gradients of more than a few percent the corrections can be quite significant. Thus for a D_{10} size of 1.0 in. (2.54 cm) and a hydraulic gradient of 0.4 the flow potential is reduced to a little less than one-quarter of that for purely laminar flow. If designers allow liberal factors of safety in establishing dimen-

TABLE 5.1 Values of Factor C Estimated from Flume Tests on Crushed American River Gravels Containing No Fines ($q = kiaC$)

	D_{10} size (in.)						
i	0.01	0.025	0.05	0.10	0.20	0.50	1.0
0.01	1.0	1.0	1.0	1.0	1.0	1.0	1.0
0.02	1.0	1.0	1.0	1.0	0.95	0.88	0.83
0.05	1.0	1.0	1.0	0.98	0.88	0.66	0.55
0.10	1.0	1.0	0.98	0.88	0.75	0.55	0.41
0.20	1.0	1.0	0.91	0.78	0.63	0.46	0.31
0.40	1.0	0.96	0.83	0.69	0.56	0.37	0.24
0.60	1.0	0.90	0.78	0.65	0.51	0.32	0.20
0.80	1.0	0.86	0.74	0.61	0.48	0.29	0.17
1.00	1.0	0.84	0.72	0.58	0.46	0.27	0.15

Note. Laminar flow is assumed at $i = 0.01$.

sions and permeabilities of drains (say 10 to 20) or because of construction feasibility a drain zone is made much thicker or wider than is theoretically needed for transmissibility, the adjustments described here to allow for turbulence would not normally be significant. It is important to recognize, however, that such conditions can exist.

Designing with Flow Nets. Flow nets for *composite sections* (Secs. 3.3 and 4.6) can be used for designing hydraulically balanced drains (Fig. 5.9) because flow channels elongate when passing from a soil of a given permeability into a more permeable material, in accordance with *the definite relationship* expressed by Eq. 3.11:

$$\frac{c}{d} = \frac{k_2}{k_1} \tag{3.11}$$

In Eq. 3.11, k_1 is the permeability of the soil, k_2 is the permeability of the drain, and c and d are the length and width of the flow net figures in the drain. The derivation of Eq. 3.11 assumes square figures in the soil.

In this chapter the subscript s designates a soil, f, a *filter* or drain; hence k_s and k_1 are identical and k_f and k_2 are identical. As used in this chapter, Eq. 3.11 becomes

$$\frac{c}{d} = \frac{k_f}{k_s} \tag{5.10}$$

In the design of drains with flow nets two procedures can be used.

1. Start with drains of assumed dimensions and determine their required permeabilities.
2. Start with assumed permeability ratios k_f/k_s and determine the required dimensions of drains.

When a flow net has been correctly drawn for a composite section, *with squares in the soil*, the ratio c/d for the flow net figures in the drain is numerically equal to the ratio k_f/k_s. If the permeability of the soil is known or can be estimated, the permeability of the drain can be estimated from a flow net by using

$$k_f = k_s \left(\frac{c}{d}\right) \tag{5.11}$$

If drains in dams or under roadbeds, and the like, are to control seepage effectively, the saturation level must generally be kept wholly within the drains; otherwise water may rise into soil layers that need to be protected from saturation. This is a basic assumption of the drain-design methods described in this section and illustrated in Secs. 5.5 and 5.6. The design of drains with flow nets has previously been described (Cedergren, 1961 and 1962), although the techniques are relatively new. The method is illustrated by an example in Sec. 5.6.

In some instances one flow net may provide enough information to design a drain; in others, however, a number of flow nets which permit a comparison of the costs of several designs may be well worth the time needed for their construction. To facilitate the study of seepage in drainage systems families of flow nets may be constructed and useful design charts developed, as described in Sec. 5.6.

In the construction of flow nets for the purposes described here the shape of the saturation line in a drain is determined simultaneously with the construction of the flow net (see also Secs. 4.5 and 4.6), for it is not known in advance.

5.5 EXAMPLES OF THE USE OF DARCY'S LAW IN THE DESIGN OF DRAINS

"Chimney" Drain in a Dam

Assume that an earth dam is to be constructed to the cross section given in Figure 5.8a, with an inclined "chimney" drain A to intercept seepage through the dam and a horizontal drainage blanket B to remove seepage through the dam and its foundation. Determine minimum thicknesses and permeabilities of parts A and B to ensure ample factors of safety with respect to discharge capacities.

First estimate the probable rate of discharge through the dam and the foundation, using Darcy's law, $Q = kiA$, or Eq. 3.18, $Q = kh(n_f/n_d)$ (Sec. 3.4). If the foundation has a permeability that is different from that of the core of the dam, seepage quantities may be estimated with composite flow nets of the general types shown in Figures 4.7 and 4.8.

In this example (Fig. 5.8) it is assumed that seepage through the dam Q_1 has been estimated by appropriate methods as 2 cu ft/day; seepage through the foundation Q_2 has been estimated as 10 cu ft/day. Accordingly, the chimney drain must be capable of discharging $Q_1 = 2$ cu ft/day and the outlet portion of the drain must discharge $Q_1 + Q_2 = 12$ cu ft/day. These quantities are the discharge rates per running foot of dam and drain.

Assuming that the chimney portion of the drain in Figure 5.8 has been designed with a horizontal width of 12 ft to permit its placement with normal earth-moving equipment, the cross-sectional area normal to the direction of seepage within the chimney is about 11 sq ft (Fig. 5.8b) and its required permeability can be approximated as

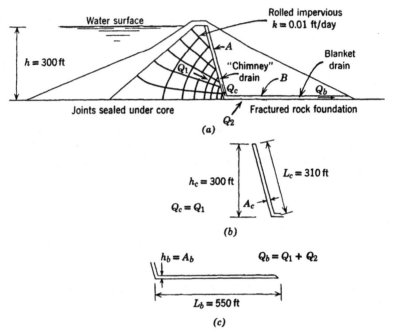

FIG. 5.8 Example of drain design procedure assuring adequate discharge capacity. (*a*) Cross section and flow net, (*b*) Key dimensions of "chimney" drain. (*c*) Key dimensions of blanket drain (outlet).

$$k = \frac{Q}{iA} = \frac{Q_1}{(h_c/L_c)(A_c)} = \frac{2 \text{ cu ft/day}}{(300 \text{ ft}/310 \text{ ft})(11 \text{ sq ft})} = 0.2 \text{ ft/day}$$

According to this approximation, a material with a permeability of about 5 ft/day would ensure reasonable discharge capacity for the chimney. Clean, washed concrete sand, low in fines, is usually about this permeable.

Outlet Blanket Drain for a Dam

With reference to Figure 5.8c, the outlet portion of the drain for this dam must be capable of discharging $Q_1 + Q_2 = 12$ cu ft/day (per foot). The relative values of Q_1 and Q_2 are fairly typical of many dams, for the amount of leakage through jointed or porous foundations is often greater than the amount of seepage through well-compacted dams.

On the assumption that the *allowable maximum head* h_b in the blanket can be no greater than its thickness A_b (Fig. 5.8c), its minimum permeability should be

$$k_b = \frac{Q}{iA} = \frac{Q_1 + Q_2}{(h_b/L_b)(A_b)} = \frac{Q_b}{(h_b/L_b)(A_b)}$$

but $h_b = A_b$. Therefore

$$k_b = \frac{Q_b}{(A_b/L_b)(A_b)} = \frac{Q_b L_b}{A_b^2}$$

Substituting numerical values, we have

$$k_b = \frac{(12 \text{ cu ft/day})(550 \text{ ft})}{A_b^2} = \frac{6600}{A_b^2} \quad \text{and} \quad A_b = \sqrt{\frac{6600}{k_b}}$$

With this relationship several theoretically adequate thickness-permeability combinations can be developed by assigning several values to A_b and calculating corresponding values for k_b or by selecting permeabilities representing several available aggregates and determining the required thickness A_b of the blanket drain for each. The second method is illustrated as follows:

Trial 1. Assume that the outlet drain in Figure 5.8 is to be constructed of washed filter aggregate with a permeability coefficient k_b of 10 ft/day:

$$A_b = \sqrt{\frac{6600}{k_b}} = \sqrt{\frac{6600}{10}} = \sqrt{660} = 25.7 \text{ ft}$$

Trial 2. Assume that pea gravel (¼ in.) with a permeability of 3000 ft/day is to be used:

$$A_b = \sqrt{\frac{6600}{k_b}} = \sqrt{\frac{6600}{3000}} = \sqrt{2.2} = 1.5 \text{ ft}$$

Trial 3. Assume that screened gravel (⅜ to ¾ in.) with a permeability of 40,000 ft/day is proposed:

$$A_b = \sqrt{\frac{6600}{k_b}} = \sqrt{\frac{6600}{40,000}} = \sqrt{0.17} = 0.4 \text{ ft}$$

Obviously the choice in trial 1 is impracticable, but either of the other two choices is entirely reasonable. When, as in this problem, substantial quantities of seepage must be removed by flat drains working under small hydraulic gradients, the least costly and most positive solution is the use of a highly permeable water conducting layer between filter layers. Because this drain is for a large dam, its discharge capacity should have a liberal factor of safety. An appropriate design would be

Material	Thickness
Upper protective filter	1 to 2 ft
⎰Inner drainage layer	3 to 4 ft
⎱⅜ to ¼ in. gravel)	
Lower protective filter	1 to 2 ft
Total thickness	5 to 8 ft

This example points up the inability of relatively low permeability aggregates (such as the washed filter aggregate in trial 1) to remove large quantities of seepage. The great necessity for designing drains as conductors of seepage is amplified by the fact that drains frequently must be capable of removing many times the rates assumed in this example.

The selection of liberal dimensions for drainage layers also safeguards against moderate reductions in the "effective" thickness caused by intermingling or mixing of materials at outer boundaries.

Whenever seepage must flow across a fine filter to enter a coarse drainage layer, as between the foundation and the outlet drain blanket in this example, the *minimum required permeability* of the filter should be estimated. Unless the filter is permeable enough to allow the water to enter the coarse drainage layer freely under only a small hydraulic gradient (0.5 or less) the effectiveness of a drain can be drastically reduced. Assume an average hydraulic gradient of 0.5 across the filter in the preceding example and that $Q_2 = 10$ cu ft/day per foot is a reasonable determination of the amount of water that will enter the first (left) 200 ft of the drain. According to Darcy's law, $k = Q/iA$, so $k = 10$ cu ft/day/$(0.5)(200) = 0.1$ ft/day. Every filter, however, must be permeable enough to provide a reasonable reserve for higher than expected flows. Even though a foundation is relatively uniform and free of important joints and cracks and its permeability has been estimated with a testing program, the specifications for the *filter* should require washed and screened aggregate with a minimum permeability *after placement and compaction* of at least 10 to 20 times that calculated theoretically. If large concentrated flows are considered possible, special measures should be taken to reduce the amount of water flowing and special drainage features should be developed to remove the water safely to prevent it from causing damage to a structure.

5.6   EXAMPLE OF THE USE OF FLOW NETS IN THE DESIGN OF DRAINS

Steeply Inclined Drains

The use of *composite flow nets* for the design of drains is illustrated with reference to Figure 5.9, which shows two flow nets for seepage into sloping embankment drains.

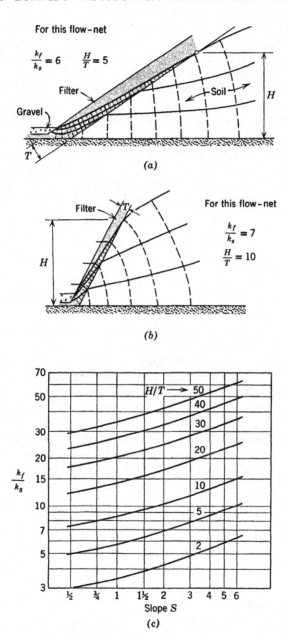

FIG. 5.9 Typical flow nets for seepage into sloping embankment drains (*a*) 1 1/2:1 discharge face (*b*) 1/2:1 discharge face (*c*) Chart developed from family of flow nets.

If drains in earth fills are to provide maximum protection against seepage and groundwater, they must have sufficient permeability to intercept and remove all the seepage that enters without allowing it to spread into the fill beyond the drains. Hence in the construction of the flow nets in Figure 5.9 all the flow channels from the soil are compressed entirely within the sloping drains.

After flow nets have been constructed (Figs. 5.9a and 5.9b) their fundamental characteristics automatically provide a means for determining the necessary permeability of drains in relation to the water-bearing soil (for the drain widths assumed). As described in Sec. 5.4, the c/d ratio for the flow net figures in a drain is numerically equal to the permeability ratio k_f/k_s and the *required permeability* of a drain of assumed dimensions is

$$k_f = k_s \left(\frac{c}{d}\right)$$

(5.11)

Thus in the flow net in Figure 5.9a each pair of flow lines in the drain encompasses six squares between full equipotentials; hence the c/d ratio in the filter is $6/1 = 6$. For the design shown in Figure 5.9b each pair of flow lines elongates to seven squares between full equipotentials; hence the c/d ratio is 7, and $k_f/k_s = 7$.

Examination of the flow nets in Figure 5.9 shows that flow in these sloping drains is fairly steep and hydraulic gradients are large. For these drains, which are provided with highly permeable gravel collector drains at the bottom, reasonable factors of safety with respect to discharge capacity often can be assured if the drain materials are 40 to 50 times more permeable than the soil; but if the formations being drained are nonuniform and the permeabilities are known only approximately it is usually necessary to use pea gravel or coarser aggregate enveloped in a suitably fine filter or filter fabric.

When a number of flow nets have been constructed in the study of a seepage or drainage problem, the value of the work can often be greatly broadened if a few additional flow nets are drawn and dimensionless design charts developed. Later, if a reasonably similar problem with different dimensions or permeabilities is to be solved, the charts may come in handy. The design chart in Figure 5.9c was developed with this thought in mind.

Summary Comments

Use of composite flow nets or Darcy's law for designing drains for discharge capacity, as described in this chapter, can secure the fullest, most permanent protection from the harmful effects of groundwater and seepage when adequate filter protection is also provided. When the *piping requirements* have not been fulfilled, serious failures have been common, as described in various parts of this book. When the *permeability requirement* has not been properly analyzed, many drains have only partly accomplished their purpose—in some cases being so ineffective that they could have been omitted with no discernible loss in benefit. Figure 5.10 illustrates a large dam designed with multilayered drains to collect seepage through and under the structure. Unfortunately the specifications permitted so many fines in all drain layers that this drain had negligible effect on underseepage. The actual pore pressures that built up under the downstream part of this dam (from reservoir seepage) agreed almost exactly with those obtained from a flow net that assumed that the drain was completely impermeable (Fig 5.10). If the permeability and discharge needs for this drainage system had been evaluated by the methods in this chapter, it is unlikely that this mistake could have been made.

5.7 USE OF SYNTHETIC MATERIALS IN DRAINS

Background

In order to have sufficient permeability to remove all the water that reaches them freely, drains for civil engineering works must contain a layer or core with high water-conducting capability, protected against intrusion and clogging from adjacent fine materials by suitable filters. As noted in the discussion of the dam shown in Figure 5.10 (Sec. 5.6), serious problems have arisen because drainage layers that were depended on for water removal contained so many fines that their *permeabilities* were too low to meet the discharge needs of the drainage systems. In part, this problem was increasing because of diminishing supplies of good quality aggregates. But even in the early 1950's I began to see that many drains for engineering works were not performing as required because of a serious flaw in filter and drain design philosophy—failure to think of these devices as conveyors of water and the necessity for making hy-

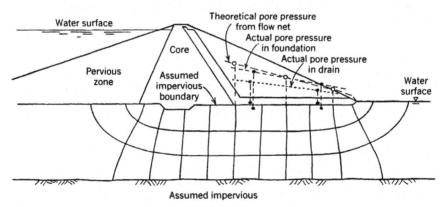

FIG. 5.10 Uplift pressures that built up under an earth dam with an expensive but ineffective drain were not measurably affected by the drain. (From *Embankment-Dam Engineering, Casagrande Volume,* Wiley, New York, 1973, p. 42).

draulic calculations to determine their discharge needs. This led to the conclusion that the mixtures of sand and gravel being used widely in drains did not have sufficient permeability to remove water fast enough in many situations (Sec. 5.4. "The Basic Problem"). The "standard" criterion for filter permeability (right half of Eq. 5.1) was not always insuring adequate permeability in drains to remove water without excessive build-up of head.

The availability of fabric filters in the late 1960's opened the door to vastly improved drainage systems, as it was now possible to take advantage of the high water-removing capabilities of open-graded aggregates (those containing no fines) by protecting them with good filter fabrics. Starting about 1967, I used fabric filters for drains for a low-lying county road in central California, a slide correction for a logging road in northern California, and so on. These systems used clean, crushed rock or gravel in the size range of 1/2 in. to 1 1/2 in. for the drainage layer protected by filter fabric (see also Cedergren, 1977). Modern synthetic filter fabrics (now called "Geotextiles") offer a way to develop drains with high discharge capacity and good resistance to clogging or piping—often at least overall cost. As described later in this section, various kinds of *composite drains* are being developed that are constructed entirely of synthetic materials. The growing interest in geotextiles is exemplified by the many national and international conferences being held—such as the International Conferences on Geotextiles, held in Paris in 1977, in Las Vegas in 1982, and in Vienna in 1986. Hundreds of workers worldwide are making tests and investigations of the properties and behavior of geotextiles, and enormous numbers of papers and reports are being published.

Various natural plastics, notably nitrocellulose, have been in use since the middle of the nineteenth century. Synthetic resin plastics, first made about 1910, have become increasingly widespread in use since the early 1960s. A number of manufacturers make fabrics or cloths from polyvinylidene chloride, polypropylene, nylon (from long-chain polyamides), polyvinylidene, or other synthetic resins, used alone or in combination. With regard to filter fabrics, Seemel (1976a) says:

> Strength and inertness of these materials vary, but they are all generally rot-proof, mildew-proof, salt water-proof, insect-proof, and rodent-proof. They are, also, not affected by hot or cold climates. Some are affected by alkalies, others by acidic material, components of asphalt, or fuels oils. Most are seriously affected by long exposure to ultra-violet components of sunlight. . . .

He also says:

> For any tendency to deteriorate, there are inevitably offsetting desirable characteristics. Do not forget that the same is true of granular materials.

When used with proper regard for their limitations, synthetic fabrics or cloths can provide a useful service in many civil engineering works. *The experi-*

*ence record of synthetic fabrics and cloths in drains is good, but verification
of the long-term effects is needed and designers should make sure that a given
material has the required properties and minimum life expectancy needed for
a given usage before allowing it to be incorporated in an important structure.*
Some of the geotextiles being promoted for drainage purposes were specifically
developed with drainage in mind. Others may look nearly identical but have
entirely different filament sizes and fabric structure and behave in an entirely
different manner.

Filter fabrics are basically of the *woven* or *nonwoven* types. They should
not be confused with degradable surface coverings used to allow seed germina-
tion for control of surface erosion or with the watertight membranes used for
reservoir liners. The woven filter fabrics are constructed with monofilament
yarns, calendered after weaving to give the filaments a distinct position in
relation to one another, and have openings of a specific controlled size. The
nonwoven fabrics are dispersed in random patterns with synthetic filaments
partly bonded by fusion. Other fabrics have the general appearance of felt and
no distinct openings except small holes that are punched while the fabric is
being made.

The use of filter fabrics has had a good deal of impetus in areas in which
problems have developed with sand and gravel filters that did not meet the
requirements or in situations in which lack of space made their use difficult
and costly. In areas in which strong currents or heavy wave action can phys-
ically remove filter aggregates under concrete slabs with wide joints, large rock
for riprap, and so on, filter cloths have successfully held materials in place,
partly by their tensile strength.

In developing designs for drains, we must not overlook the fact that high
quality natural mineral aggregates are virtually indestructable by agents of
weathering, whereas the records of synthetics are not long enough to prove
just how they will perform. Also, if the fabrics are allowed to become torn or
severely punctured or adjacent sheets are not adequately overlapped or held
against separation during construction, the best of these fabrics cannot perform as expected by the designer. Sloppy or careless construction can ruin the
best material or design, including both aggregate and fabric filters.

Applications

Within the last two decades thousands of projects in North America and Europe have made use of filter fabrics. One of the earliest applications was in
shore-erosion protection structures such as stone seawalls and jetties (Barrett,
1966). Dunham and Barrett (1974) point out that when sand and gravel filters
were installed to hold fine soils in place under erosion protection structures
scour and erosion sometimes undermined these protective devices but woven
plastic filter cloths have been used successfully.

Figure 5.11 shows the way in which filter cloths have been applied under
rubble revetments for the protection of erodible soils from heavy wave and

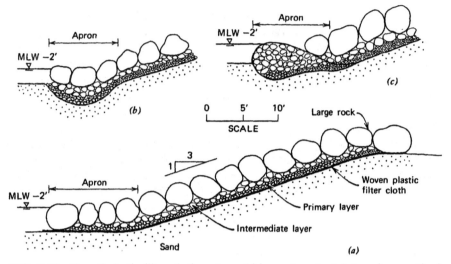

FIG. 5.11 Use of plastic filter cloth under rubble revetments. (*a*) American method. (*b*) Toe detail, Spanish method. (*c*) Toe detail, Dutch method. (After Barrett, 1966, Fig. 2.)

current actions. After the earth has been properly sloped, compacted, and fine-graded the filter cloth is spread on the soil, its edges overlapped or sewn as required, and held in place by pins or other means. A layer of fine sandy gravel (primary layer, Fig. 5.11*a*) is then placed over the filter cloth with great care to prevent tearing, puncturing, or other damage. A layer of coarse gravel (intermediate layer, Fig. 5.11*a*) is then spread over the fine gravel and large rocks are dropped by derrick or other equipment to provide a uniform rubble layer capable of withstanding wave and current action without danger of displacement. Figure 5.11*a* shows the American method of constructing rubble rivetments with woven plastic filter cloth, Figure 5.11*b* shows the Spanish method, and Figure 5.11*c* illustrates the way in which the toe is constructed by the Dutch method. *Whenever filter cloths or fabrics are to be used on slopes, designers should be sure that the coefficient of friction between the cloth or fabric and adjacent materials is sufficient to preclude the formation of a plane of weakness.* If manufacturers are unable to furnish this information, laboratory or field tests should be made to verify the coefficient that can be expected on a given project.

Several additional uses for filter cloths or fabrics are described in Figure 5.12. Figure 5.12*a* shows a stone breakwater or jetty with a core of limestone, granite, basalt, or other rock or suitable quarry waste. A woven plastic filter cloth holds the soil in place under the rock. Figure 5.12*b* is a cross section through an emergency type of road constructed over a swamp. First a filter fabric is rolled out over the area on which the road is to be built. Next a layer

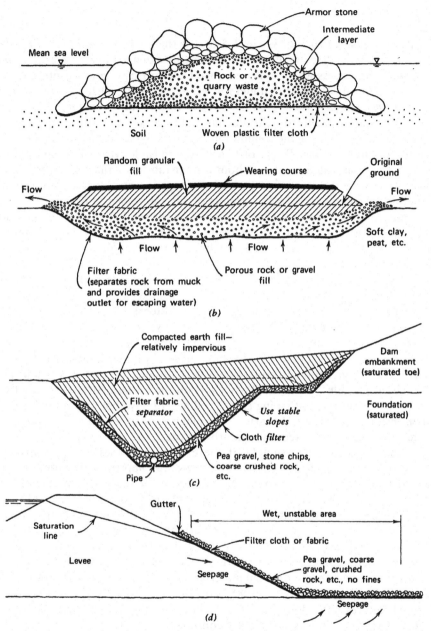

FIG. 5.12 Several uses for filter fabrics for drainage. (*a*) Under a stone breakwater or jetty. (After design in "Filter Handbook," printed with permission of Carthage Mills Incorporated, Cincinnati, Ohio.) (*b*) Under a temporary military or other access road over a swamp. (*c*) In new toe drain for earth dam with seepage problem at downstream toe. Is *accessible,* if future repairs should be needed. To be installed only when safe to do so (reservoir drawn down). (*d*) Emergency or permanent seepage protection for levees.

of porous crushed rock or gravel 1.5 to 2 ft thick is bulldozed over the area; the fabric and coarse aggregate are necessarily extended to the sides to allow the escape of excess water squeezed out of the soft clay, peat, or other muck foundation as it consolidates under the weight of the fill. Then over the porous rock a layer of any stable granular fill 1.5 to 3 ft thick is spread. The filter fabric separates the fill from the muck foundation and permits the excess water to escape, thus allowing the foundation to consolidate and stabilize in the minimum amount of time. When ordinary impermeable types of fill are placed over a weak saturated swamp without a filter fabric the fill tends to mix with the muck, with the result that vertical drainage is virtually zero, an unstable condition develops, and large amounts of fill are needed to obtain sufficient stability to allow heavy vehicles to operate on the road. The type of construction illustrated in Figure 5.12b can help to obtain a stable road, both for emergency and permanent use, with the least amount of material in the least amount of time.

This general concept has also been used for building permanent fills on unstable ground. Robison (1985) tells about the underwater site at Washington National Airport where the fill for a runway extension was built with the aid of fabric, and probably saved the lives of many passengers on a DC-10 that ran off the end of the paved runway in 1984, but stopped on the grassed extension. A geotextile mat placed underwater allowed fill to be placed on soft silty clays where conventional displacement or hydraulic fill methods would not have been feasible. She says, "Passengers on the Eastern Airlines shuttle that went off the runway last spring (1984) will never know that they probably owe their lives to a thin sheet of fabric." The plane came to rest on an embankment "that had been completed only months before."

Another remarkable example of the use of geotextiles to allow difficult earthwork on soft ground is the construction of approach fills for the new Dumbarton Bridge at the south end of San Francisco Bay. Hannon (1982) points out that the earth fills needed to be placed at both ends of the bridge in open water three to 15 feet deep on soft bay mud 10 to 40 feet deep. Fabrics made the work possible by providing separator and reinforcing functions that enabled initial fills to be placed without failures (see also Fig. 7.15).

The general concept shown in Figure 5.12b also enabled dikes to be constructed on low-strength dredged materials at the Cranes Island Disposal Area at Norfolk, Virginia. Fowler (1985) says that the strength of the fabric prevented failures and allowed the fills to be literally "floated" on fabric until the soft foundation was sufficiently consolidated.

Figure 5.12c shows a toe drain for an earth dam (or levee) which has developed undesirable underseepage and saturation at its downstream toe. After the reservoir or river has lowered sufficiently to remove any hazards, the toe drain is constructed, using slopes flat enough to be safe from slipouts during construction. It can be seen that this accessible toe drain has not been buried in the structure and can therefore be repaired in the future if the need should arise. This general method was used for repairing and upgrading the safety of the 17.4 mile long embankment surrounding the cooling water reservoir of a Florida Power & Light Co. generating plant that failed on October 30, 1979 (see *Civil Engineering Magazine*, ASCE, Jan. 1981, pp. 46-47). Heavy seepage with "pin boils" and "sand boils" and large uplift pressure under the downstream toe area had been observed regularly over the 18-month period water had been stored in the reservoir. While the precise failure mechanism could not be determined, the panel of engineers reviewing the failure recommended that the repair should include a comprehensive drainage system to protect the downstream slope and foundation from potential seepage failures after reconstruction of the breached area (see also Fig. 10.15, Sec. 10.5, "Storage Reservoirs.")

Figure 5.12*d* shows an emergency or permanent piping-prevention treatment for the downstream toes and foundation areas of levees saturated by shallow seepage. A filter fabric is rolled over the area and covered with a layer of clean permeable gravel or crushed rock. This type of treatment can help save levees or dams that show severe seepage during high water stages. The permanent stockpiling of drainage aggregates and supplies of filter fabrics, as well as the maintenance of suitable mechanical loading equipment at vital downstream areas, has been suggested (Cedergren, 1976) as a way to make emergency repairs to dams in cases of serious piping conditions until reservoirs can be lowered to allow more permanent repairs to be made.

Filter fabrics have two basic uses in drains for engineering works:

1. To serve as a true filter that must also act as a separator to hold the soil in place and allow the free escape of water for long periods of time.
2. To serve as a separator or barrier to prevent the soil from mixing with a coarse aggregate layer when there is no significant long-time flow of water.

These two uses are illustrated in Figures 5.1 and 5.12.

In all cases in which a fabric must serve as a true filter (Use 1), it must have openings small enough to prevent no more than minute amounts of the adjacent soil from passing through, but it must also have enough *open area* composed of sufficiently large openings to allow unobstructed flow of water.

In relation to *woven* fabrics, the U.S. Army (1975) says:

Several types of filter cloths have been evaluated by laboratory tests and limited observations of field performance, however, the actual life of these cloths is not known and the use of plastic filter cloth in inaccessible areas must be considered carefully. Filter cloth should not be used to wrap piezometer screens or as filter material within or on the upstream face of earth dams.

Filter design criteria for plastic filter cloths are based on the "equivalent opening size" (EOS) and "percent open area." The EOS is defined as the number of the U.S. standard sieve with openings closest in size to the filter cloth openings. The "percent open area" is defined as the summation of the open areas divided by the total area of the filter cloth.

Procedures for testing for the equivalent opening size (EOS) and the percent of open area are provided by the Corps.

Continuing with its discussion of criteria for woven fabrics, the U.S. Army (1975) says:

> It is the responsibility of the designer to specify filter cloth that retains the soil being protected, yet will have openings large enough to permit drainage and prevent clogging. The designer should select the "equivalent opening sizes" (EOS) and "percent open area" based on the following criteria:
>
> a. Filter cloth adjacent to granular materials containing 50 percent or less by weight fines (minus No. 200 materials):
>
> (1) $$\frac{85 \text{ percent size of soil}}{\text{opening size of EOS sieve}} \geq 1.$$
>
> (2)   Open area not to exceed 36 percent.
>
> b. Filter cloths adjacent to all other type soils:
>
> (1)   EOS no larger than the openings in the U.S. Standard Sieve No. 70 (0.0083 in.).
> (2)   Open area not to exceed 10 percent.

Also, in relation to *woven fabrics* intended for use in civil works construction, the U.S. Army (1975) says:

> To reduce the chance of clogging, no cloth should be specified with an open area less than 4 percent or an EOS with openings smaller than the openings of a U.S. Sieve Size No. 100 (0.0059 in.). When possible, it is preferable to specify a cloth with openings as large as allowed by the criteria. However, because of strength requirements and the limited number of cloths available, it may not always be possible to obtain a suitable cloth with maximum allowable percent open area.

Calhoun (1972) describes investigations of *woven* filter cloths carried out by the U.S. Army Waterways Experiment Station. Among the important physical properties for which tests were devised are (1) size of opening, (2) % open area, (3) strength, (4) abrasion resistance, and (5) resistance to deterioration from the elements. Chemical composition and physical properties of various fabrics were also determined by tests. Both field exposure and filtration tests were made with several soil types and several plastic cloths.

Rosen and Marks (1975) discuss a laboratory investigation of the behavior of *nonwoven* fabric filters in subdrainage applications. Tests were also made with conventional aggregate filters for a comparison with the fabric-filter performance. These investigators concluded that when water flows from a soil into a filter (either aggregate or fabric) some of the finest soil particles will move into and through the filter, some will be trapped in the filter, and others will be trapped in back of the filter in the form of a "filter cake." These actions will reduce the effective permeability of the filter below its initial value. To ensure the long-term performance of filters constructed with nonwoven fabrics, these authors make the following recommendations:

> The design criteria for fabric filter systems . . . be the same as those established for conventional aggregate filters: (a) to provide sufficient permeability so that seepage can be removed without the buildup of excessive hydrostatic pressures, and (b) to ensure that piping of soil particles does not occur during the drainage process.

Carroll (1975, 1976) says:

> The use of *nonwoven* fabrics by the drainage industry is a relatively new concept. Because of the depletion of aggregate availability and, in many cases, the inability to acquire the proper gradation of aggregates for filtration, the acceptance of subgrade fabric filters is rapidly growing.

And:

> Fabric filter cloths for subsurface drainage must be composed of strong rot-proof polymeric fibers oriented into a stable network such that the fibers retain their relative positions with respect to each other. The fabric should be free of any chemical treatment or coatings which might significantly reduce permeability, and should have no flaws or defects which could significantly alter its physical properties.

Carroll also adds:

> The permeability design criterion for fabric selection is that the fabric permeability should be five times that of the protected soil. However, for highly permeable soils which exceed this criterion designers should recognize that the fabric could become the limiting factor in the system permeability.

In relation to the nonwoven fabrics in subsurface drains Miano (1976) notes that successful installations have been made with fabrics having the following properties:

Water permeability, k: 0.05 cm/sec
Average pore size (P avg): 0.13 mm
Grab strength (ASTM D1682): 100 lb
Grab elongation (ASTM D1682): 110%
Fabric toughness (grab strength times grab elongation): 11,000 lb%

To prevent piping by soils when filter fabrics are used I suggest the following criterion:

$$\frac{85\% \text{ pore size of filter } (P_{85})}{85\% \text{ size of soil } (D_{85})} \leq 1 \tag{5.12}$$

The physical significance of the criterion represented by Eq. 5.12 is illustrated in Figure 5.13, which shows several pore-size distribution curves for filter fabrics in relation to several soil gradation curves. It is a little more conservative than the criteria suggested by others based on the equivalent opening size (EOS) sieve or the average pore size $(P\text{avg})$.

Examples of the use of filter fabrics that give filter or separator protection appear in other chapters (Figs. 7.7, 7.11, 7.13, 7.15, 8.15, 8.22, 9.10, 9.16, 9.17, 10.1d, 10.3, 10.7, 10.9, and 10.15). If filter fabrics are to provide long-lasting protection to engineering works, only the right kind for a given type of construction and use must be specified and the entire design and construction must be correctly carried out. Because of the many factors that can affect the performance of filter fabrics and the drains they protect, the preparation of adequate acceptance and performance specifications is extremely important.

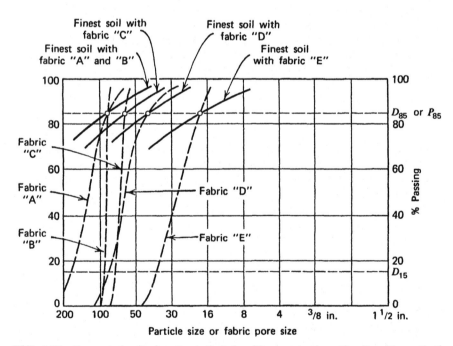

FIG. 5.13 Suggested criterion (tentative) for filter protection of soils with synthetic fabrics: D_{85} of soil $\geq P_{85}$ of fabric.

The U.S. Department of the Army, Corps of Engineers (1986*a*) again states its position about the use of geotextiles in drains for earth dams, namely that because so little is known about their long-term performance, they should not be used in inaccessible areas, on the upstream faces of earth dams, to wrap piezometer screens, or within any portion of the embankment [see also U.S. Army, Corps of Engineers (1975)].

Many investigators have been testing fabric filters for important properties relating to the proper functioning of the synthetic materials in engineering works. Bell and Hicks et al. (1978) made a thorough review (with the help of a select group of engineers serving as a task force) of all important properties of these materials for engineering uses and prepared a detailed report for the Federal Highway Administration (FHWA) on testing methods and criteria for use of fabric filters. Shrestha and Bell (1982) tested creep characteristics of six typical geotextiles under static loads to predict their long-term creep. Using a Rate Process Theory and a four-element rheological model, they predicted the time to reach failure was much shorter than the normal design life expected and shorter than indicated by actual field experience. They concluded that further study with longer duration and more accurate tests was needed to clear up this inconsistency. This is typical of essentially all accelerated types of tests, as it is very difficult to simulate real conditions with short-term tests.

Koerner and Ko (1982) describe methods used for laboratory tests from which they concluded that the long-term performance of geotextiles with freedom from filter cake formation, arching, binding, and clogging are of utmost importance and can best be verified by long-term flow tests. Halliburton and Wood (1982) conducted gradient ratio testing following U.S. Army Corps of Engineers procedures, and concluded that geotextile EOS was not related to geotextile clogging resistance, but woven geotextile open-area was found to be directly related to geotextile clogging resistance. Giroud (1982) says the simplistic filter criteria often used to select geotextile filters can be misleading and suggests other permeability and retention criteria which he says are more reliable than the others. He says that the usual permeability requirement that the fabric be more permeable than the soil "is excessively demanding and can eliminate geotextiles that are actually suitable".

Those planning to use the synthetic materials in engineering works should make sure that the ones used have sufficient strength, longevity, filter capabilities that prevent piping or clogging, and adequate permeability and water-conducting capability for the needs of the individual project usages.

Flow Capabilities of Fabric Filters

The fabric filter usages discussed in previous pages pertain to situations in which the flow of water is *through* a thin filter layer into a highly permeable conducting layer which discharges it freely to pipes or other exits. Many fabrics have good capabilities for this kind of flow. Some fabric makers, however, are suggesting that their fabrics be used for situations where the flow would be *along* the thin dimension, which is known as *in-plane flow* or *planar flow*. This kind of usage should be considered with care, because a fabric with excellent flow capability *perpendicular to* its plane may have hardly any capability for *in-plane flow*.

Raumann (1982) tested the in-plane permeability of compressed fabrics and concluded that for the thick, bulky fabrics, theoretical values of transmissibility may be reasonable, but should always be used with caution. He says the transmissibility of thin fabrics is so low as to provide negligible drainage. Koerner and Sankey (1982) also tested for in-plane permeability of geotextiles under various normal pressures and concluded that, "Typically the transmissibility up to 96 kPa (equivalent to approximately 10 m of water) for a geotextile weighing 600 gm/m^2 is equivalent in its hydraulic capacity to 2.5 cm of sand". To determine how much flow this represents, I made several estimates of flow rate, q, in sand layers 2.5 cm thick for several permeabilities representative of clean sands (see Table 5.2). As an example, if one assumes that a sand layer with a thickness of 2.5 cm or 0.08 ft has a coefficient of permeability of 50 ft/day (0.017 cm/sec), and a hydraulic gradient of 0.1 (which is probably about the maximum allowable in such a layer without excessive head build-

TABLE 5.2 In-Plane Discharge Capabilities of Fabric Filters and Equivalent Sand Layers Compared with Several Clean Aggregate Layers

Material	Sand or gravel layer analyzed	Equivalent k, ft/day	i	Discharge Capability	
				cu ft/day/ft	gpm/100 ft.
Filter fabric	Equivalent	50	0.02	0.08	0.04
with hy-	sand A		0.05	0.20	0.10
draulic ca-			0.10	0.40	0.20
pability	Equivalent	100	0.02	0.16	0.08
equal to 2.5	sand B		0.05	0.40	0.20
cm of sand			0.10	0.80	0.40
(area = 0.08	Equivalent	200	0.02	0.32	0.16
sq. ft/ft.)	sand C		0.05	0.80	0.40
			0.10	1.60	0.80
Clean gravel	Aggregate a	5000	0.02	8.0	4.2
or crushed	1/8 in. to 1/4		0.05	20	10
stone, 2.5	in. pea gravel		0.10	40	21
cm thick	Aggregate b	10,000	0.02	16	8
(area = 0.08	1/4 in. to 1/2		0.05	40	21
sq. ft/foot)	in. gravel		0.10	80	42
	Aggregate c	40,000	0.02	64	33
	1/2 in. to		0.05	160	83
	3/4 in. gravel		0.10	320	166

up), the flow quantity q = kia = 50 ft/day × 0.1 × 0.08 sq ft/ft = 0.4 cu ft/day per foot of drain. This would be 0.002 gallons per minute per linear foot or 0.2 gpm for each 100 foot of drain. For 500 feet of drainage fabric, the flow would be 1 gpm, which is actually very little water.

To carry out this example further, I also calculated the flows for sands with coefficients of permeability of 100 ft/day and 200 ft/day. Since q is proportional to k, the material with k = 100 ft/day would have twice the capability as one with k = 50 ft/day, and the one with k = 200 ft/day would have four times that capability. These volumes are still almost infinitesimal.

Then, to relate the potential flow in the fabric (or an equivalent sand layer) with the capabilities of several coarsenesses of mineral aggregates in a layer 2.5 cm thick, I calculated the capabilities of aggregate a (1/8 in. to 1/4 in. pea gravel), aggregate b (1/4 in. to 1/2 in. gravel, and aggregate c (1/2 in. to 3/4 in. gravel). The estimated flow rates for the sands and aggregates are given in Table 5.2. In Table 5.2 it may be seen that the flow capability of the most optimistic sand equivalent of the fabric, for k = 200 ft/day and i = 0.1, is less than 1 gpm for each 100 feet of fabric. In contrast, the finest of the aggregates (1/8 in. to 1/4 in.) has a flow capability of 21 gpm per 100 feet of drain, which is 25 times greater than the best sand equivalent of the fabric. So, 25 filter fabric layers would be needed to conduct the same quantity of water as 2.5 cm of the 1/8 in. to 1/4 in. pea gravel. It seems apparent that trying to use filter fabrics to conduct water along their thin dimensions (by in-plane flow) can be extremely costly in comparison with aggregate drains protected with filters.

Some engineers may wonder why a filter fabric that can conduct large volumes of water perpendicular to or *across* its plane (as through a filter enveloping a permeable aggregate or synthetic core) can conduct only minute quantities along its plane by *in-plane flow*. Applying Darcy's law to this question, it becomes instantly apparent that the typical filter fabric can conduct about 1/2000th to 1/10,000th as much water along its plane as across. According to Darcy's law, the volume of water conducted is equal to the coefficient of permeability, k, times the hydraulic gradient, i, in the direction of flow times the cross-sectional area, a, perpendicular to the direction of flow. For flow perpendicular to its plane a fabric uses its entire surface area, and the hydraulic gradient can be fairly large. For example, a head of only 1 in. on a fabric with a thickness of 1/8 in. produces a gradient = 1 in./1/8 in. = 8. And a head of 2 in. produces a gradient of 2 in./1.8 in. = 16, and so on. In most cases a head of an inch or two perpendicular to the plane of a fabric ought to be acceptable.

But for flow along the plane of a filter fabric, conditions are very much different than for flow across the fabric (perpendicular to its plane). Now, the cross-sectional area available for flow is limited to the thickness of the fabric. Thus, for a 12 in. wide strip, the cross-sectional area is 12 in. × 1/8 in. = 1.5 sq. in. or 0.01 sq. ft. And the maximum allowable hydraulic gradient for in-plane flow is very limited, and usually cannot exceed a few percent, or more than several feet in a horizontal distance of 100 ft. The comparison is made as follows:

Flow Across the Plane of a Fabric (Transverse Flow):

Assume a 12-in. × 12-in. area, for $a_1 = 1.0$ sq. ft.
Assume an allowable hydraulic gradient $i_1 = 5.0$.

$$\text{Then, } q_1 = k_1 i_1 a_1 = k_1 (5.0)(1.0) = 5.0 k_1$$

Flow Along the Plane of a Fabric (In-Plane Flow):

Assume 1/8 in. thickness and a 12-in. wide strip, for $a_2 = 0.01$ sq. ft.
Allow a hydraulic gradient of 0.05 (5 ft. of head loss/100 ft.) for $i_2 = 0.05$

$$\text{Then, } q_2 = k_2 i_2 a_2 = k_2 (0.05)(0.01) = 0.0005 \, k_2$$

Ratio of Flow Rate Along Plane of Filter Fabric to Flow Rate Across:

Assuming that $k_1 = k_2$:

$$q_2/q_1 = 0.0005/5 = 0.0001$$

So the flow rate along the plane of the fabric is 1/10,000 th of that across (for equal permeabilities in both directions).

Even if the hydraulic gradient *across* the fabric were limited to 1.0, the flow rate along the plane would be only 1/2,000th of that across. Clearly, filter fabrics that can serve well as filters to protect permeable drainage layers may be virtually useless as conveyors of water along their plane. So, too much should not be expected of filter fabrics for *in-plane* flow. Before such a usage is contemplated, the discharge needs should be calculated and no material used that cannot remove at least 4 or 5 times the estimated rate (for a theoretical factor of safety of at least 4 or 5).

Flow Capabilities of Composite Synthetic Drains

Composite Synthetic Drains (also called Prefabricated Drainage Composites) make use of high water-transmitting, high strength plastic cores enveloped within fabric filters. There are many varieties of composite drains used for a wide range of purposes. Some are for protecting walls and floors of basements from groundwater seepage. A type called *wick* drains is used as a replacement for the conventional sand drain for increasing the rate of consolidation of soft, compressible clay and peat foundations for embankments, as described in Sec. 7.2 under "Vertical Drains". Wick drains have a strong inner core with high vertical discharge capability, wrapped within a fabric filter strong enough to resist forces produced when they are driven into the ground. They are driven vertically downward with special vibrating tools into soft formations that need to be rapidly consolidated so embankments can be placed over the formations after a reasonable consolidation period under partial load, without the danger of failing.

Another very common type of composite drain, sometimes called the *fin* drain, uses a strong plastic core with high vertical transmissibility that conducts seepage to a collector-discharge pipe at the bottom for removal to safe exits. The whole assembly is wrapped within a fabric filter. One of the first of this kind—if not the first—was developed by and patented by Healy and Long (1971). They used a core of corrugated or otherwise formed plastic material to provide vertical channels for downward conduction of water to the pipe at the bottom. A fabric filter enclosed the whole assembly. They emphasize that the cross-sectional area of the channels must be larger than the openings in the fabric to allow the few particles that wash through the fabric to wash into the pipe and out of the system.

The use of two kinds of plastic materials in the composite drains offers a way to obtain high flow rates with synthetic drains. The effectiveness of the composite drains lies in the fact that they are a *two-material system,* with one material serving as a filter and another as a water-conducting element. As with mineral aggregates, it is virtually impossible to achieve both good filtering and high water-removing capability with a single material (see Sec. 5.1). Effective water removal with the composite drains depends on a core structure that has high strength and contains rather large passageways for the water to travel freely to a pipe or other exit. Because this kind of drain is relatively thin in comparison with a trench backfilled with permeable aggregate, the discharge capabilities of some of the composites are in a relatively low range. Others that have high compressive strength and flow channels with effective diameters of 0.5 in. and greater, can have large flow capabilities.

Before designing a drainage system making use of composite drains, the engineer should make his best estimates of the quantities of water that need to be removed, using Darcy's law *(q = kiA)* with realistic coefficients of permeability of the formations from which water flows, hydraulic gradient inducing the flow, and the area through which water is flowing to a pipe drainage system. Then materials for core, and size of pipe should be selected with manufacturer's ratings (or user's tests) of discharge capabilities of the materials to be used.

To illustrate the basic capabilities of vertical composite drains for removing water, I prepared Table 5.3 for several sizes of flow channels and several ranges of "effective" or "quasi" coefficients of permeability of core materials. Hydraulic gradients (in the drains) range from 0.02 to 1.0. For vertically downward flow the gradient can be 1.0 (100%), but for other flow conditions it may be as low as 1 or 2 percent. Though it is based on hypothetical conditions, I believe Table 5.3 illustrates the general order of magnitude of flow

TABLE 5.3   Estimated Flow Rates for Composite Drains with Highly Permeable Cores

Effective flow channel dia., in.	Thick- ness, in.	Estimated turbulence factor, C	Estimated "quasi" k', ft/day	Hydraulic gradient, i	Estimated flow rates* cu ft/day per foot	Estimated flow rates* gpm per foot
0.25	0.3	0.20	16,000	1.0	360	2.0
		0.27	23,000	0.5	258	1.34
		0.50	33,000	0.1	74	0.39
		0.65	50,000	0.05	56	0.29
		0.90	80,000	0.02	36	0.19
0.5	0.6	0.16	60,000	1.0	2700	14.0
		0.21	78,000	0.5	1760	9.2
		0.42	156,000	0.1	700	3.6
		0.55	206,000	0.05	464	2.4
		0.86	320,000	0.02	290	1.5
0.8	1.0	0.12	100,000	1.0	7500	39.1
		0.16	134,000	0.5	5025	26.1
		0.36	300,000	0.1	2750	11.7
		0.50	416,000	0.05	1560	8.1
		0.85	610,000	0.02	915	4.8

*Approximations, based on Quasi Darcy k values obtained for turbulent conditions (see Fig. 3.30) and $q = k'iA$, with the effective discharge area, A, equal to 90% of the cross-sectional area of a composite drain. Each composite drain type should be tested for its flow rate for specified hydraulic gradient and confining pressure.

possible in composite drains. The sketches in Figure 5.14 show basic arrangements for two kinds of prefabricated composite fabric drains. A "fin" drain with downward flow to a collector-discharge pipe is shown in Figure 5.14a, and so-called "wick" drains with upward flow are shown in Figure 5.14b. A schematic cross section for drains of these kinds is given in Figure 5.14c. The calculated flow rates, as given in Table 5.3 are for hydraulic gradients ranging from 1.0 to 0.02 and effective flow channel diameters of 0.25 in., 0.50 in., and 0.80 in.

In Table 5.3 it can be seen that for downward flow under a hydraulic gradient of 1.0 in a vertical drain with a thickness of 0.3 in. and an effective flow channel diameter of 0.25 in. the potential rate of flow is 2 gpm per foot. For a 100-ft length of drain, the flow capability would be 200 gpm, which is a substantial amount of water. If one of these drains was installed in a sandy soil with a coefficient of permeability of 50 ft/day, and a hydraulic gradient of 0.5 inducing flow to the drain, the inflow into a 20-ft deep drain would be about 5 gpm per 100 ft. from one side.

Referring again to Table 5.3, it is seen that the upward flow rate for a drain with a quasi coefficient of permeability of 80,000 ft/day and a hydraulic gradient, $i = 0.02$, would be 0.19 gpm per foot of width, or 0.063 gpm for a

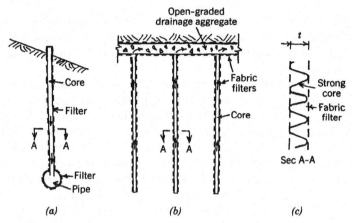

FIG. 5.14 Illustration of two kinds of composite synthetic drains, (a) Fin drain, (b) Wick drains, (c) shematic section A-A through typical prefabricated drain.

4-in wide wick drain. This would be 91 gallons per day or 12 cu ft/day for a single drain. Each cubic foot of water removed from a wick drain represents that amount of consolidation of the soil within its influence. With drains at a 4-ft spacing in both directions, 12 cu ft/day would represent (12 cu ft/day)/(16 sq. ft.) = 0.75 ft/day of consolidation, which is a very significant amount.

The flow rates in Table 5.3 are presented as rough indicators of potential rates for the conditions assumed. All prefabricated drains to be used in projects should have their capabilities tested before they are used to make sure they can perform as required.

Koerner (1986) made flow tests on a variety of "prefabricated drainage composites" using laboratory equipment that could handle up to 100 gpm per foot with applied normal pressures up to 30 psi for long periods of time. Fabrics with weak core structure had big reductions in flow under higher pressures, but three of the composites tested had large flow rates with little reduction under increasing pressure. One—measuring 2.8 cm in thickness with a core made up of columns of polyethylene 7 mm in diameter at 2 cm centers—had a flow rate of about 80 gpm under all pressures. Another—2.5 cm thick, with a "waffle" core of polystyrene—had flow rates of 12 to 11 gpm per foot under pressures from zero to 30 psi. The third—0.5 cm thick with a 0.75 cm grid of polypropylene for its core—had a flow rate of 2 gpm per foot, with no measurable reduction with increased pressure. All of the flow rates given here are for a hydraulic gradient of 1.0. Koerner concludes that more information is needed from manufacturers on flow rates of their products. He says specifications for composite drains should "list a minimum flow rate at a stipulated hydraulic gradient and applied pressure". Also, the protective filter "cannot intrude (much less fail) between the flow channels of the core material." His work is very timely as it gives answers to the key question about prefabricated drains, "What is their capability for discharging water under field conditions?" As discussed at length in this chapter and elsewhere in this book, the discharge capabilities of all kinds of drains for engineering works is of prime importance.

Chen and Chen (1986) made tests on permeability characteristics of five prefabricated vertical drains and found that drain materials with thicker and harder cores had higher permeabilities than thinner and weaker core materials. Also, the large flow channels gave higher permeabilities than smaller ones. They found that Calhoun's criterion, $0_{95}/d_{85} \leq 2$ or 3, was better than another criterion that had been suggested. With soil confinement they obtained permeability values in the range of 11–21 cm/sec at 0.2 Kg/cm^2 pressure to 8–15 at 3.0 Kg/cm^2 confining pressure. Without soil confinement they obtained permeabilities in the range of 18–30 cm/sec at 0.2 Kg/cm^2 pressure to 10–24 cm/sec at 3.0 Kg/cm^2 confining pressure.

Suits (1986) describes procedures used by the New York State Department of Transportation (NYSDOT) to determine the acceptability of prefabricated wick drains. He concludes that rigid core drains lose very litle flow capacity under confining pressures, but soft core prefabricated drains lost 50% to 90% of their flow capacity under the same loads applied to those with rigid cores. Flows of 30 cc/sec to 210 cc/sec were obtained through rigid core drains. Like California Dept. of Transportation, New York State DOT requires that wick drains be installed with a protective mandrel. Spacings have been between 4 and 7 feet for three projects that all gave excellent performance.

Prysock and Walsh (1988) describe methods used by the California Department of Transportation (CALTRANS) to test prefabricated vertical drains for flow capacity characteristics to determine suitability for use. Sections of drain are immersed in water in a testing device that forces water to flow through the fabric and out through the core. A limiting pressure is set up for acceptance of any specimen proposed for use in a state project.

5.8 SPECIFICATIONS FOR AGGREGATE FILTERS AND DRAINS

Although naturally occurring gravels and sands can sometimes be used for filters and drains with little or no processing, drainage aggregates usually need to be treated. Efforts to save money by using untreated local materials almost always produce marginal or unsatisfactory filters and drains. Most natural sand and gravel deposits are highly variable in grading from point to point in a borrow area or are covered or interbedded with silt and clay that is difficult to remove. Rarely, natural deposits of clean washed beach sands are potential sources of permeable materials for highway drainage. Other relatively uniform natural deposits which are located near projects can be looked on as possible sources. When feasible, local deposits should be explored with drill holes, test pits, and trenches and tested for grain-size distribution. Usually, however, *nat-*

ural deposits cannot be depended on for consistent, adequate sources of drainage materials without washing and screening.

Specifications for permeable aggregates for filters and drains should avoid the use of general terms such as *pervious* and *free draining* unless used in conjunction with specific requirements for grading, soundness, and permeability that ensure the necessary physical properties. When grading requirements are part of a specification, the *time and place* of the sampling for testing should be clearly stated; otherwise confusion arises, and enforcement of designers' intentions is difficult. Many aggregates break down and develop a greater proportion of fines during handling, placement, and compaction. Such materials may meet a grading specification at the plant or when they arrive at the job but may fail if tested *after* compaction. The important criteria are the grading and permeability of the material as it will be in the finished work; hence samples for testing should be taken of the materials *after they have been compacted.*

Frequently specifications for permeable aggregates are unenforceable because of their vagueness. A typical unsatisfactory specification is the following, which was used in the construction of an earth dam.

A gravel drain shall be constructed in the downstream portion of the embankment to the lines, grades, and dimensions shown on the plans, or as directed by the Engineer in the field. the gravel shall be *clean, well-graded, free-draining* (our emphasis), and contain no cobbles greater than 6 inches in diameter. Suitable gravel is available in the banks of the river from _____ to _____ feet downstream from the center of the dam, but some care in selecting the gravel may be required to insure its being clean and free-draining.

Although the writers of this specification thought they were giving the engineer a means of controlling the quality of the drain material, the only specific control he was able to enforce was that limiting the maximum size to 6 in. All of the other requirements were matters of opinion. As a consequence the contractor was careless in stripping overburden from the borrow pit and permitted excessive amounts of silt and clay to be included in the gravel. The drain material contained 8 to 10% of minus 200-mesh material and produced an impervious, entirely unsatisfactory drain that failed to fulfill its intended purpose. Subsequently the owners had to spend substantial additional funds to have other drains installed to make the dam safe for use, but it will never be drained so well as the designers intended, largely because of a meaningless, unenforceable specification.

It is extremely difficult to write specifications that are completely enforceable if the contractor is constantly looking for loopholes; a reasonable level of integrity on the part of the contractor must be assumed. Some descriptive terms cannot be avoided, but when used with specific test values the strength of the specification is greatly enhanced. A sample of a type of specification that produced a satisfactory job is the following.

The aggregate used shall be composed of hard, durable mineral particles free from organic matter, clay balls, soft particles and other impurities or foreign matter. When tested by Test No. _____, when sampled *after being compacted in the work,* the material shall conform to the following grading requirements:

Sieve No. or size	Percent passing by weight
1½ in.	100
¾ in.	50 to 100
No. 4	20 to 40
No. 16	7 to 20
No. 50	0 to 5
No. 100	0 to 2

Drain gravel shall be placed with spreader boxes or other approved equipment in horizontal lifts not over 12 inches in thickness before compaction. To minimize segregation and to facilitate its compaction, the material shall be thoroughly saturated at the time of its placement and compaction. Each lift will be compacted with _____ passes of a _____ roller weighing _____ lbs per linear foot.

The material as placed and compacted in the work shall be free of segregation and free of all contaminating materials. If unsatisfactory materials or contamination are permitted in the work, the unsatisfactory materials shall be removed to the satisfaction of the Engineer and replaced with acceptable materials at no additional cost to the owner.

The drain shall always be maintained at least 12 inches above adjacent embankment zones. At no place shall the dimensions be smaller than those given in the drawings or stated in the specifications. Equipment crossovers shall be limited to not more than two at any given level of the embankment. Each crossover shall be cleaned of all contaminating materials to the satisfaction of the Engineer and approved by the Engineer before additional drain materials are placed in these areas.

In cases of dispute over the acceptability of any portion of the placed materials, referee samples weighing 100 lbs shall be secured by the Engineer for testing. If such a sample fails to meet the specification requirements, all of the material represented by the sample shall be considered unacceptable and shall be removed to the complete satisfaction of the Engineer.

No other single feature of many civil engineering works is more vital to long, trouble-free performance than the drainage features. The need for high quality workmanship in the construction of drains cannot be overemphasized. Well-written, enforceable specifications are a prerequisite for good quality construction.

REFERENCES

Babbitt, H. R. and J. J. Doland (1939), *Water Supply Engineering*, McGraw-Hill Book Co., p. 6.

Barrett, Robert J. (1966), "Use of Plastic Filters in Coastal Structures," *Proceedings, 10th International Conference on Coastal Engineering,* Tokyo, September 1966, pp. 1048-1067.

Bell, J. R. and R. G. Hicks, et al. (1978), "Test Methods and Use Criteria for Filter Fabrics," *Interim Report,* prepared for the U.S. Dept. of Transportation, Federal Highway Administration (FHWA), Oregon State University, Sept. 1978.

Bertram, G. E. (1940), "An Experimental Investigation of Protective Filters," Publications of the Graduate School of Engineering, Harvard University, No. 267, January 1940.

Calhoun, C. C., Jr. (1972), "Development of Design Criteria and Acceptance Specifications for Plastic Filter Cloths," Technical Report S-72-7, U.S. Army Engineer Waterways Experiment Station, Vicksburg, Miss., June 1972.

Carroll, Robert G., Jr. (1975), "Mirafi 140 Filter Fabric For Subsurface Drains," *Drainage Contractor,* Ontario, Canada, Vol. 1, No. 2, December 1975, pp. 133-134.

Carroll, Robert G., Jr. (1976), "Design Considerations for Specifying Fabrics Used in Subsurface Drainage Structures," Internal publication, Celanese Fibers Marketing Co., July 1976 (unpublished).

Cedergren, Harry R. (1961), "Discussion of 'Design of Control Measures for Dams and Levees,' by W. J. Turnbull and C. I. Mansur," *Transactions,* A.S.C.E., Vol. 126, Part I, pp. 1531-1534.

Cedergren, Harry R. (1962). "Seepage Requirements of Filters and Pervious Bases," *Transactions,* A.S.C.E., Vol. 127, Part I, pp. 1090-1113.

Cedergren, Harry R. (1976), "Evaluation of Seepage Stability," Paper presented at Engineering Foundation Conference on "Evaluation of Dam Safety," Pacific Grove, Calif., November 28-December 3, 1976; pp. 195-218 of *Proceedings,* A.S.C.E., published in 1977.

Cedergren, Harry R. (1977), The Need for Fabrics in Hydraulic Engineering," *Proceedings;* International Conference on the Use of Fabrics in Geotechnics, Paris, France, April 20-22, 1977; Vol. II 249-254.

Cedergren, Harry R. (1987), "Drainage of Highway and Airfield Pavements," Updated printing by Robert E. Krieger Pub. Co., Malabar, Florida, 1987; originally published by John Wiley & Sons, Inc., N.Y. 1974, pp. 174-180; (1987a), pp. 136-141.

Chen, R. H. and C. N. Chen (1986), "Permeability Characteristics of Prefabricated Vertical Drains," *Proceedings,* 3rd Int. Conf. on Geotextiles, Vienna, Austria, April 7-11, 1986, Vol. III, pp. 785-790.

Civil Engineering Magazine, ASCE (1981), "Dike Safety Upgraded with Millions of Square Feet of Fabric," January, 1981, pp. 46-47.

Creager, W. P., J. D. Justin, and Julian Hinds (1950), *Engineering for Dams,* John Wiley and Sons, New York, Vol. 3, pp. 682-3.

Dunham, James W., and Robert J. Barrett (1974), "Woven Plastic Cloth Filters for

Stone Seawalls," *Journal,* Waterways, Harbors, and Coastal Engineering Division, A.S.C.E., Vol. 100, No. WW1, February 1974.

Engineering News-Record (1976), "Preliminary Report Faults Grout Curtain in Teton Dam Failure," June 17, 1976, pp. 9–10.

Fowler, Jack (1985), "Building on Muck," *Civil Engineering* (New York), Vol. 55, No. 5, May, 1985, pp. 67–69.

Giroud, J. P. (1982), "Filter Criteria for Geotextiles," Proc. 2nd Int. Conf. on Geotextiles, Las Vegas, Nevada, August 1–6, 1982, Vol. I, pp. 103–108.

Grass, L. B., and A. J. MacKenzie (1972), "Restoring Subsurface Drain Performance," *Journal,* Irrigation and Drainage Division, *Proceedings,* A.S.C.E., Vol. 98, No. IR1, pp. 97–106.

Halliburton, T. Allen, Thomas M. Petry, and Myron L. Laydon (1975), "Identification and Treatment of Dispersive Clay Soils," Report to the U.S. Department of the Interior, Bureau of Reclamation, Denver, Research Contract 14-06-D-7535 with Oklahoma State University, July 1975, p. 142; (1975a), p. 143.

Halliburton, T. A., and P. G. Wood (1982), "Evaluation of the U.S. Army Corps of Engineer Gradient Ratio Test for Geotextile Performance," Proc. 2nd Int. Conference on Geotextiles, Las Vegas, Nevada, Aug. 1–6, 1982, Vol. I, pp. 97–101.

Hannon, J. (1982), "Fabrics Support Embankment Construction Over Bay Mud," Proc. 2nd Int. Conf. on Geotextiles, Las Vegas, Nevada, Aug. 1–6, 1982, Vol. III, pp. 653–658.

Healy, K. A. and R. P. Long (1971), "Prefabricated Subsurface Drains," *Highway Research Record* 360, 1971, p. 57.

Karpoff, K. P. (1955), "The Use of Laboratory Tests to Develop Design Criteria for Protective Filters," *Proceedings,* American Society for Testing Materials, Vol. 55, p. 1183, 1955.

Koerner, R. M., and F. K. Ko (1982), "Laboratory Studies of Long-Term Drainage Capability of Geotextiles," Proc. 2nd Int. Conf. on Geotextiles, Las Vegas, Nevada, Aug. 1–6, 1982, Vol. I, pp. 91–95.

Koerner, R. M., and J. E. Sankey (1982), "Transmissivity of Geotextiles and Geotextile/Soil Systems," Proc. 2nd Int. Conf. on Geotextiles, Las Vegas, Nevada, Aug. 1–6, 1982, Vol. I, pp. 173–176.

Koerner, Robert M. (1986), "Designing for Flow," *Civil Engineering,* ASCE, Oct. 1986, pp. 60–62.

Miano, Ralph R. (1976) (private communication citing test report for Celanese Fibers Marketing Company, prepared by Law Engineering Testing Company. Ore Transfer Site, Chesapeake, Virginia. Letco Project No. NK-5-137, March 1976).

Prysock, R. H., and Tom Walsh (1988), "California's Rapid Test for Predicting Relative Field Performance of Wick Drains," prepared for presentation at 67th Annual Meeting, Transportation Research Board (TRB), Wash. D.C., Jan. 11–14, 1988.

Raumann, G. (1982), "Inplane Permeability of Compressed Geotextiles," Proc. 2nd Int. Conf. on Geotextiles, Las Vegas, Nevada, Aug. 1–6, 1982, Vol. I, pp. 55–60.

Ripley, C. F. (1983), Discussion of ASCE Paper No. 16807, "Design of Filters for Clay Cores of Dams," By Peter R. Vaughan and Hermusia F. Soares, Jan. 1982; discussion in vol. 109, No. 9, *Journal of the Geotechnical Engineering Div.*, ASCE., Sept. 1983, pp. 1193–1195.

Robison, Rita (1985), "Engineering With Fabric," *Civil Engineering*, ASCE, Dec. 1985, pp. 52–55.

Rosen, William J., and B. Dan Marks (1975), "Investigation of Filtration Characteristics of a Nonwoven Fabric Filter," Transportation Research *Record* 532, 54th Annual Meeting of the TRB, pp. 87–93.

Seemel, Richard N. (1976), "Filter Fabrics," *Civil Engineering*, A.S.C.E., April 1976, pp. 57–59; (1976*a*), p. 58.

Sherard, J. L., R. S. Decker, and N. L. Ryker (1972*a*), "Piping in Earth Dams of Dispersive Clay," *Proceedings*, A.S.C.E. Specialty Conference on the Performance of Earth and Earth-Supported Structures, Purdue University, June 1972, Vol. 1, Part 1.

Sherard, J. L., R. S. Decker, and N. L. Ryker (1972b), "Hydraulic Fracturing in Low Dams of Dispersive Clay," *Proceedings*, A.S.C.E. Specialty Conference on the Performance of Earth and Earth-Supported Structures, Purdue University, June 1972, Vol. 1, Part 1.

Sherard, James L., Lorn P. Dunnigan, and Rey S. Decker (1976), "Identification and Nature of Dispersive Soils," *Journal*, Geotechnical Engineering Division, A.S.C.E., Vol. 102, No. GT4, April 1976, pp. 287–301.

Sherard, James L., Richard J. Woodward, Stanley F. Gizienski, and W. A. Clevenger (1963), *Earth and Earth-Rock Dams*, Wiley, New York, p. 84.

Sherard, James L. (1986), "Hydraulic Fracturing in Embankment Dams," *Journal*, Geotechnical Engineering Division, A.S.C.E., Vol. 112, No. 10, Paper 20963, October, 1986, pp. 905–927.

Shrestha, S. C., and J. R. Bell (1982), "Creep Behavior of Geotextiles Under Sustained Loads," Proc. 2nd Int. Conf. on Geotextiles, Las Vegas, Nevada, Aug. 1-6, 1982, Vol. III, pp. 769–774.

Suits, L. David (1986), "New York State Department of Transportation's (NYSDOT) Basis of Acceptance and Specifications for Prefabricated Wick Drains," *Transportation Research Circular*, TRB, No. 309, Sept. 1986, pp. 7–9.

Taylor D. W. (1948), *Fundamentals of Soil Mechanics*, Wiley, New York, p. 134 (Fig. 7.4).

U.S. Army Corps of Engineers, Waterways Experiment Station (1941), "Investigation of Filter Requirements for Underdrains," Technical Memorandum No. 183-1, December 1941.

U.S. Army Corps of Engineers (1955), "Drainage and Erosion Control—Subsurface Drainage Facilities for Airfields," Part XIII, Chapter 2, Engineering Manual, Military Construction, Washington, D.C., June 1955, p. 15; (1955a) p. 13.

U.S. Army Corps of Engineers (1975), "Guide Specification for Plastic Filter Cloth," CW 02215, October 1975, pp. i–iv.

U.S. Army Engineer Waterways Experiment Station (WES) (1980), EM 1110-2-1906, 30 Nov 1970, *Laboratory Soils Testing;* Appendix XIII, "Pinhole Erosion Test for Identification of Dispersive Clays," 1 May 1980.

U.S. Army, U.S. Navy, and U.S. Air Force (1971), "Dewatering and Groundwater

Control for Deep Excavations," TM 5-818-5, NAVFAC P-418, AFM 88-5, Chapter 6. April 1971, p. 39.

U.S. Bureau of Reclamation (1973), *Design of Small Dams,* U.S. Government Printing Office, Washington, D.C., 2nd ed.

U.S. Department of the Army, Corps of Engineers, Office of the Chief of Engineers (1986), "Seepage Analysis And Control for Dams," Engineering Manual EM 1110-2-1901, 30 September, 1986, pp. D-1 to D-5; (1986*a*), p. D-6.

CHAPTER SIX

SEEPAGE CONTROL IN EARTH DAMS AND LEVEES

6.1 GENERAL

A prominent engineer once said that any well drained structure is safer and more economical than its poorly drained counterpart. Because the water held back by dams and levees can pose serious hazards to the people living below or behind these structures, every effort should be made to make them as safe as possible. To achieve this objective, controlling seepage is a top priority; therefore, the development of good systems for controlling water is one of the most important responsibilities of civil engineers. Sound methods for designing features that can eliminate or at least minimize problems with water in dams and levees are presented in this chapter.

Historical

Levees or dikes have protected lands since primitive times and earth dams have been used for the storage of water for human needs and protection for more than 2000 years. In the year 504 B.C. an earth dam containing nearly 20 million cu. yd. of earth was completed in Ceylon (now Sri Lanka).

Early dams and levees were constructed simply by heaping earthen materials across an area to be blocked, human traffic often producing all the compacting effort. Many of the early efforts were washed out by overtopping, underseepage, or other destructive forces, but eventually standards of practice emerged that can be called "rules of thumb." These practices often had no real basis, except that something had worked at a number of locations; hence it might work elsewhere.

Even into the twentieth century dams and levees were being designed largely by empirical methods. Wegmann (1922) states:

> The design of such works (earth dams) should not be based upon mathematical calculations of equilibrium and safe pressure, as in the case of masonry dams, but upon results found by experience. Most of the earth dams constructed within the last century have had a large margin of safety in resisting the water-pressure, both as regards overturning and sliding, and yet frightful disasters, such as the rupture of the Dale Dyke and the Johnstown dams, have resulted from faults in *designing some details* or from *neglect in the construction* of the work.

From about 1930 to the present time (1988), analytical and experimental methods have had an increasingly important part in the design and construction of earthworks. They will continue to play an important role in the design of dams and levees, but experience will also have a dominant place. As the weaknesses of modern practices come to light in occasional failures new standards will emerge. These standards will continue to improve because they will be based on fundamental principles and broad experience.

Requirements for Safety

Earth dams and levees must be safe against overtopping, their slopes must be stable under all conditions, their foundations must not be overstressed, and they must be safe against internal erosion and water forces and pressures. Justin (1936) has tabulated earth dam failures. Among those that did not fail from overtopping about 80% failed because of piping, sloughing at the toe, or other results of uncontrolled seepage. Horsky (1969) tells about a reservoir that was constructed on a limestone foundation which leaked so badly that the reservoir would not hold water. A large percentage of earth dam failures reported by Sherard et al. (1963) likewise were seepage failures. The greatest need, in addition to the prevention of overtopping, is to prevent internal damage from seeping water; therefore methods of analyzing the influences of seepage will always be among the most important aspects of designing and building safe dams and levees.

An ASCE/USCOLD (American Society of Civil Engineers/ United States Committee on Large Dams) committee on failures and accidents to large dams (ASCE/USCOLD, 1975) thoroughly studied reports of 349 dams with problems. Of those studied, 100 had failed from uncontrolled seepage or had leaks of varying degrees of severity. Jansen (1980), in a thorough review of problems with dams worldwide, emphasizes that control over seepage is one of the most important requirements for safe dams. He says, "Any leakage at an earth embankment may be potentially dangerous, since rapid erosion may quickly enlarge an initially minor defect." He urges dam owners to have thorough surveillance and monitoring programs that can forewarn of impending problems from seepage and other factors influencing the safety of each dam.

Most of the foundations of dams and levees are covered with gravelly or sandy soils and often with a thin cover of fine sand, silt, or clay. The underlying bedrock is often relatively tight, but it may be highly weathered, jointed, and fissured and it may contain crevices, permeable zones, fault planes or other hidden inconsistencies. Minor soils and geological details which exist in the foundations of many dams and levees can cause dangerous seepage conditions if not detected and controlled.

Many of the serious dam and levee failures of our times have occurred because of highly erodible or permeable formations that were not detected before or during construction (see Figs. 1.1 and 10.15). Figure 6.1 shows a dam in an arid region of the United States which was built to store water, but leaked so badly that almost no water remained in the reservoir more than a few days after a storm (as seen here). After one heavy storm had filled the reservoir about half full, boils appeared at the downstream toe, and the downstream slope was on the verge of a collapse. Investigations made to determine the reasons for the problems revealed a layer of extremely permeable open-work gravel under a thin cover of silty clay. Had the open-work layer been detected in advance of the construction, a shallow cut-off trench through this layer into tight underlying materials could have prevented the trouble. This dam illustrates the importance of having adequate investigations of foundations of all water-impounding facilities made prior to the design and construction of projects.

FIG. 6.1 A reservoir in an arid area which cannot hold water because of open-work gravel layer not detected before or during construction.

Earth dams and levees should be designed to utilize available materials to the best advantage and to conform to actual conditions at sites. Sherard et al. (1963a) say, " . . . the characteristics of the particular site have a greater influence on the design of an earth dam then they do on many other engineering structures." Design details sometimes will be influenced heavily by the strengths of foundations and construction materials, but the basic features are usually dictated by seepage considerations. The more common methods for controlling seepage in earth dams and levees and in their foundations are described in subsequent sections, in which examples are given that not only point up the advantages and weaknesses of various methods but also illustrate the use of analytical methods in the design of safe dams and levees. Properly designed filters and drains that allow free discharge of seepage through and under dams, while preventing the washing out of erodible materials, are essential features of every important dam.

Not only must dams and levees be *designed* for safety, they must be *constructed* with safety in mind, for unsafe construction practices can lead to failures. Adequate stripping of unsatisfactory materials from foundations and abutments is essential to the security of earth dams and levees. Impervious cores should be in intimate contact with nonerodible watertight soil or rock formations. Loose materials should be removed from all exposed fissures or joints and such areas backfilled with "dental" concrete or slush grout that is protected from damage and sealed with an impervious membrane curing compound. Pneumatically applied mortar (gunnite) is often used for the protection of impervious cores against piping along contacts with jointed, fissured, sheared, and erodible rocks.

Impervious fill in direct contact with rough or uneven surfaces should be placed slightly wet of optimum to ensure its molding to the shape of the surface. Care must be taken to avoid the use of excessively wet impervious fill at the bottom of cutoff trenches or other locations in which seepage or groundwater may be encountered.

Many of the seepage problems and failures of earth dams and levees have occurred because of careless or incomplete cleanup and preparation of their foundations and abutments. Realizing the importance of good preparations, an entire Journal of the S. M. & Foundations Division (now Geotechnical Engineering Division), ASCE, was devoted to this topic. Acker and Jones (1972) describe procedures used for several high dams in foreign countries: Ambukloa Dam—a 430-ft high rockfill dam in the Philippines; Derbendi Khan—a 442-ft high rockfill dam in Iraq; Angat—a 430-ft high rockfill dam in the Philippines; and Guri—a 279-ft high rockfill dam in Venezuela. Barron (1972) describes procedures used by the U.S. Corps of Engineers in preparing foundations and abutments of high dams on rock foundations to ensure watertight contacts and control over seepage. Burke et al. (1972) and Pratt et al. (1972) describe procedures used in constructing dams from 580 feet high up to 800 feet high in the U.S. and Canada. Stropini et al. (1972) tell about foundation treatment methods for the State of California's Oroville Dam, Cedar

Springs Dam, and Pyramid Dam, which range in height from 380 feet up to 770 feet. Practices in Australia were described by Wallace and Hilton (1972), and U.S. Bureau of Reclamation practices were described by Walker and Bock (1972) in relation to 538-ft high Trinity Dam, 322-ft high Reudi Dam, and 390-ft high Blue Mesa Dam. Some of the authors also give detailed descriptions of instrumentation systems designed to forewarn of any undesirable conditions occurring while reservoirs are filling, so that remedial measures can be taken before serious problems develop.

In the construction of impervious cutoffs into pervious waterbearing soils excavations can become badly flooded unless adequate dewatering measures are employed to remove the incoming seepage (sec. 7.1). Sometimes in an effort to reduce foundation dewatering costs impervious backfill for cutoffs is bulldozed into the water until sufficient thickness has been placed to support construction equipment. Then the surface is rolled lightly and additional fill is placed in layers and *thoroughly compacted* to produce dense fill over loose fill. This practice is an invitation to trouble because well-compacted fill tends to arch over cavities that form in loose underlying material because of piping or reduction in volume produced by saturation. When cavities or pipes erode to the reservoir side, a sudden rush of water may cause rapid failure of the structure. I examined one earth dam that washed out in 1965 because of this condition and another that leaked so badly that it could not serve its intended purpose of storing water. No poorly compacted materials should be permitted in any earth embankment.

When all seepage discharge areas are covered with well-designed filters and drains (as in Figs. 6.7 and 6.14*d*), piping problems can be virtually eliminated. Nevertheless, good construction practices should always be mandatory.

An important safeguard to the safety of a dam or reservoir is a good observation and monitoring program, in which periodic reading of piezometers, settlement monuments, horizontal movement stakes, and seepage quantity measurements are requirements. Calibrated weirs are used for measuring the quantity of seepage below dams and reservoirs. Whenever possible, drainage systems for dams and reservoirs should be separated into several distinct parts and the discharge from each should be isolated for separate measurement. Then, if changed conditions occur, they can at least be partly narrowed down. It is important that seepage observation and measurement be made a permanent part of the operation of all important dams and reservoirs, for changes that may become dangerous if not detected can go on for years. Marsal and Pohlenz (1972) tell about the failure of the Laguna Dam in Mexico from underground piping after 60 years of operation. They maintain that the cause of this failure would have been difficult to anticipate as one of the many unknown erosion channels or "pipes" in the embankment foundation " . . . may shortcircuit the reservoir in a matter of hours and cause disaster." Even though it may not be possible to avert every failure, thorough monitoring of dams and reservoirs certainly should help to keep the number of failures down to the lowest possible minimum.

Before modern experimental and rational design methods were developed, hardly any earth dams were provided with more than minimal instruments for recording significant behavioral patterns. But, essentially all dams of importance built in the past several decades have instruments for measuring pore pressures and other important indicators of behavior. Pratt et al. (1972a) describe elaborate systems of motion indicators and piezometers for Bennett Dam and Mica Dam in Canada, completed in 1967 and 1973, respectively. O'Rourke (1974) discusses extensive systems for Oroville Dam in California, for recording various kinds of motion, and pore pressures. Stanage (1982) tells of the use of 3-dimensional finite element analysis for the layout of instrumentation for measuring motions and pore pressures in the 640-ft high New Melones Dam in California, completed in 1978. Simmons (1982), in a review of the seepage problems and remedial measures for Wolf Creek Dam on limestone formations in Kentucky, tells about the use of more than 300 piezometers for studying seepage in and under this dam. He also tells about dye tracer tests, and temperature surveys of subsurface water that aided in understanding seepage behavior. Other techniques were also employed in the studies leading to the design and construction of a deep concrete diaphragm wall through this dam and into its limestone foundation to cut off all significant seepage, to prevent further problems with this dam. Londe (1982) describes a "new generation of instruments" using constant tension invar wires for best accuracy of distance measurements, and continuous piezometers for boreholes in rock formations. A mobile torpedo with packers is lowered to any desired point where pore pressure readings are desired. This instrument not only gives pore pressures but allows rough estimates to be made of permeability from the time required for equilibrium to be reached. Cooke and Sherard (1987) point out that since the embankments of Concrete-Face Rockfill Dams (CFRD) are dry, with only a thin layer of water flowing over the rock foundations, "instrumentation is not a requirement for safety monitoring." And, "Crest settlement movements and a leakage weir are adequate" for these dams. Clearly, monitoring systems must be designed for individual structures and site conditions, and the more difficult the conditions the more detailed the system must be.

A variety of devices are used for measuring or recording pore pressures or hydrostatic head in and under earth dams and levees and other water-impounding structures. Figure 6.2 shows a "pop-on" pressure gage in place for a reading of pore pressure in the foundation beneath the downstream toe of an earth dam in California. It can be attached without releasing any water, so gives instantaneous readings of the correct pressure head. Older types of gages that required removing a cap from the piezometer pipe to install the gage or adjusting a valve, which allowed water to escape, needed a good deal of time for the head to build back up to the correct value. Permanently installed gages are subject to vandalism unless housed in boxes or large-diameter pipe risers with locks for doors or covers. A simple riser pipe, as shown here, is less likely to be damaged by vandals, so this type of pressure gage is very

FIG. 6.2 A "pop-on" pressure gage in position for pressure reading in a piezometer at downstream toe of an earth dam (used with permission of the Sacramento Dist., U.S. Army Corps of Engineers).

convenient at locations where an attendant can make readings of pressures as frequently as desired.

At locations where frequent readings are needed, but access is difficult or time-consuming, automatic recording devices such as shown in Figure 6.3 are often used. A float on a flexible cable actuates a needle that makes an ink imprint on a paper on a slowly moving drum, registering water level at any desired time interval, such as 15 minutes or more, thus providing a relatively continuous record of pressure head. At suitable intervals, such as 30 days or so, an attendant visits the site and removes the chart, replacing it with a fresh one.

Some projects have piezometers and water-level instruments equipped with recording devices that send a signal by radio or direct wire to operations buildings that may be long distances from projects, and where attendants are on duty 24 hours a day. Any sudden increase in pressure or water level triggers a signal to alert operators to a possible problem before it becomes serious.

The quantity of seepage through or under dams or levees can be of great

FIG. 6.3   An automatic device for recording the level of water in piezometer tube in an earth levee in California (used with permission of the Sacramento Dist., U.S. Army Corps of Engineers).

significance, particularly if any sudden increase takes place. Where attendants can make inspections one or more times a day, calibrated "V"-notch or flat weirs offer a simple way to obtain measurements of seepage quantity. A calibrated Parshall Flume, such as shown in Figure 6.4, is somewhat more accurate than the weirs, and is preferred by many engineers.

In addition to the use of seepage quantities and piezometer measurements to monitor seepage behavior, various types of seepage tracer have helped engineers to locate areas that were causing increases in seepage. Peter et al. (1970) tell about the construction of dams on the Danube, where spots liable to piping failures by subsurface erosion are localized by the use of one-hole or two-hole methods to observe seepage flow by radionuclid methods. Birman et al. (1971) describe the use of thermal monitoring to detect leakage and changes in leakage. Thermister probes calibrated to 0.1°C were read periodically. According to these authors, the observed data could be correlated with observed visible leakage zones, which demonstrated that the method is reliable. Sentürk and Sayman (1970) say that the monitoring of an instrumented dam revealed a sudden 300% increase in seepage. Tests with fluorescein dye and radioisotope

FIG. 6.4 A Parshall Flume for measuring quantity of seepage at the downstream toe of an earth dam (used with permission of the Sacramento Dist., U.S. Army Corps of Engineers).

tracers indicated that the increase was caused by a crack that developed in an impervious blanket upstream from the dam. Piezometer and seepage quantity measurements for a small reservoir in southern California revealed that five times the predicted underseepage developed during the first filling of the reservoir and excess head built up under the downstream toe in amounts much larger than predicted from the flow-net studies for its design. After the reservoir was emptied numerous sinkholes were found in the clay lining just upstream from the dam, with evidence that a large amount of water had leaked through them. After they were repaired and other drainage improvements had been made the reservoir was refilled and water losses dropped to less than 20% of those in the first filling and agreed almost exactly with those originally predicted with the flow nets. A good monitoring program for this reservoir had quickly revealed the presence of the sinkholes, and verified the benefits of the repairs.

Detrimental Actions of Seeping Waters

Although this book is limited almost entirely to the physical factors that influence seepage and its control, other factors such as changes in the chemical composition of soils and the resistance of soils and rocks to dissolution by seeping waters can also be important. Water is generally considered to be an inert substance, yet seeping waters can decrease or increase soil permeabilities (see *Test Precautions and Details,* Sec. 2.4). In situations in which changes in permeability under the influence of seepage could be detrimental, chemical analyses should be made of reservoir waters and soils and soft rocks.

Waters high in sodium tend to flocculate soils high in alumina, thereby reducing the permeability. On the other hand, fresh water may deflocculate clay soils and render them more permeable. Aitchison and Wood (1965) attribute the failure of a number of small dams in Australia to postconstruction deflocculation that led to piping when fresh river water was suddenly turned into reservoirs that had been kept filled with saline water. These authors recommend that deflocculation-susceptible compacted earth fill have an upper limiting value of permeability of 10^{-5} cm/sec to 10^{-7} cm/sec as a means of avoiding piping. They also recommend compaction at optimum or slightly wet of optimum as further means of avoiding piping failures in clay soils. In extremely arid climates in which little or no water is available for compaction they recommend the use of chemical aids. They also state: "Maintenance of a low sodium adsorption ratio in the percolating water is essential to prevent undesired changes in the soil Sodium Adsorption Ratio following leaching."

To safeguard dams and reservoirs against failures caused by the loss of suspended or dissolved matter by seeping water, seepage quantities should be measured periodically and samples tested for suspended solids and dissolved matter. Large amounts of matter lost by leaching may cause the formation of channels or cavities, which if uncorrected can lead to piping failures. Sometimes to safeguard dams and reservoirs from failure it is necessary to close off these channels and to fill cavities with chemical or cement grout. If unexplained increases in rates of seepage occur in the foundations or abutments of dams or reservoirs, thorough investigations should be made to determine the causes and develop remedies.

If all seepage exits are protected with filters (Chapter 5), the dangers of piping caused by increased permeability are virtually eliminated.

Progressive reductions in the permeabilities of reservoir and canal linings and the cores of earth dams can be beneficial, but reductions in the permeabilities of filters and drains under the actions of percolating waters can be detrimental and should be guarded against. A fine-grained drainage layer which had ceased to function as a drain several years after the construction of a highway, was dug into and found to have become cemented into a hard, impervious mass. Such detrimental actions can largely be minimized by setting high standards of cleanness for drainage aggregates and using fundamental principles in designing seepage control systems for all important engineering works.

Methods for Controlling Seepage

As in other engineering works, earth dams, levees, and their foundations can be protected from seepage by two fundamental processes (Sec. 1.1): (1) those that keep the water out or reduce the seepage quantities and (2) those that use drainage methods to control the water entering. Methods commonly used under (1):

Cutoff trenches
Grout curtains
Sheet-pile walls and other thin cutoffs
Impermeable upstream blankets
Thin sloping membranes

Methods used under (2):

Embankment zoning (with filters and drains)
Longitudinal drains and blankets
Chimney drains extending upward into embankments
Partially penetrating toe drains
Relief wells

In general, combinations of methods are used; however, no attempt is made in this chapter to cover all possible combinations.

Two or more methods for the control of seepage are employed in most large dams, particularly when they are built on deep alluvial formations or badly jointed rock foundations. Nawaz and Nagvi (1970), for example, tell about the Tarbella Dam on the River Indus in Pakistan, where thick alluvial deposits 60 to 120 m deep constitute the dam's foundation. Seepage control measures include an inclined impervious core, an impervious earth blanket extending 1432 m upstream from the toe of the main dam, grout and drainage curtains, a drainage blanket, a collector gallery, and drainage wells. Each project must be provided with sufficient safeguards to provide a high level of protection against all potential seepage hazards. Even though jointed rocks in the foundation of Teton Dam were extensively grouted, failure to provide filters and drains between its highly erodible silt core and unsealed joints in the rock allowed this dam to fail by piping (E. N.-R., 1977).

6.2 SEEPAGE CONTROL BY METHODS THAT REDUCE QUANTITY

Cutoff Trenches

Frequently dams and levees are constructed on foundations composed of various kinds of alluvial materials underlaid by some form of bedrock. A cross section is shown in Figure 6.5 in which a foundation of permeable soils of moderate thickness covers an impermeable bedrock formation. If the alluvium is not excessively thick, a cutoff trench excavated without bracing offers a means for completely controlling underseepage. In deep alluvium other meth-

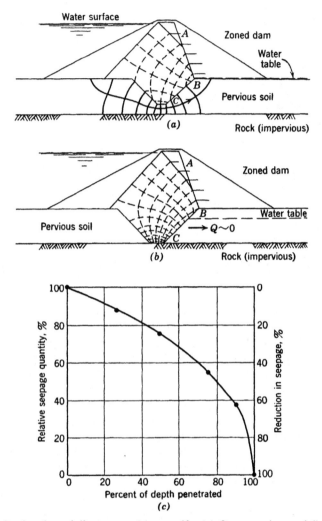

FIG. 6.5 Study of partially penetrating cutoffs, (*a*) Cross section and flow net for a partial cutoff. (*b*) Complete cutoff (minute flow through dam). (*c*) Relationship between depth of cutoff and seepage quantity. Note that suitable filters must be provided to prevent piping of soil at faces *A-B-C* in (*a*) and (*b*).

ods are used; for example, sheet-pile walls and narrow trenches held open with a thick slurry of clay and bentonite and backfilled with various soil, cement, and bentonite mixtures or unreinforced or reinforced concrete.

Open trench cutoffs offer a number of advantages because they provide full-scale exploration trenches that expose all soil strata, permit high quality backfilling operations to be carried out and treatment of exposed bedrock when necessary, and the installation of filters and drainage blankets to control seepage and prevent piping into coarse foundation strata. Also, in the process, a large mass of foundation soil is removed and replaced with stronger material, which often helps to stabilize the dam.

The construction of an open-trench cutoff can be a large operation, for the excavation may need to be quite deep and excess groundwater can create major problems. To ensure that all permeable strata in a foundation will be intercepted by a cutoff and that a good watertight joint will be made between the impermeable core of a dam and sound rock an entire excavation should be carried out "in the dry." The control of seepage into excavations can be difficult, particularly if the water sources are not known. Methods of selecting and designing dewatering systems for below-the-water excavations are described in Sec. 7.1.

Success in the use of cutoff trenches requires the interception of all important sources of inflowing water, and the use of dependable pumps, other equipment and power supplies that allow the work to go on without interruption until all crucial aspects of the work have been completed. Sometimes delays and even inferior quality fill and drains have been caused by a lack of understanding of the sources of the water, or equipment or power failure. Figure 6.6a shows a cutoff trench excavation being made under very adverse conditions. The sloppy condition was caused by the shutdown of dewatering pumps when the gasoline-powered generators for the pumps ran out of fuel. Unbelievable as it seems, it occurred almost every night! Since I had developed the cross-section and drainage features for the dam, I was invited to look at the job and offer recommendations to cure the problem. To relieve the power-failure problem, I suggested that dependable electric power lines be run to the spot. This was done, and that ended the pump shut-off problem.

But, an equally serious problem still remained. Those responsible for the construction believed that the sumping method was not working because of heavy inflows through the valley alluvium from the reservoir area upstream from the dam. They were convinced that a second cutoff through the alluvium at the upstream edge of the dam area was needed to control the inflows. It would have cost a great deal because it would all be done as "extra work" under a change order. To try to understand where the water was coming from, I sat on the spot where the photo in Fig. 6.6a was taken and carefully studied the area, looking for sources of water. At first, nothing was obvious. But after a time, I began to detect numerous seeps coming out of the rock formations at the bottom of the far slope. Suddenly I realized that the inflows were from the fractured rock formations at the far end of the excavation. Also, I realized that the proposed second cutoff would be worthless and a complete waste of money. So I suggested that a number of small sump drains with pumps be installed into the rock near the bottom of the excavation wherever noticeable seeps were flowing into the excavation. At first this suggestion was resisted, but it was eventually done. Fortunately it completely controlled the inflow problem, as may be seen in the next photo (Fig. 6.6b) taken a few weeks after the visit to the site when I took the photo in Figure 6.6a. This example is

FIG. 6.6   Cutoff trench for a 90-ft high earth dam in California. (*a*) Sloppy conditions caused by lack of steady power supply for sumping pumps and failure to discern sources of inflowing water; (*b*) Same trench a few weeks later after dependable power supply had been installed, and inflowing water was controlled by sumps drilled into primary wet spots.

described in this detail because it points up the necessity of "knowing where the water is coming from", as well as the need for dependable dewatering systems.

Because a deep, open cutoff trench can be an involved and expensive procedure, the trend has been toward the use of the slurry-trench method for seepage control for earth dams and many other kinds of structure with foundations below the water table (see also Sec. 6.6).

If deep cutoffs with slurry trenches are to be used for permanent dams or levees, the backfill must not pipe into openwork gravel or open rock seams which may exist in the foundations. Because this work cannot be inspected, its resistance to piping may be uncertain; however, if the trench is backfilled with concrete, piping problems can be eliminated.

Considerable practical information about the design and construction of slurry-trench cutoffs (using soil-bentonite or soil-cement backfill, or concrete membrane walls) is given in an updated engineering manual prepared by a U.S. Governmental agency (U.S. Department of the Army, Corps of Engineers, 1986).

To be effective, cutoffs must penetrate pervious strata thoroughly. In this respect they closely parallel other systems that depend on keeping the water out of foundations or embankments. A partly penetrating cutoff trench and typical flow net are shown on the cross section in Figure 6.5a, a completely penetrating cutoff in Figure 6.5b, and the relation between depth of cutoff and seepage quantity in Figure 6.5c. This chart reveals, for example, that a 90% complete cutoff trench of the proportions shown reduces seepage only 61%.

In Figure 6.5a it is assumed that the downstream zone of the dam is highly permeable in relation to the foundation; consequently, there is no appreciable rise in the saturation level in the downstream part of the dam. Filters and drains are not shown in this simplified cross section.

Table 6.1 provides a great deal of useful information about the installation of various kinds of cutoffs, with notes about their applicability (see U.S. Navy, Naval Facilities Engineering Command, 1982, NAVFAC Design Manual 7.1).

Grout Curtains

For many decades fluid cement pastes pumped through small-diameter drilled holes into crevices and joints in rocks have been used for tightening dam foundations. This process, *grouting*, is standard practice for treating the foundations of nearly all important modern concrete dams. It is also used for tightening foundations beneath the impervious cores of earth and rockfill dams. After the cement paste has set it forms a barrier to the passage of water. Grouting is not entirely foolproof because cement grout cannot be forced into pores smaller than about 1.0 mm and it is not always possible to seal all important joints. To overcome the difficulty of sealing small cracks which are too fine to be entered by cement particles but can still pass considerable water, various chemical grouts have been used. Though chemical grouts can be highly effective in sealing small rock joints, they cannot *always* be depended on to control seepage through soils. Chemical grouts with a fluidity approaching that of water usually can penetrate silts or fine sands to control seepage. To achieve increased penetration of cement grout into coarser soil deposits and rock formations, some investigators advocate using very thin, fluid mixtures with high water contents. But Warner (1978) says "There can be no argument that very thin grouts are easier to mix and pump and perhaps less costly than their thicker counterparts." But, he says that the final quality is the most important

TABLE 6.1 Cutoff Methods for Seepage Control

Method	Applicability	Characteristics and requirements
Sheet pile cut-off wall	Suited especially for stratified soils with high horizontal and low vertical permeability or pervious hydraulic fill materials. May be easily damaged by boulders or buried obstructions. Tongue and groove wood sheeting utilized for shallow excavation in soft-to-medium soils. Interlocking steel sheet piling is utilized for deeper cutoff.	Steel sheeting must be carefully driven to maintain interlocks tight. Steel H-pile soldier beams may be used to minimize deviation of sheeting in driving. Some deviation of sheeting from plumb toward the side with least horizontal pressure should be expected. Seepage through interlocks is minimized where tensile force acts across interlocks. For straight wall sheeting an appreciable flow may pass through interlocks. Decrease interlock leakage by filling locks with sawdust, bentonite, cement grout, or similar material.
Compacted barrier of impervious soil	Formed by compacted backfill in a cutoff trench carried down to impervious material or as a core section in earth dams.	Layers or streaks of pervious material in the impervious zone must be avoided by careful selection and mixing of borrow materials, scarifying lifts, aided by sheepsfoot rolling. A drainage zone downstream of an impervious section of the embankment is necessary where the compacted cutoff may be imperfect or cracking of cutoff material is likely.
Grouted or injected cutoff	Applicable where depth or character of foundation materials make sheet pile wall or cutoff trench impractical. Utilized extensively in major hydraulic structures. May be used as a supplement below cutoff sheeting or trenches.	A complete positive grouted cutoff is often difficult and costly to attain, requiring a pattern of holes staggered in rows with carefully planned injection sequence and pressure control.

Slurry trench method	Suited for construction of impervious cutoff trench below groundwater or for stabilizing trench excavation. Applicable whenever cutoff walls in earth are required. Is replacing sheetpile cutoff walls.	Vertical-sided trench is excavated below ground-water as slurry with specific gravity between 1.2 and 1.8 is pumped back into the trench. Slurry may be formed by mixture of powdered bentonite with fine-grained material removed from the excavation. For a permanent cutoff trench, such as a foundation wall or other diaphragm wall, concrete is tremied to bottom of trench, displacing slurry upwards. Alternatively, well graded backfill material is dropped through the slurry in the trench to form a dense mixture that is essentially an incompressible mixture; use cement-bentonite mix when using coarser gravels (which may settle out).
Impervious wall of mixed in-place piles	Method may be suitable to form cofferdam wall where sheet pile cofferdam is expensive or cannot be driven to suitable depths, has insufficient rigidity, or requires excessive bracing.	For a cofferdam surrounding an excavation, a line of overlapping mixed in-place piles are formed by a hollow shaft auger or mixing head rotated into the soil while cement grout is pumped through the shaft. Where piles cannot be advanced because of obstructions or boulders, supplementary grouting or injection may be necessary.
Freezing-ammonium brine or liquid nitrogen	All types of saturated soils and rock. Forms ice in voids to stop water. Ammonium brine is better for large applications of long duration, liquid nitrogen for small applications of short duration where quick freezing is needed.	Gives temporary mechanical strength to soil. Installation costs are high and refrigeration plant is expensive. Some ground heave occurs.

After: (U.S. Navy, Naval Facilities Engineering Command, 1982).

criterion and questions the propriety of using grouts with water-cement ratios up to 20:1. He comments that "the grout used should always contain the least amount of water possible. In cases where large voids or caverns exist such as in limestone, grout of a thick mortar-like consistency would be preferrable." And, "although the equipment commonly used in dam foundation grouting is not capable of either mixing or pumping such consistency material, suitable equipment is available."

Grouting operations should always be carried out by highly experienced people. Cambefort (1987) emphasizes that grouting operations can be carried out successfully "only if the characteristics of grouts and their injection methods are fully understood." And "a great deal of investigation is necessary if a grouting program is to be satisfactory. In addition, great skill is required by the grouting operatives because of the variable nature of the ground which can only be revealed at the time of the excavation. This is why grouting is still a skilled operation and the mixing of grouts almost constitutes a science in itself." He says that sodium silicate gels can be quite effective in grouting most formations that cannot be penetrated by cement grouts, but "unfortunately are very costly." In relation to bitumen grouts, he says "bitument emulsions and hot bitumens have been tried as grouts, but with little success. Injection is a very delicate operation."

Grout curtains may substantially reduce seepage quantities, but their effect on hydrostatic pressures is often less than desired. To be effective grouting must be exceedingly thorough, because only a small percentage of ungrouted joints in permeable rock formations can render them highly ineffective in controlling hydrostatic pressures.

The U.S. Department of the Army, Corps of Engineers (1986a) says that the effectiveness of grouting depends upon "being able to rather specifically locate the leaking areas and fill the culprit openings without damage to the embankment" (without causing cracking of impermeable cores, foundations, or abutments, or clogging drains). In another publication of the U.S. Army Corps of Engineers, Karol (1988) emphasizes that for most purposes grouts must be permanent and have adequate strength and imperviousness and not deteriorate with age or by contact with ground or ground water.

The effectiveness of foundation grouting can be examined by taking readings on piezometers installed at pertinent locations in dam foundations. Such measurements have been made in the foundations of many concrete and earth dams. Typical of an instrumented foundation under an earth dam is the section shown in Figure 6.7 for a zoned dam with a highly impervious clay core extending down to jointed bedrock. A single row of grout holes 60 ft deep was drilled into the rock, and a pervious blanket drain was located downstream from the cutoff trench to collect seepage and relieve hydrostatic pressures in the foundation. Typical pressure measurements are shown for several piezometers that were drilled a few feet into the rock. The dashed line in Figure 6.7 shows that the maximum loss in head that can be attributed to the grout curtain is one-third of the differential head acting on the dam at the time of the

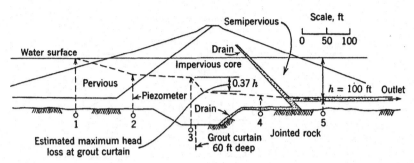

FIG. 6.7 Hydrostatic pressures in jointed rock foundation with grout curtain.

readings. The drainage blanket is effectively draining the foundation. A flow-net study would have shown that a single shallow row of grout holes could not have had a major influence on the seepage pressure distribution in this foundation, but a good drainage system was essential to the dam's safety.

The usefulness of flow nets for studying the potential benefits of grout curtains is illustrated in Figure 6.8. A cross section through a zoned earth dam 200 ft high on a pervious rock foundation and the flow net that would exist without grouting are given in Fig. 6.8a. The shaded area in the foundation beneath the core of the dam represents a grouted zone 30 ft wide. In studying the benefits of grouting, several degrees of tightness in the grouted zone were assumed and head losses were estimated for each of them. If grouting is 90% effective in reducing permeability, the grouted zone is assumed to have a mass permeability one-tenth that of the ungrouted rock; if grouting is 95% effective, the permeability of the grouted zone is one-twentieth that of the un-grouted rock, and so on.

In the flow net in Figure 6.8a, which has 11 equipotential drops, one equipotential drop just coincides with the width of the grouted zone; hence without grouting one-eleventh, or 9%, of the head h is used up in this zone. If the effective permeability of the grouted (shaded) zone were reduced to one-fifth that of the untreated rock, five equipotential drops would occur in the grouted zone; if its effective permeability were reduced to one-tenth that of the un-grouted rock, 10 equipotential drops would be in this zone, and so on. By keeping in mind the number of equipotential drops that occur in the grouted zone hydrostatic pressures at the bottom of the dam are readily determined with the aid of the flow net in Figure 6.8a. Hydrostatic pressures for the three conditions just described are given in Figures 6.8b, 6.8c, and 6.8d. It is shown (Fig. 6.8b) that if the grouting is 80% effective, as defined here, the head losses in the grouted zone will increase to one-third of the total head difference h; if the grouting is 95% effective (Fig. 6.8d), they will increase to two-thirds of h. It will be noted that the grouted zone in this study has been assumed to penetrate the jointed rock formation completely, whereas in many cases only partial penetration is achieved by grout curtains. Even under the highly favor-

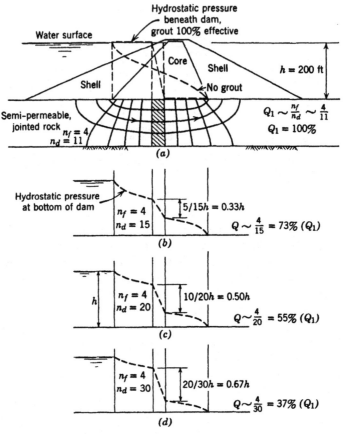

FIG. 6.8 Flow net study of effectiveness of grout curtain under core of earth dam. (a) Cross section and typical flow net. (b) Grout 80% effective ($k_g = 0.2k$). (c) Grout 90% effective ($k_g = 0.1k$). (d) Grout 95% effective (k_g 0.05k). k = permeability of ungrouted rock.

able conditions assumed in this study grouting must be substantially perfect to influence hydrostatic pressures noticeably. To achieve 90% reduction in the permeability of a grouted zone more than 99% of the cracks must be grouted.

The flow net in Figure 6.8a also permits the *relative seepage quantities* for various grouting efficiencies to be estimated, for $Q \sim n_f/n_d$ (from Eq. 3.18). For the ungrouted foundation this ratio is 4/11 for 0.364, which may be called 100%. If the grouted zone is 95% effective (Fig. 6.8d), the ratio $n_f/n_d = 4/30 = 0.133$, which is 0.133/0.364 (100%) = 37% of the rate of flow without grouting.

Descriptions of a number of grouting processes, procedures recommended by the U.S. Navy, Naval Facilities Engineering Command (1982), and notes on their applicability are given in one of its design manuals.

Thin Cutoffs

Interlocking steel sheet piling, interlocking wood, reinforced and nonreinforced concrete, and other materials have been used extensively for constructing thin cutoffs through deep alluvial materials in the foundations of dams and other structures. The success of such cutoffs depends on obtaining a high degree of perfection in the cutoff, a good connection into an impermeable formation at its bottom, and a permanently tight connection with an impervious zone at its top. Relatively small openings in cutoffs and gaps at the bottom or top can let large amounts of water through.

Casagrande (1961) presents a mathematical analysis of the efficiency of imperfect cutoffs. He demonstrates that a steel membrane which has 1/16-in. slits spaced every 5 ft with an open-space ratio of 0.1% would have a theoretical cutoff efficiency of 29%. A sheet pile or other cutoff with this amount of open space would thus allow 71% of the normal flow to pass through.

Flow nets enable studies to be made of the general nature of seepage through imperfect cutoffs. The curves in Figure 6.9b were developed with the aid of a group of flow nets similar to the one in Figure 6.9a to show the relationship between the percentage of open area remaining in a cutoff and the percentage of reduction in seepage. It is seen that the efficiency of cutoffs for the conditions shown in Figure 6.9a depends on the total amount of open area and its distribution. If it occurs at one point only, the efficiency will be considerably greater than if it were distributed among several openings. Thus a cutoff with 5% open area at one point reduces the seepage by 60%, whereas a cutoff with the same amount of open space equally divided among eight openings is less than 20% efficient.

Impermeable Upstream Blankets

Levee systems often extend many miles along one or both sides of rivers, frequently on foundations with natural impermeable covers over pervious sands and gravels. The impermeable cover may be thin or substantially missing on the river side, as shown in Figure 6.10a. Under these conditions levees may successfully withstand a number of low flood stages, only to fail suddenly when a large flood comes along. Because extensive blankets or other effective seepage control measures can be very costly, designers may prefer to construct levee systems with reasonably good seepage control and determine the need for any additional measures by observation of the behavior at moderate river stages. When this can be done safely, considerable savings in cost may be possible; however, such levees may be potentially unsafe if subjected to a high river stage before the necessary seepage control measures are constructed.

Observation wells or piezometers installed at the landward toe and at one or more additional points, as shown in Figure 6.10a, permit uplift pressures to be measured and the degree of security evaluated. If readings can be obtained at several low-to-moderate river stages, such as 1, 2, and 3 in Figure

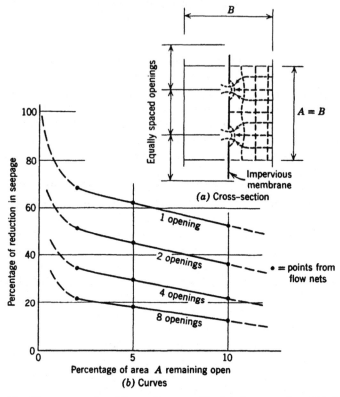

FIG. 6.9 Flow net study of imperfect cutoffs. (a) Cross section. (b) Curves.

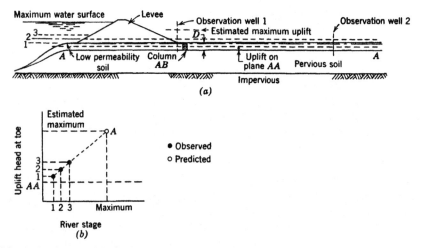

FIG. 6.10 Levee with dangerous uplift condition at landward toe. (a) Cross section. (1), (2), (3) are low and intermediate levels at which readings have been taken in observation wells. (b) River stage versus head at toe.

6.10*a*, a plot of "river stage" versus "uplift head" at the landward toe can be plotted (Fig. 6.10*b*) and the uplift at maximum river stage (point *A*) can be estimated by extrapolation. The degree of security against blowouts and sand boils depends on the magnitude of the maximum *uplift gradient* (Sec. 3.5) at the landward side of the levee. The factor of safety G_s against uplift failures can be calculated as the ratio of the downward forces acting on column of soil *AB* to the upward forces. Assuming that column *AB* has an area of 1 sq ft and a height *D*, the downward force due to its submerged weight is (D) (62.5 lb/cu ft)$(G - 1)/(1 + e)$, and the upward seepage force is (D) (62.5 lb/cu ft) (i). The factor of safety is therefore

$$G_s = \frac{(D)(62.5)(G - 1)/(1 + e)}{(D)(62.5)i} = \frac{(G - 1)/(1 + e)}{i}$$

Because the hydraulic gradient *i* is equal to the uplift head h_t on plane *AA* divided by the height *D* of column *AB*,

$$G_s = \frac{(D)(G - 1)/(1 + e)}{h_t}$$

and the maximum safe uplift head at the toe is

$$h_t = \frac{(D)(G - 1)/(1 + e)}{G_s}$$

To be reasonably secure against blowouts and boils the factor of safety in the last expression should be at least 2 to 2.5. If the predicted uplift h_t is greater than a safe value, control measures should be provided. Turnbull and Mansure (1961), in describing underseepage under the Mississippi River levees, conclude that ". . . it appears that heavy seepage and sand boils should be anticipated whenever estimated upward gradients exceed 0.5 to 0.8, depending on site conditions." Seepage conditions related to upward gradients through the top stratum measured by piezometers during the 1950 high water are summarized in Table 6.2.

Several kinds of remedial measures are available for upgrading the safety of levees against seepage failures during high water stages, such as (1) improving upstream blankets, (2) installing relief wells along downstream toe areas (see Fig. 6.20), and (3) installing slurry wall cutoffs at upstream toe areas to tie into impervious parts of embankments (see Fig. 6.26).

Thin Sloping Membranes

A fundamental principle in the design of any dam or levee is that the energy of the water pressing against the structure must be safely consumed. A theoret-

TABLE 6.2 Seepage Conditions and Measured Upward Gradients, Mississippi River Levees

Seepage conditions	i
Light to none	0 to 0.5
Medium seepage	0.2 to 0.6
Heavy seepage	0.4 to 0.7
Sand boils	0.5 to 0.8

(After Turnbull and Mansur, 1961)

ically ideal design is furnished by a thin, highly impervious sloping membrane, such as steel or reinforced concrete, on the upstream face of a rock or gravel embankment (Fig. 6.11). The resultant hydrostatic pressure P (Fig. 6.11a) presses downward into the foundation, thus increasing frictional resistance to sliding on the base. Also, when dams are constructed this way, their entire embankments are in a "dry" state and earthquakes cannot cause pore pressures in the rockfill voids; so they are virtually immune to damage by earthquake shaking.

Thin impervious membranes are susceptible to cracking due to differential settlements caused by high pressures (Fig. 6.11b) and are rather costly. Tight cutoffs and well-anchored surface slabs are essential to the safety of membrane dams (Fig. 6.11c). At high altitudes, where impermeable soils are scarce, rockfill dams are sometimes built with thin sloping membranes; but when ample

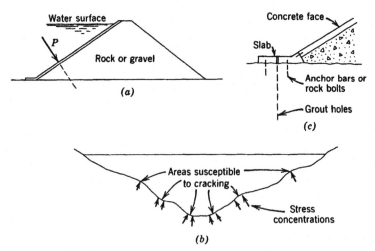

FIG. 6.11 Dam with thin impervious membrane. (a) Cross section. (b) Profile. (c) Toe detail.

supplies of impermeable earth are available near damsites *zoned* earth dams with internal drains almost always are more economical than any other kind.

When thin sloping membranes are used on dams, it is important that the selected materials serve the intended purpose for many years without excessive repairs. Detailed studies may be required to make sure that no important factor has been overlooked and that the design selected will be satisfactory for the specific project. Herreras (1973) tells about the selection of reinforced concrete for the upstream impervious membrane for the Pozo de Los Ramos Dam in Spain. This rockfill dam, which supplies water to Madrid, the country's capital, was faced with a reinforced concrete membrane after detailed study of various alternatives. The New Spicer Meadow Dam in California (265-ft high and 2,150-ft long) used a rockfill with a reinforced concrete face slab 12 to 20 in. thick, placed in 50-ft wide sloping strips with a modified bridge deck paver (see *ENR,* Aug. 4, 1988, pp. 52–54).

Asphalt concrete surfacings have been used on a number of dams and reservoirs throughout the world in areas in which good impervious materials were scarce or uneconomical or this type of lining was considered advantageous for specific operating conditions (see also Chapter 10). Koenig and Idel (1973), for example, describe the behavior record of refined asphalt concrete surfacings in West Germany. After 16 years of service no significant repair work had been required and the linings had remained watertight. In some projects, however, two courses of dense asphalt concrete (with 1 to 2% pores) had been used, and in some of them water that could not escape from between the two layers had caused blisters. These authors recommend that the lower layer (of a two-layer surface) have larger pore volume (3½ to 4%) to prevent such occurrences.

Sembenelli and Fagiolo (1974) describe the steel facing used on the Aquada Blanca Rockfill Dam on the Chile River in southern Peru. Restricted working space in the narrow gorge and a scarcity of good impervious material led to the selection of a rockfill dam with a steel facing. These authors point out that two delicate problems had to be solved: (1) a tight fit between the membrane and the abutment rock and (2) the proper grouting of the highly jointed rock. They also emphasize that the prevention of corrosion is a major problem with steel facings. In addition, the prevention of buckling under large temperature fluctuations had to be considered. After a thorough analysis of all important problems the design was developed with a lining of high quality steel, coated on top and bottom, and resting on a sand asphalt cushion course. Sliding joints in the steel were designed to allow contraction and expansion without the buildup of serious stresses.

The preceding cases represent seepage control by methods that depend on the principle of keeping water out or reducing seepage quantities. Almost always such methods must be used in combination with drainage facilities. In the next paragraphs methods of draining dams and levees and their foundations are illustrated.

6.3   SEEPAGE CONTROL BY DRAINAGE METHODS

Embankment Zoning

Zoned earth dams have an internal impervious section called a *core,* which furnishes watertightness, and outer sections on both sides of the core called *shells,* which furnish strength. Depending on the availability of materials and personal preference, dam designers vary the location and thickness of impervious cores in zoned dams (Fig. 6.12). Some designers prefer extremely thin sloping cores (Fig. 6.12*a*), sometimes called "Growdon" dams after J. P. Growdon, the originator of this type of structure. Others prefer the somewhat thicker, moderately sloping core shown in Figure 6.12*b*; still others, the thick, centrally located core illustrated in Figure 6.12*c*.

Filters must always be used when required for the protection of cores. Thin inclined cores dissipate hydrostatic energy quickly but are subjected to large hydraulic gradients that increase the dangers of piping if filter protection is not extremely good. Thin cores offer less protection than thick cores against offset displacements due to earthquakes or other earth movements. In the past they have been used in rockfill dams in which much of the rock was placed by dumping methods; however, they are also adaptable to rolled rockfill methods.

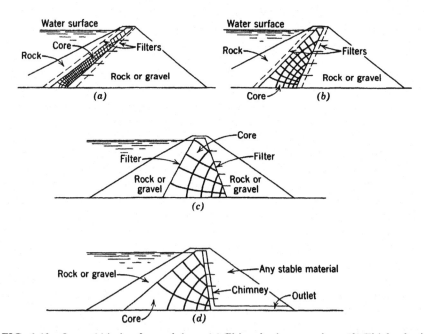

FIG. 6.12   Several kinds of zoned dam. (*a*) Thin, sloping core dam. (*b*) Thick, sloping core dam. (*c*) Central core dam. (*d*) Dam with chimney drain and outlet blanket drain.

When a thin sloping core is used, most dam designers insist on a thick rolled rock section between the core and dumped rock fill. Some designers try to avoid the use of dumped rock in important dams because of its high compressibility.

Some dam designers prefer the thick central core typified by Figure 6.12c because it provides a wide zone for the dissipation of energy, relatively low hydraulic gradients, and long contacts with foundations. A long seepage path along the foundation is particularly desirable on foundations of badly jointed or soft rocks. This design offers savings in cost when liberal quantities of highly impermeable soils are readily available. A disadvantage is that large volumes of fairly permeable granular materials are required for the outer shells (Figs. 6.17 and 6.18).

Figure 6.12d shows a modified homogeneous dam with a granular zone on the upstream slope for drawdown protection and an internal "chimney" drain for the control of seepage. The downstream zone is kept free of saturation by a properly designed and constructed vertical drain and outlet drain; hence the downstream shell can be constructed of any stable material, regardless of its permeability. This design offers great economic savings because it often permits the utilization of materials in the downstream zone that ordinarily might be wasted. Also, a properly designed outlet blanket drain under the entire downstream half of the dam protects the dam from being damaged by excessive seepage through a jointed or porous foundation. Of course, highly plastic clays and other low-strength materials should not be used in large dams, and stability analyses should be made when necessary to establish safe slopes for the materials to be used.

By means of flow nets for the cross sections just described (Fig. 6.12) differences in hydraulic gradients are strikingly evident, for they are inversely proportional to the sizes of the squares in the flow nets (for other examples see Fig. 3.21).

In general, wide cores and wide transition zones and drain zones are better able to conform to nonuniform deformation caused by settlement of foundations, consolidation of fill materials, and ground movements than thin zones. Nichiporovich and Teitelbaum (1973) who tell about the occurrence of fissures in cores of rockfill-earth dams say that if fissures are to be minimized in high dams in narrow canyons transition zones must be sufficiently wide and well-graded and that construction should proceed uniformly to minimize nonuniform deformation.

Dams with extremely narrow cores induce large hydraulic gradients not only within themselves but also in their foundations. Grouting is often employed to reduce seepage through foundations (Sec. 6.2). An interesting use of flow nets to study the influence of grouting on seepage gradients beneath impervious cores in dams is described in Figure 6.13. In this simplified study it is assumed that a thin sloping core rests on permeable rock and that a partly penetrating grout zone is located beneath the center of the core (Fig. 6.13a).

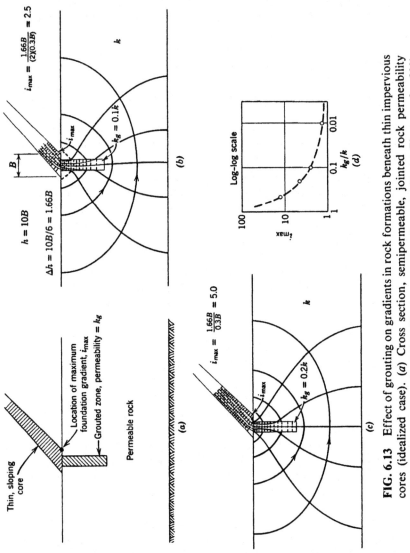

FIG. 6.13 Effect of grouting on gradients in rock formations beneath thin impervious cores (idealized case). (a) Cross section, semipermeable, jointed rock permeability equals k (b) Flow net for 90% effective grouting ($k_g = 0.1k$). (c) Flow net for 80% effective grouting ($k_g = 0.2k$). (d) Maximum escape gradient, i_{max} versus grouting efficiency.

Flow nets were constructed on the assumption that the grouted zone would behave as a homogeneous mass. Various degrees of efficiency of the grouting are expressed in terms of relative permeability (Fig. 6.13*b* and 6.13*c*). As in another grout efficiency study (Sec. 6.2), 90% efficiency is taken to mean that the permeability of a grouted zone has been reduced to one-tenth that of the ungrouted rock. By reference to the summary curve in Figure 6.13*d* we can see that critical hydraulic gradients beneath the cores of dams can be reduced substantially by thorough grouting, although they are improved very little by poor grouting. Joints, fissures, and other openings in rock or soil formations in direct contact with earth cores should be sealed with grout or dental concrete, and filters and drains should always be provided to prevent erosion and piping of fine material through unguarded openings.

Dams should be zoned to take advantage of the most economical combinations of readily available materials at sites, and each project should be designed individually to fill the needs at a given site safely and economically. No single type of design can be said to be most suited to all locations, for the final selection is influenced largely by site conditions and the relative costs of materials available for construction. As previously noted, however, vertical "chimney" and horizontal blanket drains should nearly always be provided as measures for controlling excessive leakage. Typical cross sections of some existing earth and rockfill dams appear in Figure 6.14.

The zoning of dams must be developed in conjunction with foundation seepage-control requirements because their performance and that of levees depends heavily on these structures functioning as complete embankment-foundation systems. This point is illustrated by a flow net study for a zoned dam on a permeable foundation (Fig. 6.15). The core is assumed to be impervious, and the downstream shell is assumed to have several degrees of permeability. In Figure 6.15*a* the downstream shell is assumed to be extremely pervious in relation to the foundation and the saturation line is low. In Figure 6.15*b* the downstream shell is assumed to be 10 times as permeable as the foundation. Although this ratio may seem favorable, it is noted that saturation rises substantially in the downstream zone. In Figure 6.15*c* the downstream zone has the same permeability as the foundation, and the saturation rises dangerously high. In both Figures 6.15*b* and 6.15*c* a permeable blanket drain is needed under the downstream zone to ensure good drainage, even for the favorable assumption that the embankment is nonstratified. A wide variety of conditions occur in dam foundations. Embankment designs must be adapted to site conditions to ensure safe operation under unfavorable conditions. The use of well-designed vertical and horizontal drains, as shown in Figures 6.7, 6.12*d*, and 6.14*d*, is one of the best safeguards against erratic or unknown conditions affecting seepage through dams. If foundations are highly stratified, it may be necessary to reduce the flows by the use of various kinds of cutoffs (see Sec. 6.2) or to install relief wells along the downstream toe.

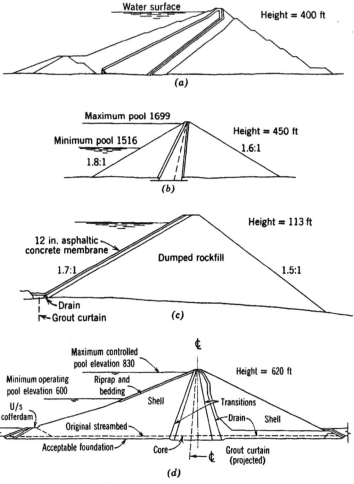

FIG. 6.14 Cross sections of some typical earth and rock dams. (*a*) Trangslet Dam. (From *Transactions,* A.S.C.E., Vol. 125, 1960, part II, p. 568.) (*b*) Cougar Dam. (From *Transactions,* A.S.C.E., Vol. 125, 1960, Part II, p. 673.) (*c*) Montgomery Dam. (From *Transactions,* A.S.C.E., Vol. 125, 1960, Part II, p. 435.) (*d*) New Don Pedro Dam, section at streambed. (From *Proceedings,* A.S.C.E., Vol. 98, No. SM10, Oct. 1972, p. 1035).

Longitudinal Drains and Blankets

One of the most economical types of drain for dams and levees is a longitudinal aggregate drain with perforated or jointed pipe, designed according to criteria in Chapter 5. Hundreds of miles of levees and some dams have been constructed with drains of this type. Narrow longitudinal drains are usually

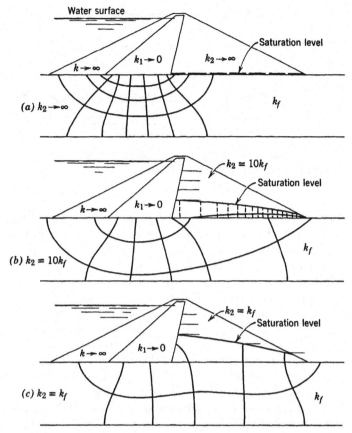

FIG. 6.15 Zoned dam study. Relationship between downstream shell permeability (relative to foundation permeability) and saturation level in shell.

located about midway between centerline and the downstream toe. Unfortunately, they provide little or no protection against flaws in embankment construction (see Fig. 6.16.) Even very wide blanket drains placed under downstream portions of dams or levees can do little to protect against the kind of condition illustrated in Figure 6.16 if a vertical "chimney" drain has not been provided. Under severe conditions of stratification a properly designed and constructed chimney drain and outlet drain can furnish complete control over seepage through embankments.

Chimney Drains

Borrow pits for embankments are never completely uniform; therefore successive loads of material, which are placed in horizontal lifts, vary somewhat

"Homogeneous" cross section

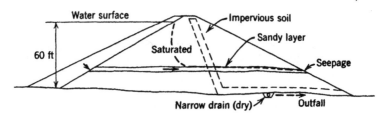

FIG. 6.16 Dam with a construction flaw that could have been controlled by a chimney drain with an outlet blanket. (See dashed lines.)

from load to load. As a result, the average horizontal permeabilities of constructed embankments are almost always somewhat greater than the vertical. Also at times, extreme variations in permeabilities of embankment materials do occur, as illustrated in Figure 6.16, which shows a 60-ft-high earth dam that was designed as a homogeneous embankment with a very narrow longitudinal drain located midway between the center of the dam and the downstream toe. Although a liberal supply of impervious clay soil was available for its construction, the equipment operators inadvertently dug into a sandy layer under the clay, which went unnoticed for a few days by the engineer in charge. The sand was spread out over most of the area and was soon covered (and hidden) with clay, which was used for the balance of the dam. Shortly after the reservoir was filled, considerable leakage appeared on the downstream slope and a boggy condition developed on the slope and at both abutment groins. Borings made in the dam revealed the sand layer. The narrow longitudinal drain could not possibly collect all the seepage, but if a chimney drain and outlet blanket drain of suitable filter and drainage aggregates had been constructed, as shown by the dashed lines, this accidental error would have been of no consequence. Other dams constructed with good internal drains are shown in Figs. 6.7, 6.12*d*, 6.14*d*, and 6.23*a* and *b*.

Gravel and rock shells often have an extreme degree of horizontal stratification produced by the way in which these zones are constructed. Frequently the top of a lift (sometimes several feet thick) degrades under the compaction and hauling equipment and becomes impermeable, although the lower part of the lift may be thousands of times more permeable. This layering can cause concentrations of seepage at the bottoms of lifts, much as the sandy layer in the dam in Figure 6.16 had a concentrated flow. Peter (1971) describes the failure of a 41 m high rockfill dam, with a sloping core. The rock fill of weathered granodiorite was crushed at the lift tops by the compaction equipment, which caused the trapping of seepage and high seepage reservoirs (mounds) to build up in the downstream part of the dam. Water springs and small slides appeared on the downstream slope. Provision of a good internal drainage system (as shown in Figs. 6.12*d*, 6.14*d*, and by dashed lines in Figure 6.16) is *almost* *always* needed to prevent the problems that occur when zoning is depended on for control of seepage in dams.

It is often thought that if the outer shells of zoned dams are several times more permeable than the cores, seepage in the downstream shells surely will be well controlled. The flow-net study in Figures 6.17 and 6.18 shows the influence of permeability and stratification on the position of the saturation line in a zoned dam. The study is for a central core dam on an impervious foundation, with the permeability of the core designated as k_1 and the permeability of the downstream shell as k_2. The ratio k_2/k_1 varies from more than 1000:1 to 1:1. Figure 6.17 gives flow nets for nonstratified soil conditions ($k_h = k_v$) and Figure 6.18 for stratified conditions ($k_h = 16k_v$). The saturation lines for both cases are summarized in Figure 6.17.

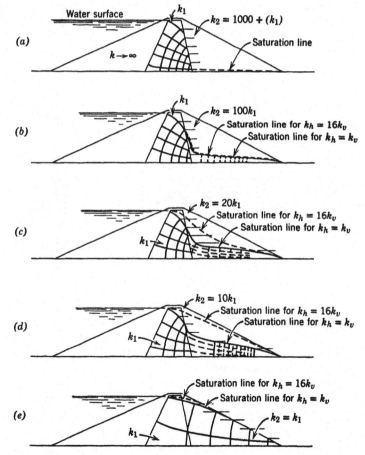

FIG. 6.17 Flow nets for $k_h = k_v$. Study of seepage through a dam with a central core (summary sheet showing effect of stratification on saturation line).

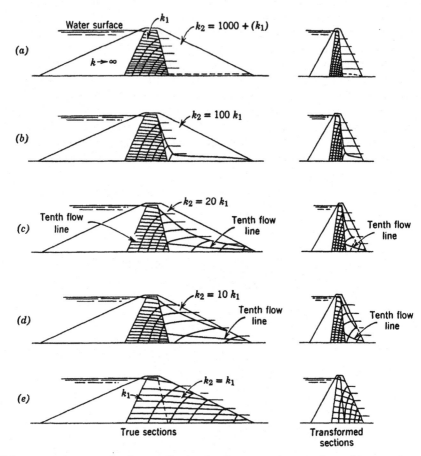

FIG. 6.18   Flow nets for $k_h = 16k_v$. Study of seepage through dam with central core.

This study shows that the saturation level in zoned dams can be expected to rise substantially unless the downstream shells are *many* times more pervious than the cores. It also shows that stratification has a marked influence on the saturation level in shells, particularly when their permeabilities are only moderately greater than their cores.

A complete safeguard against the uncertain saturation levels that can develop in the shells of zoned dams can be obtained by the use of chimney drains. Properly designed and built chimney drains with adequate outlets frequently save many times their cost in lowered permeability requirements for large volumes of shell materials. When good internal drainage is provided, the downstream shells can be constructed of any materials of adequate strength with no restrictions on permeability and often at substantial cost reduction.

To function effectively chimney drains and their outlets must have sufficient permeability to remove seepage without excessive buildup of head. It is not safe to assume that control of seepage is automatically assured because a dam contains a chimney drain. Such drains have been constructed of very dirty, low-permeability aggregates, little more permeable than the soils they are supposed to drain (Sec. 1.3). In such cases they cannot serve their purpose (see Fig. 5.10). The design of drains with adequate discharge capacity to meet any specified rate of flow is described in Chapter 5.

Partly Penetrating Toe Drains

As already noted, earth dams and levees should be designed and constructed with drainage systems that will control seepage through their embankments and foundations. Even though some drainage is provided in an original design, additional drainage may be needed after a project has been completed if undesirable seepage conditions develop (Cedergren, 1961). A partly penetrating toe drain of the type shown in Figure 6.19 offers a means of improving the seepage

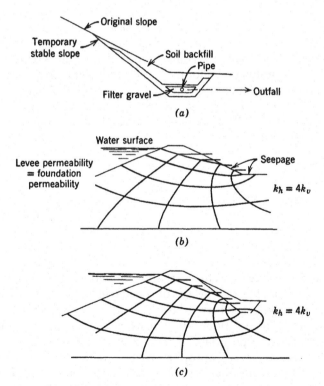

FIG. 6.19 Partly penetrating toe drain. (*a*) Cross section through partly penetrating toe drain. (*b*) Flow net without toe drain. (*c*) Flow net with toe drain (From *Transactions,* A.S.C.E., Vol. 126, 1961, Part I, pp. 1532–33.)

conditions in some existing levees and small dams. This drawing shows a toe drain that penetrates a moderate depth into a pervious foundation. It can be installed by excavating into the landward toe of an existing levee or dam when the water level is low and the ground is well drained. These drains should be provided with perforated or slotted pipes connected to gravity outlets. They are best suited to levees or small dams on relatively homogeneous foundations or on pervious foundations covered by shallow topsoil layers which can be penetrated by the drainage trench. If this type of drain is to improve uplift conditions at the landward side of a levee or small dam, it must be in contact with water-bearing strata. The flow net in Figure 6.19b shows potentially unstable conditions that existed before the installation of a toe drain. The flow net in Figure 6.19c shows the seepage pattern after the installation of such a drain.

When foundation conditions are such that a toe drain would be separated from underlying pervious strata by impervious layers, only a portion of the seepage could be removed by the drain and only limited control, obtained. In such cases a toe drain can be combined with relief wells or replaced by a system of relief wells.

At the present time (1988) a drain of the type shown in Figure 6.19 could be constructed with highly permeable drainage aggregate protected with fabric filters, rather than to try to depend on clean, washed filter aggregate to provide adequate discharge capacity. A comprehensive toe drain system was used in the repair and upgrading of Florida Power and Light Co's Martin County storage reservoir that failed by piping on Oct. 30, 1979. That system used a layer of 3/8 in. to 1 in. aggregate protected with fabric filters to prevent piping of the cohesionless foundation and embankment materials into the highly permeable drainage layer (see Fig. 10.15).

Relief Wells

When pervious or jointed strata beneath dams or levees are too deep to be penetrated by cutoffs or shallow drains, relief wells offer a means of relieving uplift pressures because they can penetrate the most pervious waterbearing strata in a foundation. Their discharge resistance must be small, and they should be spaced sufficiently close together to lower the adjacent water pressures to a safe level. Relief wells in erodible formations must be designed with screens or filters that prevent the loss of soil and must be resistant to corrosion and to deterioration caused by bacteria. To ensure permanent performance without movement of soil filter criterion 1 (Sec. 5.2) must be satisfied.

Figure 6.20 shows a portion of a relief well system at the downstream toe of an earth dam in central California. Wells were installed to relieve moderately high uplift pressures that developed at high reservoir levels. Even though calculated factors of safety were above allowable limits without the wells, this well system provided an extra margin of safety against the possibility of problems with this dam.

FIG. 6.20 View along downstream toe of an earth dam in Central California, where relief wells were installed to upgrade stability of the dam (with permission of Sacramento District, U.S. Army Corps of Engineers).

Relief wells must be able to retain their discharge capabilities for long periods of time or be restored or replaced. A comprehensive five-year investigation by the U.S. Army Corps of Engineers (1972) of relief wells protecting Mississippi River leveees in Illinois showed that the specific yield of 24 test wells decreased 33% in a 15-year period. Incrustation on well screens and in gravel filters was believed to be a major factor in the reduction of the specific yield of the wells (iron bacteria growth on screens and in filters or from precipitation of iron oxides and hydroxides and calcium carbonates in gravel filters). The relief wells—constructed of wood-stave pipe risers of 8 in. inside diameter and slotted wood-stave screens wound with galvanized or stainless steel wire—were surrounded by a graded gravel filter 6 in. thick. They were installed to depths of 70 to 100 ft (21 to 30 m) on 75-to-300-ft (23-to-91-m) centers parallel to the levee system. A check valve and metal well guard prevented backflow into each well. The aquifer consists almost entirely of sand and gravel. Small amounts of iron bacteria were found in water samples taken from all wells. None of the wells had piping problems.

By periodic maintenance the Corps found that the original efficiency of most wells can be recovered. Some treatments that the Corps has tried are listed in Table 6.3.

When well systems are being designed, the expected long-time efficiency (capacity or specific yield) should be used in determining the number and spac-

TABLE 6.3 Treatments Used to Restore Efficiency of Relief Wells[a]

Incrustation Cause	Treatment
Incrustation of gravel filter particles by calcium carbonate	Hydrochloric acid (acid attacks metal and should be used with care)
Iron oxides	Chemical treatment with polyphosphates and surging
Iron bacteria	Small amount of calcium hypochlorite with phosphates
Bacterial slime and iron oxide deposits	Chlorine (deteriorates wooden well screens)
Iron bacteria and incrustation	Acids and chlorine (deteriorate well screens and should be used with care)

[a]From U.S. Department of the Army, Corps of Engineers (1972).

ing of wells required. Further research is needed in the prevention of serious reduction in well capacities over long periods of time.

An advantage of relief-well systems is the ease with which they can be expanded if an initial installation fails to furnish the needed control. An initial system based on the best available knowledge of soil conditions can be installed, and if this system does not furnish the degree of control desired additional wells can be added until an adequate system is obtained. This procedure is most suitable in situations in which the level of the water surface behind a structure can be controlled. If there can be no control, an initial system should be at least the minimum judged adequate to prevent failures.

Middlebrooks and Jervis (1947) developed formulas for the design of fully penetrating relief-well systems based on seepage theory and model studies. Turnbull and Mansur (1961) describe methods developed by the U.S. Army Corps of Engineers for designing partly penetrating wells.

The design of a relief-well system requires determination of the most economical spacing and penetration of wells that will lower the uplift pressures to a safe level. The U.S. Army Corps of Engineers' method starts with an infinite line of wells, and the spacing is reduced when necessary to allow for the lower efficiency of a finite line of wells compared with an infinite line.

6.4 PROTECTING FROM EARTHQUAKES AND EARTH MOVEMENTS

Basic Considerations

Although relatively few dam and reservoir failures have been caused by earthquakes and earth movements, those that have occurred have made designers eminently aware of the need for constant vigilance to avoid them. The most obvious are those caused by severe ground movements that literally shake structures to pieces. Possibly less obvious, but of equal importance, are those caused by the splitting of foundations and dams produced by fault movements or deep subsidence resulting from extraction of oil or water.

Earthquakes can cause failure of earth dams in a number of ways, of which the following are considered the most important (Sherard et al., 1963b; Seed, 1973):

1. Piping failure through cracks formed in the dam, its foundation, or its abutments.
2. Overtopping caused by loss of freeboard from compression of the foundation or the embankment induced by shaking.
3. Collapse of the embankment because of liquefaction of loose overburden in the foundation or loose embankment material caused by the shaking.
4. Shear slides in the embankment caused by shaking.
5. Overtopping caused by large waves (seiches) in the reservoir, uplifting of the reservoir bottom, major reservoir slope collapses, or blockage of the spillway or outlet works.

Among these five ways cracking is the most dangerous from the standpoint of piping failures and the most directly controllable by designing "self-healing" drainage features and zones that will slough into any openings and slow down any through-seepage and possibly prevent piping failures. Collapse failures from liquefaction are also at least partly controllable by drainage systems that keep saturated zones to the minimum sizes possible.

In regard to failure by erosion caused by water flowing through cracks produced by earthquake deformations of an embankment, Castro (1976) says, "The deformations to be expected from an earthquake cannot be reasonably predicted with our present state of knowledge. Regardless of what the results of a sophisticated seismic stress strain analysis might indicate, one should provide a good drainage system with high water-removing capabilities that would minimize damages resulting from seismic deformations of the embankment."

Protecting from Ground Shaking

The massive landslides of recorded history nearly always occurred in saturated earth. Many landslides take place during or immediately after earthquakes (Sec. 8.1). Sherard et al. (1963c) report that the only dam known to have failed completely during an earthquake (Sheffield Dam, California) was unzoned and loosely compacted; the foundation and lower part of the embankment were saturated at the time the earthquake occurred.

The mechanics of failure of earth masses under earthquake shocks are discussed in Chapter 8. The conclusion is reached (Sec. 8.2) that good drainage is one of the most effective means of improving slope stability during earthquakes. Also strong, dense materials are considerably more resistant to damage by earthquakes than loose, weak materials. During construction any materials in the foundations of dams that are likely to liquefy under severe shocks should be removed and replaced with embankment materials thoroughly compacted.

At one time it was thought that if sands and gravels were well compacted they could not liquefy or be otherwise substantially weakened by earthquake shocks. Evidence now points to the probability that even moderately well compacted gravels may liquefy under severe shocks if they are under the great pressures the sometimes develop deep within the massive earth dams that are now being constructed and planned. One of the best ways of preventing liquefaction failures in dams of major proportions is to require thorough compaction and the best possible drainage.

If the saturation level is permitted to rise substantially in the downstream shells of major dams, all such saturated zones must be considered potentially susceptible to liquefaction under severe earthquakes. Unless watertight upstream membranes are used, it is impossible to keep water out of the upstream portions of zoned dams; however, the pressure of the water helps to counteract outward earthquake forces. High compaction, the use of strong permeable rock or gravel fill, and sufficiently flat slopes must be depended on for the stability of saturated upstream shells.

To ensure the maximum stability of major zoned dams during earthquake shocks the saturation level in downstream shells should be kept as low as possible. If the volume of saturated shell material is small in relation to the size of a dam, it represents little hazard; however, if a large volume is filled with free water the hazard is greatly increased. Figure 6.21a shows an earth dam with graded filter drains that virtually eliminate free water in the downstream shell. The volume of shell susceptible to liquefaction is very small; hence the downstream slope is extremely resistant to damage. A similar dam, but without positive drainage, is shown in Figure 6.21b. This dam has a downstream sandy gravel shell which is several times more permeable than the core, but the saturation level is high, and a substantial volume of shell material is filled with free water. This dam is considerably more susceptible to damage during severe shocks than the dam in Figure 6.21a. *Every important dam should be well drained to ensure maximum stability during earthquakes.*

Superficial Drainage Gives False Security Superficial drainage features such as small rock toes, small line drains, and thin layers of gravel or boulders on slopes may impart an external appearance of rugged stability without adding a great deal of security from deep-seated failures. The downstream slope of an earth dam may show no signs of seepage, yet be potentially unstable. The dam in Figure 6.22b, for example, which has a small rock toe, would appear from the surface to be considerably safer than the dam in Figure 6.22a because the slop would be dry. Yet the dam in Figure 6.22b is almost as saturated as that in Figure 6.22a and would be almost as vulnerable to failure along a deep-seated surface such as *ABC* (Fig. 6.22c).

Good, deep-seated drainage is beyond question one of the best ways to

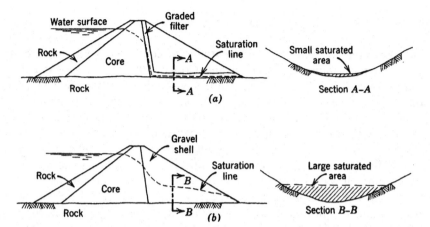

FIG. 6.21 Good drainage increases earthquake stability of dams. (*a*) A large zoned dam *with* graded filters. (*b*) A large zoned dam without graded filters.

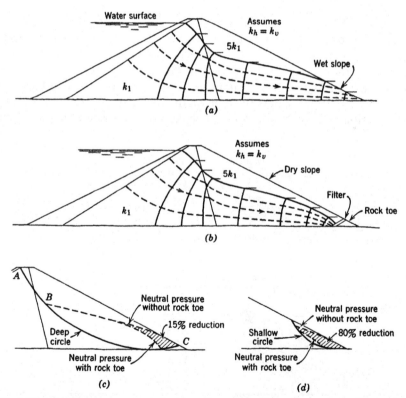

FIG. 6.22 Study showing deceptive benefits of small rock toes in earth dams. (*a*) Seepage in zoned dam with no toe drain (poor appearance). (*b*) The same dam with small rock toe (good external appearance but little improvement in stability). (*c*) Neutral pressures on deep failure circle reduced only slightly by small rock toe. (*d*) Neutral pressures on shallow failure circle reduced substantially by small rock toe.

assure the safety of dams under essentially all conditions including ground shaking from earthquakes. Sherard and Cooke (1987) in an assessment of Concrete-Face Rockfill Dams (CFRD), say—under Earthquake Considerations—"Since the entire CFRD embankment is dry, earthquake shaking cannot cause pore pressures in the rockfill voids." Also, "Earthquakes can only cause small deformations during the short period of strong shaking. After the earthquake is over the CFRD is as stable as before." I wonder how many readers fully comprehend the enormous importance of these statements. In my view, keeping all water-impounding structures in as "dry" a state as possible is the key to high stability and high security against failure. The closer these structures can be kept in the "dry" state (and the better they are drained), the safer they will be. This is why I have been emphasizing for years the importance of good internal drainage for dams, as illustrated in Figure 6.21a and elsewhere in this book and other publications.

Protecting from Offsets and Tension Cracks

Figure 6.23 shows some of the design features that can be incorporated in embankments to increase resistance to damage from offsets and tension cracks caused by earthquakes or other actions in the earth's crust, whether natural or man-made. In Figure 6.23a a dam on firm rock with little overburden is shown. Shallow stripping puts the dam on the rock and the need is to develop a cross section with high resistance to shakes and offsets. A strong, highly

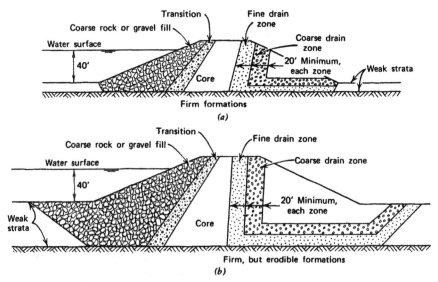

FIG. 6.23 Design features minimizing dangers from offsets and tension cracks include nonrigid cores, coarse upstream zones, and thick drain zones. (a) Site with shallow overburden. (b) Site requiring deep stripping.

permeable rock or gravel upstream zone can help to reduce the dangers of slipouts of the upstream slope. An impervious core flexible enough to adjust to earth offsets without opening wide cracks will help to reduce the dangers of through-seepage and piping failures. Wide downstream drain zones can reduce the risks of complete rupturing of drains.

Figure 6.23b shows the kind of design that can be used for dams on thick beds of questionable alluvial material. Deep stripping of this foundation to firm nonliquefiable rock is combined with the basic features of Figure 6.23a to safeguard the dam from severe shakes and large offsets or cracks.

Whenever possible dams and reservoirs should not be located on or near known faults or known areas of large oil or groundwater extractions that could have an adverse effect on these structures. When these sites cannot be avoided, special measures (see Fig. 6.23a and b) can be taken to reduce the dangers of serious damage:

1. Use impervious cores and blankets that are sufficiently flexible to adapt without cracking to anticipated earth movements.
2. Use wide transitions and drains constructed of materials of suitable ranges of particle sizes that will not rapidly pipe out through cracks or joints but will allow these zones to slough into any cracks or crevices that start to form. If these zones are noncohesive, it is more likely that they will behave as "self-healing" zones that will not allow rapid erosion through dams or foundations and thus will at least delay through-seepage and piping failures that have been known to occur when cohesive zones were used.

Rigid cores and cohesive drains should be avoided in all areas in which cracking or offsetting is likely to occur. The behavior of the Baldwin Hills Reservoir (Leps, 1972) bears out this statement. In discussing the failure of this reservoir in 1963, Leps says that it had a rigid pea-gravel drainage blanket with a pervious gunite topcoat and an impervious asphalt undercoat and lead-off pipes in rigid concrete cradles. Progressively increasing tensile strains and vertical offsets that occurred over a period of 13 years led to serious damage to the lining and drainage system. According to Leps, open-through-going planar voids created in fault planes by horizontal tensions provided an immediate and highly erodible escape avenue for reservoir water. Open fractures in the underdrain system permitted piping through the clay lining and the erodible foundation.

In discussing the failure of the Baldwin Hills Reservoir, Casagrande et al. (1972) conclude that the failure was caused by foundation strata "which are highly sensitive to erosion and are crossed by faults." They say that if the brittle drainage system beneath the clay lining had not been used it might have taken years longer to produce the degree of underground erosion that led to failure, although the reservoir might eventually have failed in the same manner. Making the drain cohesionless—not rigid—while taking care to fulfill piping criteria would have given it a better opportunity to adjust to the earth movements and might have forestalled the failure by allowing the reservoir to be emptied and corrective measures taken.

When dams and reservoirs are constructed in areas and on foundations that might be susceptible to the formation of cracks, it is particularly important that thorough instrumentation be employed. Automatic recording and transmitting devices should signal an alarm in the dam attendants' quarters if any unusual increase in piezometric head or flow quantity occurs. If such changes should take place, the instrumentation should be sufficient to alert operators to the fact that potentially unsafe changes are imminent and an immediate investigation should be made. Even a few hours forewarning of an impending problem might be enough to alert downstream people of the potential danger and allow effective remedial measures to be started in time to avert a total failure.

Sherard (1973) points out that cracks can form in dams from a number of causes:

> One of the most probable and hazardous effects of earthquake shaking on embankment dams is settlement and cracking, very similar in pattern and location to the cracking commonly caused by differential settlement. Design measures to increase the safety of dams against concentrated leaks in differential settlement cracks also act to make the dam safer against cracking due to earthquake shaking, and vice versa.

He also says:

> Complete failures, in which the entire reservoir released through a crack or erosion tunnel, have usually occurred only in homogenous relatively low dams which have no internal vertical drain (chimney drain) or zoning to control concentrated leaks.

Another hazard to dams and reservoirs are the cracks that might form when land subsidence from extraction of oil or water place the dam and its foundation in tension. Jansen et al. (1967) point out that the failure of a reservoir on an erodible foundation was traced to the cracking of its rigid lining along the line of a fault that was opened by deep-seated subsidence.

6.5   NONSTEADY SEEPAGE IN DAMS AND LEVEES

Designing "Free-draining" Upstream Shells

The upstream slopes of earth dams and levees must be stable during *rapid drawdown* of the body of water retained by these structures. A thickness of a few feet of rock riprap on a bedding of rock spalls or sand and gravel selected to prevent the washing out of soil (Sec. 5.3) is usually sufficient to protect levees from waves and currents and to prevent sloughing under fluctuating river stages. Similar slope protection is often used for earth dams in areas in which sound rock and gravel are scarce; but when thin slope protections are used it is necessary to flatten slopes to obtain stability under saturation and reservoir drawdown.

When liberal quantities of natural sandy gravels are available near dam sites, upstream *shells* of these materials are placed to improve slope stability. If these materials are so "free-draining" that the saturation level falls rapidly, they add substantially to the stability of slopes subjected to rapid drawdown.

If upstream gravel zones are not sufficiently permeable to drain rapidly during drawdown, they improve slope stability much less than if they are highly permeable. To establish a criterion for the required permeability of upstream "free-draining" gravel zones in dams assume that "free draining" means sufficient permeability to allow the saturation level in the zone to follow a lowering reservoir with a small lag (Fig. 6.24). The amount of lag to be permitted is represented by the height h of the sloping saturation line at the back edge of the zone. It can be defined by a ratio h/L, which is not to exceed a specified small value, say 0.05 or 0.10.

For an assumed position of the saturation line in the gravel zone (line AB, Fig. 6.24) the *required* permeability of the zone can be approximated from the relationship

$$k_{min} = \frac{v_{dd}Ln_e}{\sin \alpha h} \tag{6.1}$$

In Eq. 6.1 the term v_{dd} is the velocity of drawdown of the water surface in the reservoir, n_e is the *effective porosity* of the gravel, and α is the angle between the median flow line CD and a horizontal line; for instance, assume that $v_{dd} = 5$ ft/day, $\alpha = 23°$, and $\sin \alpha = 0.39$:

$$n_e = 0.25; \; h = 10 \text{ ft}; \; L = 100 \text{ ft}$$

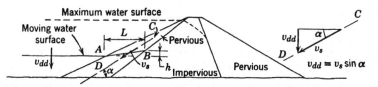

FIG. 6.24 Conditions compatible with rapid drainage of "pervious" zones on upstream faces of dams.

For these conditions, using Eq. 6.1,

$$k_{min} = \frac{5 \text{ ft/day } (100)(0.25)}{0.39(10)} = 32 \text{ ft/day}$$

Studying Progress of Saturation in Dams and Levees

It is sometimes useful to know approximately the length of time required for saturation to develop fully in earth dams and levees. If the in-place permeabilities are known, the time for saturation can be approximated by using the succession of transient flow nets described in Sec. 3.6. A study of the progress of saturation through a homogeneous dam or levee is given in Figure 3.25 and 3.26. By using this method the graph in Figure 6.25 was developed for approximating the time of saturation of earth dams with large impervious cores of the general shape shown. To apply the chart the ratio h/L is determined for a given cross section, for which the average permeability is known. The chart is

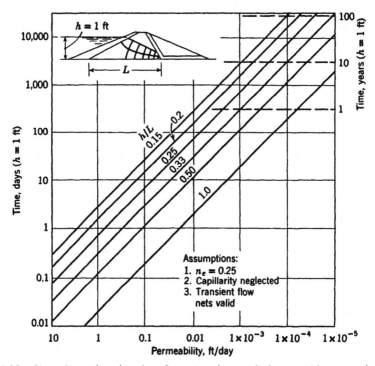

FIG. 6.25 Chart for estimating time for saturating earth dams and levees on impervious foundations. To apply (1) determine h/L for a dam or levee section similar to diagram; (2) use chart to determine time for saturation to develop in section having h = 1 ft; (3) multiply time in chart by actual h in feet.

FIG. 6.26 Trench for slurry wall cutoff for dam in California being excavated by dragline; bentonite slurry entering in foreground provides pressure against trench walls to prevent their collapse.

based on an *effective* porosity of 0.25 and $h = 1$ ft and neglects the effects of capillarity. To estimate the time of saturation of an actual dam the time determined from the chart must be multiplied by the actual height h in feet from the base of the dam to the water surface; for instance, assume that $h/L = 0.5$, $k = 1 \times 10^{-3}$ ft/day, and $h = 50$ ft. From the chart (Fig. 6.25) the time is 100 days. Multiplying by 50 ($h = 50$), the actual time is $50(100) = 5000$ days or 14 years.

6.6 REHABILITATION OF WATER-ENDANGERED DAMS

Properly designed structures that are capable of withstanding the effects of water or which have built-in seepage-reducing and drainage measures capable of controlling the inflows are usually the most economical and satisfactory over the long term. *Stage construction* to provide needed water control later on is often much more costly, except for relief wells and other measures that can be installed outside the main structures. If adequate *internal* measures are not provided in new dams and troubles develop, rehabilitation and restoration measures can be extremely costly.

When instability problems develop because of improper embankment design that did not provide for good internal drainage, some dams have been strengthened by the construction of large berms against upstream and downstream slopes. When the problems are caused by severe seepage through foundations and abutments, upstream areas are sometimes blanketed with impermeable materials, or relief wells are installed along downstream toe areas. In some cases where problems were severe, concrete diaphragm walls have been constructed from the crest down through the embankment and well into sound, non-erodible rock formations in the foundation and abutments as required. Slurry trench cutoff walls using either impervious soil-bentonite backfill or concrete diaphragms have been used to upgrade the security of many dams and large dikes. Figure 6.26 shows a slurry wall under construction in a western state for a large reservoir where leakage through a permeable sand layer was causing nuisance problems to properties below the dam. A bentonite slurry is being fed into the trench being excavated by dragline, to maintain a sufficiently high level of the fluid to prevent collapse of the walls of the trench before it can be filled with a special mixture of soil, sand, gravel, and bentonite. A well planned and constructed slurry wall filled with earth and bentonite can virtually eliminate seepage in treated areas; however, the longevity of its effectiveness cannot always be assured, as lapses can develop over a period of time. Concrete diaphragm walls should be virtually free of such problems.

With the use of draglines, the maximum practical depth of slurry trenches is around 100 feet or a little deeper, but various techniques are being used that allow much greater depths for cutoffs (e.g., see Fig. 7.1). Fairweather (1987) describes slurrywall construction for a concrete cutoff at the left abutment of the U.S. Bureau of Reclamation's Navajo Dam in New Mexico. A customized hydrofraise excavator cuts primary panels for the slurry wall construction. After filling with tremie concrete and allowing it to cure, the process is repeated for intervening secondary panels. At its maximum depth it will reach down to a record-breaking 398 feet, according to Fairweather. Michael Thomas, the Bureau's regional engineer at Salt Lake City, said the U.S. Army Corps of Engineers is considering a 420-ft deep cutoff wall at the Mud Mountain Dam near Seattle. Increasing seepage rates in the abutments of the Navajo Dam led the Bureau to install corrective measures to safeguard the dam from piping failures.

Apparently, these record-breaking depths will be exceeded in fairly short order. Using new technology, Japanese engineers have developed techniques they say will allow slurry-wall cutoffs to be constructed to depths of over 500 feet in soft strata (see *Engineering News-Record,* Sept, 3, 1987, p. 23.) They believe their system, which revolves around a double-unit rotary cutter and a new mix of concrete can be used to sink deep cutoff walls needed for a Trans Tokyo Bay Bridge and Tunnel in Japan.

Another example of restoration by diaphragm wall is the U.S. Army Corps of Engineer's Wolf Creek Dam in Kentucky. This 26-year-old earthfill dam began to leak after 17 years of service. To control seepage through the cavities and solution channels that had developed in its limestone foundation, a thin concrete diaphragm wall was installed from the crest through the fill and as deep as 70 ft into the rock (E.N.-R., 1976). Held open by the slurry method, the trench was excavated to depths as great as 268 ft at an estimated cost of $110 million; the dam cost $80 million when built in 1950.

Various kinds of grouting treatment have been used to reduce the amount of damage that occurs in projects as a result of leaks or channels that have developed from the action of seeping water. Cases have been reported of dams on foundations containing soluble limestone (or other minerals) in which cavernous openings had formed after a number of years of seepage. One such dam had been treated every 10 years or so with thousands of bags of cement grout each time the erosion had progressed to a seemingly dangerous point.

The U.S. Department of the Army, Corps of Engineers (1988), presents written lectures by recognized experts on modern "Remedial Seepage Control Methods for Embankment-Dams and Soil Foundations." Included in the Proceedings of the workshop are papers on grouting, flexible membrane linings, drainage measures, jet grouted cutoff walls, reinforced downstream berms, plastic concrete cutoff walls, and ground freezing. A video tape of the workshop, including the panel discussion, is available from the Waterways Experiment Station (WES) library, the report says.

Maintaining permanent stockpiles of fine and coarse filter aggregates downstream from dams and keeping mechanical loaders always on hand for quick placement on suspicious seepage areas is one way of retarding piping until permanent repairs can be made. Also (see Fig. 5.12d) a filter fabric can be spread on a dangerous seepage area and covered in turn with a layer of coarse crushed rock, railroad ballast, coarse gravel, pea gravel, or similar *highly permeable* material to allow water to escape while holding fines in place.

Thorough monitoring of piezometers and flow measuring devices both before and after rehabilitation measures have been provided helps to evaluate the benefits of corrective measures and is an essential part of any important repair (see also "Requirements for Safety," Sec. 6.1).

6.7 SUMMARY

It is sometimes thought that if seepage control measures of some kind are installed in dams and levees seepage will surely be controlled. The actual success of any seepage control system depends on how well it conforms to existing conditions. Measures that attempt to control seepage by keeping water out depend for their success on a high degree of perfection. Slight lapses in blankets, cutoffs, or grout curtains can drastically reduce their effectiveness. Unsealed porous joints or strata of high permeability can allow water to bypass cutoffs and grout curtains. Failure of the 305-ft Teton Dam in Idaho on June 5, 1976, was blamed on a lack of filters and drains to protect its highly erodible silt core from piping through unsealed rock joints (*E.N.-R.,* 1977). A high level of protection against damage from seepage through dams or through jointed or porous foundations can usually be obtained by the use of well-designed vertical "chimney" drains, filters, and outlet blanket drains, as outlined in this chapter. A good drainage system can provide an important "second line of defense" (even when an extensive grout curtain or other seepage-reducing system is used) to protect a dam from harmful effects of excessive seepage.

Success in seepage control in dams and levees depends on *designing and building* systems capable of coping with conditions as they really exist. Adequate exploration and testing programs, judicious application of rational design methods, and the observation of completed works are the cornerstones of successful engineering projects.

SUPPLEMENTAL READING

Sherard, James L., Richard J. Woodward, Stanley F. Gizienski, and W. A. Clevenger, *Earth and Earth-Rock Dams,* Wiley, New York (707 pp), 1963.

Independent Panel to Review Cause of Teton Dam Failure (1976), "Report to U.S. Department of the Interior and State of Idaho on *Failure of Teton Dam,*" Idaho Falls, December, 1976.

REFERENCES

ASCE/USCOLD Committee on Failures and Accidents to Large Dams of the U.S. Committee on Large Dams (1975), "Lessons from Dam Incidents, USA," Published by the American Society of Civil Engineers, New York, N. Y., 1975.

Acker, Richard C., and Jack C. Jones (1972), "Foundation and Abutment Treatment for Rockfill Dams," *Journal of the S. M. & Found. Div.,* ASCE, Vol. 98, No. SM10, October, 1972, pp. 995–1015.

Aitchison, G. D., and C. C. Wood (1965), "Some Interactions of Compaction, Permeability and Post-Construction Deflocculation Affecting the Probability of Piping Failure in Small Earth Dams," *Proceedings,* 6th International Conference on Soil Mechanics and Foundation Engineering, Montreal, Vol. II, pp. 442–446.

Barron, Reginald A. (1972), "Abutment and Foundation Treatment for High Embankment Dams on Rock," *Journal of the S.M. & Found. Div.,* ASCE, Vol. 98, No. SM10, Oct. 1972, pp. 1017–1032.

Birman, J. H., A. B. Esmilla, and J. B. Indreland (1971), "Thermal Monitoring of

Leakage Through Dams," *Bulletin,* Geological Society of America, Vol. 82, No. 8, pp. 2261-2284.

Burke, Harris H., Charles S. Content, and Richard L. Kulesza (1972), "Current Practice in Abutment and Foundation Treatment," *Journal of the S. M. & Found. Div.,* ASCE, Vol. 98, No. SM10, pp. 1033-1052.

Cambefort, H. (1987), "Grouts and Grouting," Chapter 32, *Ground Engineer's Reference Book,* edited by F. G. Bell, Butterworth and Co. (Pub.), Ltd., London, p. 32/3.

Casagrande, A., S. D. Wilson, and E. D. Schwantes (1972), A.S.C.E. Specialty Conference on Performance of Earth and Earth-Supported Structures, Purdue University, June 11-14, 1972, Vol. 1, Part 1, pp. 551-588.

Castro, Gonzalo (1976), "Seismic Stability Evaluation of Embankment Dams," presented at Engineering Foundation Conference on "The Evaluation of Dam Safety," Pacific Grove, Calif., November 18-December 3, 1976; pp. 377-390 of *Proceedings,* A.S.C.E., published in 1977.

Cedergren, Harry R. (1940), "Utility of the Flow Net in Stability Analyses," *Proceedings,* Purdue Conference on Soil Mechanics and Its Applications, Sponsored by the Society for the Promotion of Engineering Education, Purdue University, September 2-6, 1940, pp. 277-283.

Cedergren, Harry R. (1961), "Discussion of 'Design of Control Measures for Dams and Levees,' by W. J. Turnbull and C. I. Mansur," *Transactions,* A.S.C.E., Vol. 126, Part I, pp. 1531-1534.

Cooke, J. Barry, and James L. Sherard (1987), "Concrete-Face Rockfill Dam: II. Design," *Journal of the Geotechnical Engr. Div.,* ASCE, Vol. 113, No. 10, Oct., 1987, p. 1129.

Engineering News-Record (1976), "Concrete Cutoff Extends 278 ft to Seal Old Dam's Foundation," July 22, 1976, pp. 14-15.

Engineering News-Record (1977), "Teton Dam Failure is Blamed on BuRec Design Deficiencies," January 13, 1977, pp. 8-9.

Fairweather, Virginia (1987), "Stopping Seepage," *Civil Engineering,* ASCE, March, 1987, pp. 44-46.

Herreras, J. A. (1973), "The Membrane of the Pozo de Los Ramos Dam, Spain," *Transactions,* 11th Congress on Large Dams, Madrid, June 1973, Vol. 3, pp. 843-859.

Horsky, O. (1969), "Seepage from the Dam Foix in Calcareous Karst," *Vodni hospodárstvi,* Prague, Vol. 7, No. 7, pp. 201-203.

Jansen, Robert B., Gordon W. Dukleth, Bernard B. Gordon, Laurence B. James, and Clyde E. Shields (1967), "Earth Movement at Baldwin Hills Reservoir," *Journal,* Soil Mechanics and Foundations Division, *Proceedings,* A.S.C.E., Vol. 93, No. SM4, July 1967, pp. 551-575.

Jansen, Robert B. (1980), "Dams and Public Safety," A Water Resources Technical Publication, U.S. Dept. of the Interior, Water and Power Resources Service, Denver, Colorado, 1980.

Justin, J. D. (1936), *Earth Dam Projects,* Wiley, New York, p. 1.

Karol, Reuben H. (1988), "Grouting For Ground-Water Control," *Proceedings,* REMR Workshop on New Remedial Seepage Control Methods for Embankment-

Dams and Soil Foundations, 21–22 Oct., 1986, Dept. of the Army, Waterways Experiment Station, Corps of Engineers, *Final Report,* Jan. 1988, pp. 9–33.

Koenig, H. W., and K. H. Idel (1973), "Report on the Behavior of Impervious Surfaces of Asphalt," *Transactions,* 11th Congress on Large Dams, Madrid, June 1973, Vol. 3, pp. 359–367.

Leps, T. M..(1972), A.S.C.E. Specialty Conference on Performance of Earth and Earth-Supported Structures, Purdue University, June 11–14, 1972, Vol. 1, Part 1, pp. 507–550.

Londe, Pierre (1982), "Concepts and Instruments for Improved Monitoring," *Journal of the Geotechnical Engr. Div.,* ASCE, Vol. 108, No. GT6, June, 1982, pp. 820–834.

Marsal, R. J., and W. Pohlenz (1972), "The Failure of Laguna Dam, Mexico," A.S.C.E. Specialty Conference on Performance of Earth and Earth-Supported Structures, Purdue University, June 11–14, 1972, Vol. 1, pp. 489–505.

Middlebrooks, T. A., and W. H. Jervis (1947), "Relief Wells for Dams and Levees," *Transactions,* A.S.C.E., Vol. 112, pp. 1321–1338.

Nawaz, Khan S., and S. Ali Nagvi (1970), "Foundation Treatment for Underseepage Control at Tarbella Dam Project (Pakistan)," *Transactions,* 10th International Congress on Large Dams, Montreal, Vol. 2, pp. 1167–1193.

Nichiporovich, N. N., and A. Teitelbaum (1973), "Estimation of Fissure Occurrence in the Coures of Rockfill-Earth Dams," Gidrotekhnicheskoe Stoitel'stvo, Moscow, No. 4, pp. 10–15.

O'Rouke, John E. (1974), "Performance Instrumentation Installed in Oroville Dam," *Journal of the Geotechnical Engr. Div.,* ASCE, Vol. 100, No. GT2, Feb. 1974, pp. 157–174.

Peter, P. (1971). "Damage and Erosion Phenomena on an Earth Dam at Hriňva (CSSR)," *European Civil Engineering,* Vol. 2, No. 1, pp. 8–16.

Peter, P., J. Hulla, and R. Ravinger (1970), "Investigation of Failures Due to Seeping Water Using Radionuclids," Vodahospodarsky Casopis, Bratislava, Czechoslovakia, No. 5, pp. 560–578.

Pratt, Harold K., Robert C. McMordie, and Robert M. Dundas (1972), "Foundations and Abutments—Bennett and Mica Dams," *Journal of the S. M. & Found. Div.,* ASCE, Vol. 98, No. SM10, Oct., 1972, pp. 1053–1072; (1972a), pp. 1060–1063, 1069–1071.

Seed, H. Bolton (1973), "Stability of Earth and Rockfill Dams During Earthquakes," *Embankment-Dam Engineering, Casagrande Volume,* Chapter 8, Wiley, New York, 1973, p. 246.

Sembenelli, P., and M. Fagiolo (1974) "Aquada Blanca Rockfill Dam with Metal Facing," *Journal,* Geotechnical Engineering Division, A.S.C.E., Vol. 100, No. GT1, January 1974, pp. 33–51.

Sentürk, F., and Y. Sayman (1970), "Interpretation of Piezometric Indication in a Dam Resting on Permeable Foundation," *Transactions,* 10th Congress on Large Dams, Montreal, 1970, Vol. 3, pp. 845–855.

Sherard, James L. (1973), "Embankment Dam Cracking," *Embankment-Dam Engineering, Casagrande Volume,* Chapter 9, Wiley, New York, p. 343.

Sherard, James L., Richard J. Woodward, Stanley F. Gizienski, and W. A. Clevenger

(1963), *Earth and Earth-Rock Dams,* Wiley, New York, pp. 116–123; (1963*a*) p. 1; (1963*b*) p. 408; (1963*c*) pp. 159–161.

Sherard, James L., and J. Barry Cooke (1987), "Concrete-Face Rockfill Dam: I. Assessment," *Journal of the Geotechnical Div.,* ASCE, Vol. 113, No. GT10, Oct., 1987, p. 1106.

Simmons, Marvin D. (1982), "Remedial Treatment Exploration, Wolf Creek Dam, Ky," *Journal of the Geotechnical Engr. Div.,* ASCE, Vol. 108, No. GT7, July, 1982, pp. 966–981.

Stanage, Clark E. (1982), "Embankment Construction: New Melones Dam," *Journal of the Geotechnical Engr. Div.,* ASCE, Vol. 108, No. GT4, April, 1982, pp. 621–636.

Stropini, Elmer W., Donald H. Babbitt, and Henry E. Struckemeyer (1972), "Foundation Treatment for Embankment Dams on Rock," *Journal of the S. M. & Found. Engr. Div.,* ASCE, Vol. 98, No. SM10, pp. 1073–1079.

Turnbull, W. J., and C. I. Mansur (1961), "Investigation of Underseepage—Mississippi River Levees," *Transactions,* A.S.C.E., Vol. 126, Part I, pp. 1429–1539.

U.S. Department of the Army, Corps of Engineers (1972), "Investigation of Relief Wells, Mississippi River Levees, Alton to Gale, Illinois," Misc. Paper 5-72-21, Waterways Experiment Station, Vicksburg, Miss., June 1972.

U.S. Department of the Army, Corps of Engineers, Office of the Chief of Engineers (1986), "Seepage Analysis and Control For Dams," Engineering Manual EM 1110-2-1901, 30 September 1986, pp. 9-11 to 9-46, 12-6 to 12-12; (1986*a*), pp. 9-84 to 9-86, 11-2 to 11-4, 12-2 to 12-4.

U.S. Department of the Army, Corps of Engineers, Waterways Experiment Station (1988), "Proceedings of Remr Workshop on New Remedial Seepage Control Methods for Embankment-Dams and Soil Foundations," 21–22 Oct. 1986, Compiled by E. B. Perry, *Final Report,* January, 1988, 174 pp.

U.S. Department of the Navy, Naval Facilities Engineering Command (1982), NAVFAC Design Manual 7.1, "Soil Mechanics;" NAVFAC Design Manual 7.2, *Foundations and Earth Structures,* May, 1982.

Walker, Fred C., and R. W. Bock (1972), "Treatment of High Embankment Dam Foundations," *Journal of the S. M. & Foundations Div.,* ASCE, Vol. 98, No. SM10, Oct., 1972, pp. 1099–1113.

Wallace, B. J. and J. I. Hilton (1972), "Foundation Practices for Talbingo Dam, Australia," *Journal of the S. M. & Foundations Div.,* ASCE, Vol. 98, No. SM10, Oct., 1972, pp. 1081–1098.

Warner, James (1978), *Discussion* of ASCE Paper 13214, "Engineering of Grout Curtains to Standards," by Adam Clive Houlsby, Sept., 1977; discussion in *Journal of the Geotechnical Engineering Div.,* ASCE, Vol. 104, No. GT3, March, 1978, pp. 401–402.

Wegmann, Edward (1922), *The Design and Construction of Dams,* Wiley, New York, 7th ed., p. 221.

CHAPTER SEVEN

FOUNDATION DEWATERING AND DRAINAGE

Whereas the water held back by dams and levees creates most of the problems treated in Chapter 6, naturally occurring groundwaters are the source of the problems pertaining to foundations as discussed in this chapter. The temporary control over groundwater and seepage is required for many excavations made below the level of the groundwater at construction sites. Controlling water at sites where excavations are needed is discussed in Sec. 7.1, "Construction Dewatering." Permanent control over seepage and groundwater for some facilities and permanent improvement in the strengths of foundations for others are discussed in Sec. 7.2, "Foundation Improvement by Drainage."

7.1 CONSTRUCTION DEWATERING

General

Many types of engineering construction require the excavation of soil and rock below the natural water table. If the formations are naturally well cemented, water control may be simply a matter of allowing the water to seep down the excavation slopes into shallow ditches or sumps from which it is removed by pumping. On the other hand, if the waterbearing materials have low strengths, extensive dewatering systems may be required.

Either of the two fundamental methods of controlling seepage can be used for the control of groundwater during construction: (1) those that keep the water out or (2) those that depend on its control by drainage processes. Many important foundation excavations are carried out with combinations of the two basic methods. Chemical grout, cement grout, slurry trench cutoff walls, steel sheet pile walls, and caissons are means that often serve to keep out most of the water. Usually when these methods are used, pumps are required to maintain dry conditions in excavations. Frequently compressed air is used in caissons to equalize the hydrostatic head and to reduce the risks of "running" ground and blowouts. In this chapter control of groundwater, by drainage methods is the primary interest, although seepage-reducing methods are also discussed.

Some important structures with foundations below the normal water table must have water control both during the construction and after the project is completed and put in operation. If water in large volume needs to be pumped during a long construction period and after a project is completed, it may be more economical to cut off the flow than to try to drain it, particularly if a cutoff for construction can be partly utilized for permanent control of seepage. One major project in which this was the case is the second powerhouse at Bonneville Dam on the Columbia River near Portland, Oregon (*E.N.-R.* 1975).

The excavation for the Bonneville Second Powerhouse (constructed for the Portland, Oregon, District of the U.S. Army Corps of Engineers) was completely enclosed in a new type of thin concrete cutoff wall. A trench 24 in. wide was excavated through gravels, boulders, and other slide debris to depths as great as 150 ft by the slurry trench method (see also Secs. 6.2, 6.6, and 10.3). Short segments of the trench were excavated with a special clamshell bucket (Fig. 7.1) while the sides were supported with a thick bentonite slurry. After excavating 3 ft into bedrock the bottom was cleaned of sand and rock and the segment was then filled with low-strength concrete through a tremie pipe lowered to the bottom. An alternate to the slurry-trench cutoff wall for construction water control would have been pumping with deep gravel-packed wells, but the cutoff wall was considered safer and more positive and was therefore selected.

Dewatering systems have the following main purposes.

1. Intercepting seepage which otherwise would enter excavations and interfere with the work.
2. Improving the stability of slopes, thus preventing sloughing or slope failures.
3. Preventing the bottoms of excavations from heaving because of excessive hydrostatic pressure.
4. Improving the compaction characteristics of soils in the bottoms of excavations for basements, freeways, and so on.
5. Drying up borrow pits so that excavated materials can be properly compacted in embankments (a special application).
6. Reducing earth pressures on temporary supports and sheeting.

FIG. 7.1 Special narrow clamshell bucket being used for excavating trench for thin concrete cutoff wall for Second Powerhouse at Bonneville Dam, Oregon. (Reprinted with the permission of *Engineering News-Record,* November 13, 1975 issue; photo by Ray Bloomberg.)

Leonards (1962) presents a concise historical review of construction dewatering in the United States and several other countries.

Many shallow foundations are excavated below the groundwater level by the use of slopes flat enough to resist failures, together with shallow ditches and sumps to collect the seepage that enters the excavation. These simple inexpensive methods are most effective in dense somewhat cemented soil formations. Most excavations in waterbearing formations such as gravels, sands, silts, and stratified clays are stabilized by wellpoints, deep pumped wells, or other groundwater control systems. To provide a running record of the effectiveness of dewatering systems for important projects observation wells should be installed at key locations and water levels observed periodically. Water-level readings can give a forewarning of deficiencies that might be developing before they become serious. As safeguards against breakdowns, standby power facilities and standby pumps and other vital equipment should be on hand for all important projects.

Sump Pumping

When the soil or rock formations being excavated are strong enough to stand on the excavated slopes without sloughing, erosion, or other problems, the inflowing groundwater is sometimes allowed to flow into ditches on one or more sides of an excavation or into sumps at selected locations from which it is removed by pumps. When this method, called *sump pumping,* can be used for controlling water, the cost of dewatering can be minimal, but the handling of power cables, pumps, and suction hoses in an excavation and on its slopes is a nuisance and interferes with the placement of drains and the operation of compaction equipment, and thus may add indirectly to the overall cost of a project (note that Figure 6.6 shows a sumping job that was very successfully completed after a couple of early problems were solved).

Sump pumping is almost never satisfactory for deep excavations below the water table in sand, silt, silty clay, stratified alluvial soils, or any situation in which inflowing water causes "quicksand," boils, piping, or other unstable conditions. Under favorable conditions, however, it is sometimes the primary if not the only method of controlling water in major excavations.

Wellpoints

For the temporary lowering of groundwater during the construction of buildings, depressed freeways, and other works, wellpoints have been widely used for many years. A wellpoint is a small-diameter tube or pipe fitted with plastic or metal screens to permit water to enter without the loss of adjacent soil. It is often equipped with metal points that allow it to be driven or jetted into soil formations. Wellpoints have been most successfully used in gravelly sands, sands, silty sands, and similar soils. If the material is dense or bouldery, wellpoints must be installed in drilled holes which greatly increases the cost of an installation.

Wellpoints are arranged in a line or ring surrounding an excavation area and connected through a manifold to a pump that extracts seepage to lower the water table in the area to be excavated. The required spacing, usually between 3 and 12 ft, depends on the type of soil and the desired amount of groundwater lowering as well as on the type and size of wellpoint. Powers (1976) says:

> The optimum spacing for wellpoints is determined by a bewildering variety of factors and except in the simplest situations, even experts disagree. Spacing can range from two feet or less in very difficult conditions to over forty feet in free draining soils.

He also points out that although nomographs were at one time used for determining wellpoint spacings (see Chapter 7 in first edition, "Seepage, Drainage, and Flow Nets") the current practice is to use larger diameter wellpoints for high volume flows and increase the spacing. Powers emphasizes that the spacing of wellpoints may be adjusted because of a number of practical considerations; for example, if the installation is costly—as in formations containing many large rocks and boulders—it may be more economical to "open up the spacing, even if it means accepting more water in the excavation." He adds that if the installation is "quick and inexpensive, but maintenance labor is unskilled, then it may be wise to close up the spacing so the wellpoint adjustment is less critical."

One of the advantages of wellpoint systems is the flexibility of the method. If an estimated spacing does not produce the desired groundwater lowering, additional wells can be installed between the initial wells until the desired lowering is obtained.

Wellpoints are most often used for dewatering excavations that do not require deep lowering of the water table. If the water table must be lowered more than 15 to 18 ft, the maximum effective lift of suction pumps, two or more *stages* of wellpoints (or deep pumped wells) can be used. Figure 7.2

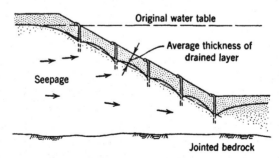

FIG. 7.2 Excavation stabilized with five-stage well point system. (After Terzaghi and Peck, Wiley, New York, 1948.)

shows a section through a deep excavation below the water table that is stabilized with a five-stage wellpoint system.

Powers (1981) suggests that if the total required drawdown is slightly greater than can be effectively handled by a one-stage system, it may be advisable to install a temporary wellpoint stage (as shown in Fig. 7.3) so that the main system can be installed deeper, and the required lift reduced to a "manageable 15 ft." If the total required drawdown is substantially more than 22 ft he says it is usually necessary to use a multistage wellpoint system.

According to Stang (1976), plastic pipes and screens have been a boon to the wellpoint industry. He says:

> The use of plastics within the wellpoint industry has not only helped in reducing installation costs but has also, because it is virtually inert, eliminated many corrosion problems that were encountered in areas where alkali, sulfur and other toxics were destructive to steel.

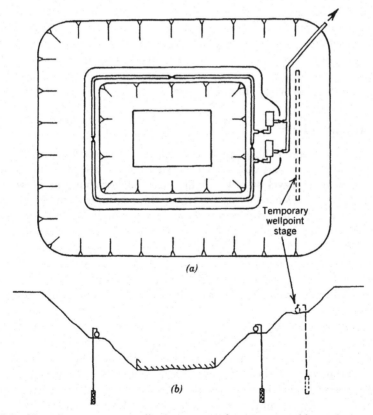

Temporary
wellpoint
stage

(a)

(b)

FIG. 7.3 Use of a temporary wellpoint stage to lower water table so the permanent single-stage wellpoint system can be installed at a lower level for better performance (Fig. 17.2 of "Construction Dewatering: A Guide to Theory and Practice," J. Patrick Powers, John Wiley & Sons, Inc., N. Y., 1981, p. 294.)

He also says:

> Wellpoint systems are probably the most flexible of any of the dewatering systems on the market today. By simply adding or reducing the number of wellpoints or pumps in a properly designed system, an increase or reduction in capacity can be obtained. Wellpoint assemblies of varying lengths or depths can be connected to a common header manifold. This design flexibility permits excavation side slope and subgrade stability as well as pressure relief of lower artesian aquifers. The wellpoint system operates most effectively in soils ranging from coarse silt to fine gravel. It is also effective in stratified soils when installed in a properly designed filter column.

Jet-Eductor Wellpoint Systems

The U.S. Army et al. (1971) point out that the depth of a wellpoint can often be greatly increased by inserting an aspirator or eductor pump near the bottom. Water is forced under pressure through an internal riser pipe to activate the eductor and pump water up to the surface and away through a return header. Jet-eductor wellpoints are most advantageously used for dewatering deep excavations in relatively low permeability soils that produce small volumes of water to be pumped. They are not very efficient; hence deep pumped wells should be used when flows are large.

Deep Pumped Wells

When the water table is deep, pumped wells are used for lowering the water level. They have been particularly effective in highly stratified formations that contain openwork gravel or other highly pervious horizontal layers that feed water to the wells. When highly pervious, thin, natural drainage layers are penetrated by large-diameter pumped wells; a few wells at wide spacing are often sufficient. Deep wells are usually installed at relatively wide initial intervals, and if sufficient groundwater lowering is not obtained additional wells are drilled. Many foundations have been dewatered exclusively with deep pumped wells. If highly pervious water-bearing formations exist at some depth, deep pumped wells may be the only practical means of controlling uplift pressures.

Deep wells are usually located around an excavation's outside edges. If an excavation is narrow, no other dewatering facilities may be needed, but if it is wide and underlaid with pervious formations a line of wellpoints at the inside toe of the slope (Fig. 7.4) or additional deep wells within the foundation area may be required. Wells drilled through impervious layers into pervious underlying strata to relieve uplift pressures are called *bleeder wells* (Terzaghi and Peck, 1948).

Although a substantial number of deep pumped wells is sometimes required, a 600-ft-long excavation for an earth dam cutoff trench was success-

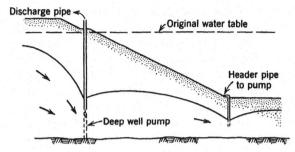

FIG. 7.4 Excavation stabilized with deep wells and well points. (After Terzaghi and Peck, Wiley, New York, 1948).

fully dewatered by using only three wells equipped with deepwell turbines, together with shallow ditches and sump pumps within the excavation.

Deep wells and wellpoints must usually be pumped for extensive periods of time. Because pumping failures can cause soils to blow up into excavations or slopes to fail, standby pumps and auxiliary power supplies are kept on hand for emergency use in the event of equipment or power failure.

The most suitable type of dewatering system for a given location depends in large measure on the soil being dewatered. Soils that are too coarse for drainage by wells or wellpoints are usually drained by gravity methods, supplemented when necessary by grout curtains or other cutoffs. Some soils that are too fine to be drained by wells and wellpoints are effectively drained by the vacuum method or electroosmosis.

Large arrays of deep pumped wells can lower a water table to much greater distances than individual or small groups of wells. A system of 24 deep gravel-packed wells which operated for more than 3 years to dewater alluvial formations at the construction site of a heavy "boat section" for a depressed freeway in a West Coast city caused so much lowering of the water table that the increased unit weight of the dewatered soil produced several tenths of a foot of land settlement as far as one-half mile away. Sidewalks and other facilities around buildings (which rested on friction piles) required many thousands of dollars in repairs and this led to serious legal questions of responsibility. Such possibilities should be considered when dewatering systems are being designed. An ample settlement monument system and detailed notations on preexisting conditions adjacent to proposed dewatering projects may prove to be helpful in combatting unjustified claims of damage. Photographs showing cracks in structures, sidewalks, etc. before dewatering was begun can also be useful.

Vacuum Method

According to Terzaghi and Peck (1948a), if the average effective grain size of a soil D_{10} is less than about 0.05 mm, gravity drainage methods (including wells) are ineffective for lowering the groundwater level because capillary suction prevents the release of water. Glossop and Skempton (1945–1946) suggest that a total silt and clay content of about 60% is the borderline. Soils that are too fine grained to be stabilized by gravity drainage frequently can be improved by the application of a vacuum to filters surrounding wellpoints (Fig. 7.5). Water is squeezed toward the evacuated filters under the pressure of the atmosphere, which is about 1 ton/sq ft. As a result, the effective pressures in the soil near the wells is substantially increased. The shear strength is thereby increased by an amount equal to $p_a \tan \phi$, where p_a is the atmospheric pressure and ϕ is *the* angle of internal friction of the soil.

The U.S. Army et al. (1971a) say that silts and sandy silts which have a low coefficient of permeability ($k = 0.1 \times 10^{-4}$ to 10×10^{-4} cm/sec) and do not drain readily by gravity methods can be drained much more rapidly with the vacuum method.

The vacuum method has been successfully used for stabilizing steep cut slopes for important excavations in soils that were too impervious to be improved by gravity drainage methods. Terzaghi and Peck (1948a) cite as examples of the use of this method an excavation in organic silt with an average effective grain size less than 0.01 mm and another in silt with an effective size of about 0.015 mm. Although the quantity of water pumped from the wells may be extremely small, a few weeks of pumping often result in remarkable improvement in the strength of weak soils. Soils that are stabilized by the vacuum method remain completely filled with water; hence reasonable precautions should be taken to avoid severe jarring of the soil structure. If the soil

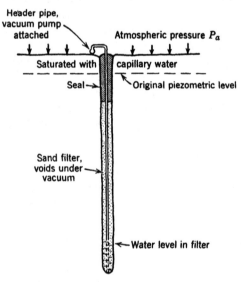

FIG. 7.5 Diagram illustrating the vacuum method of drainage. (Terzaghi and Peck, Wiley, New York, 1948.)

has a loose structure, it is possible that severe shocks could produce a collapse of the soil structure leading to liquefaction.

Electroosmosis

When an electrical potential is applied to a saturated fine-grained soil, the flow of water that results can be beneficial by reducing the water content or by producing favorable seepage forces. Reuss (1809) discovered the movement of liquids in porous media under the influence of an externally applied electric potential. An explanation of this phenomenon, which had been examined experimentally by numerous investigators, was the electric double-layer concept of Helmholtz (1879), developed further by Freundlich (1926). According to this theory, the water near the soil particles is made up of two layers. One is assumed to be bonded to the particles; the other is free moisture. The layer adjacent to the particles has an excess of anions, the outer layer an excess of cations. As a result, when an external voltage is applied, the unattached cations are free to migrate toward a negative cathode. Esrig and Majtenyi (1965) reviewed theories of osmotic flow and developed an equation that suggests that the electroosmotic velocity of water flow in porous media such as soils is related to pore ion conductivity, permeability, porosity, and soil plasticity. Others who have developed theoretical concepts and suggested practical applications for electroosmosis are Gray and Mitchell (1967) and Winterkorn and Schmid (1971).

Dr. Leo Casagrande (1947) made use of the principle of electroosmosis and put it to work in stabilizing waterlogged slopes in fine-grained soils.

To apply the electroosmosis principle to soils wellpoints are installed in a line to the depth required to lower the water table the desired amount. A metal pipe in each well serves as a suction pipe for the removal of water and a negative electrode to attract it. Rods are driven into the soil to act as anodes. When an electrical current flows between anodes and cathodes, water flows out of the surrounding soil toward the cathode well from which it can be removed by suction pumps or, if necessary, by deep well pumps. If the anodes are placed nearer the excavation or on the low side of a slope from the wells, the water flow induced by electrical current will oppose the natural hydraulic gradient and neutralize the seepage forces that would be contributing to the instability of the slope. Thus electroosmosis improves stability by lowering the water table and neutralizing harmful seepage forces. This method can be effective even though the total amount of water removed from a soil may not be great.

An example of a construction project which was carried out with the help of electroosmosis is the 6,000-ft-long cut for a double track railway north of Salzgitter near Brunswick.* The top 4 ft of the cut was in sandy soil over a

---

*See "Soil Mechanics for Road Engineers," Road Research Laboratory, Department of Scientific and Industrial Research, London, 1952, p. 338 (H. M. Stationery Office).

bed of soft silt that could not be excavated with power shovels and was too weak to stand without excessive sloughing.

A row of wellpoints was installed to a depth of about 25 ft on each side of the proposed excavation, spaced at intervals of about 33 ft. Gas tubing was inserted into the ground between the wellpoints to serve as anodes. The wellpoints consisted of perforated pipes 4 in. in diameter, surrounded by selected sand and gravel. Plain pipes 1 in. in diameter were inserted inside the filter pipes to permit the pumping of water out of each wellpoint.

During the excavating a potential of 180 V was applied to the system (in sections a few hundred feet long). The current consumption was about 19 amp per wellpoint. After the excavation was bottomed in a given section the potential was lowered to 90 V to reduce current consumption. Intermittent pumping sufficed to keep the wellpoints pumped down. After the excavation had been completed on cut slopes of 1:1 sand backfill was placed as shown in Figure 7.6. The backfull slope of 1:2, with drainage as shown, provided sufficient surcharge over the weak silt to provide a stable slope after the current was turned off.

Although the total quantity of water removed was not large, the discharge rate during current application was 150 times that without, which indicates the profound influence of electroosmosis on the seepage patterns in the soil. Without current application the 20 wellpoints produced about 90 gal in 24 hours; with current application it rose to more than 13,000 gal in 24 hours.

Casagrande et al. (1961) describe a successful electroosmotic stabilization of a slope in organic silt which had caused serious problems during the construction of the Little Pic River Bridge near Marathon, Ontario. After a serious slide in the saturated silt had jeopardized the project the designers considered a number of methods, including freezing of the slope, chemical stabilization, flattening of the slope, the use of caissons, a design change, and electroosmosis. Studies indicated that electroosmosis would be the most economical method. Three 70-kw direct current, diesel-powered generators that produced 1200 amp at 150 V furnished the necessary power. A water yield of about 1200 gal/hr was removed from deep cathodes by submersible pumps with capacities of 1 gpm at a head of 170 ft. Ordinary wellpoint pumps and headers removed the water that entered shallow cathodes. Despite the severe

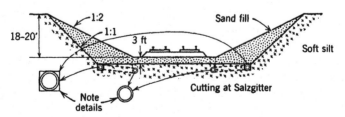

FIG. 7.6   Excavation stabilized by electroosmosis. (From *Soil Mechanics for Road Engineers*, H. M. Stationery Office, London, 1952.)

vibrations produced by the pile-driving equipment used to pound foundation piles into rock, the slope remained perfectly stable.

Electroosmosis has been employed fairly extensively in Europe but to only a limited extent in the United States. For some fine-grained soils that cannot be effectively drained by gravity methods it offers outstanding advantages. In some soils the anodes deteriorate so rapidly that the method is not practical. It is effective in only a narrow range of soils; hence it should be used only by persons experienced in the method.

Miscellaneous Methods

The groundwater control methods described in preceding paragraphs depend on natural or forced drainage to improve soil stability. If the quantity of water removed is likely to be extremely large, it may be advisable to cut off a large part of the flow either with sheet pile walls, thin slurry-trench cutoffs, or cement or chemical grouts. (see Sec. 6.2, Chapter 6).

When excavations are required in highly developed areas for large water mains, sewers, etc. braced interlocking steel sheet piling is often used to facilitate the work and prevent subsidence of ground outside the excavation with potential damage to buildings and other facilities. Frequently, any large drawdown of groundwater levels under adjacent improvements can cause harmful subsidences. Figure 7.7 shows the way modern synthetic filter fabrics in combination with high-permeability bedding material can be used to minimize

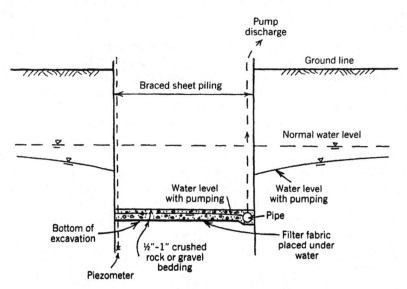

FIG. 7.7 Suggested method for dewatering excavations for large sewer lines and water mains, with aid of fabrics and open-graded aggregates to minimize drawdowns in properties adjacent to excavations.

groundwater lowerings during the construction of water mains and sewers. The excavation is made to the required depth without any removaľ of water from within the trench, a fabric filter is placed on the bottom (with the help of divers if necessary), a drainage pipe is placed in a shallow trough at one side, as shown, and the gravel bedding material is placed over the fabric. While water is being pumped from the excavation through the collector pipes and risers, uplift pressures are read in piezometers installed at the bottom edge of the piling to make sure that the uplift pressures do not exceed previously calculated safe amounts. If an initial trial does not meet requirements, the sheet piling can be driven to a greater depth, or the thickness of the bedding material can be increased to the amounts needed to prevent heave of the bottom of the excavation.

Underground ice cofferdams are a means of control of groundwater in the construction of deep cutoffs for dams and other difficult water problems. The formation of an impervious barrier of ice by freezing the water in the soil pores can be very effective; however, the installed freezing plant must have adequate capacity or there may be serious problems. If even a small trickle is permitted to flow through an ice cofferdam, it may get progressively larger and jeopardize an entire installation.

Selection and Design of Dewatering Systems

General. Groundwater control for foundation excavations may be accomplished in a number of different ways. The most appropriate method for a given job should be determined by adequate soil surveys and test borings to delineate important soil strata and locate sources of water. On important projects the permeability of the formations should be determined by field pumping tests (Sec. 2.6) or other adequate methods. Records should be developed of seasonal fluctuations in the natural groundwater level; and if a river, lake, or other body of water is near a proposed excavation, fluctuations in the level of the body of water and in the groundwater table should be recorded. All pertinent available data bearing on the dewatering problem should be studied. For any dewatering project in which failures could lead to extensive structural damage or serious flooding the design and installation of water-control systems should be carried out by higthly experienced persons.

Most dewatering systems are flexible with respect to discharge capacity and can be enlarged in capacity to take care of unexpectedly large rates of flow. Nevertheless, the approximate rate of discharge should be known in advance so that approximate power requirements will be known. The design of dewatering systems involves two important steps.

1. Evaluation of the magnitude of the dewatering project, including an estimate of the probable rate of inflow and power consumption.
2. Design of a system capable of providing the required groundwater lower-

ing for the length of time needed for the construction that is to be carried out below the natural groundwater level.

The magnitude of a dewatering project depends largely on the quantity of water that must be removed and the length of time the system must be operated. The detailed techniques employed in any system must be modified or expanded as required to produce the necessary control of inflowing seepage and uplift pressures to allow the work to be carried out safely. Sometimes major *sources* of water are not known until the work is well underway. In Sec. 6.2, under *Cutoff Trenches,* difficulties arose in the construction of a cutoff trench for a small dam, partly because the engineers thought the valley alluvium was the only important source of water that needed to be removed by sumping (see Fig. 6.6). In many projects requiring cutoff trenches in narrow valleys, the valley alluvium is the only significant source of water entering the excavations, with almost no water entering from formations at the sides of the valley; so the work is successfully carried out by systems that dewater the alluvium. But, in some cases, highly fractured and permeable rock formations in abutments add considerably to the inflows and dewatering difficulties. Such was the case with the dam illustrated in Figure 6.6. When this system was adjusted to take care of flows from the rock formations, the work was completed very successfully.

The probable inflow rates to dewatered excavations frequently can be estimated by the use of the following:

1. Formulas developed for multiple-well systems.
2. Simple well or artesian well formulas that treat the excavation as a single large well or sink.
3. Darcy's law, using reasonable values for i and A.
4. Two-dimensional flow nets, making an approximate correction for three-dimensional flow.

Seepage systems analyzed in the other chapters are predominately two-dimensional flow. As pointed out by Muskat (1937), all fluid systems must necessarily extend in three dimensions, but the significant characteristic of two-dimensional systems is that the motion in parallel planes is identical; therefore all features of the fluid motion can be observed in a single plane. Examples of two-dimensional flow are the flow nets in other chapters and the artesian flow toward wells that fully penetrate sand beds completely filled with water. Examples of three-dimensional flow are seepage around the ends of dams and the flow toward wells and foundation excavations in which a phreatic line drops as seepage approaches the hole.

Mathematical relationships pertaining to wells for dewatering and other purposes have been developed and described by a great many writers; for example, Avery (1951) presents an analysis of wells adjacent to open bodies of water. His solutions make it possible to determine the following:

1. The rate of pumping required to maintain any desired drawdown with various arrangements of wells.
2. The change in rate of pumping required for any change in drawdown, arrangement of wells, or water stage.

Peterson (1961) presents a method of designing interceptor drainage wells for artesian aquifers underlying irrigated lands. Although his theory does not pertain to foundation dewatering, the paper will be of value to those interested in well drawndown theory.

An interesting study of well theory is that developed by Hantush (1963) for the flow of groundwater through a water-table aquifer resting on a sloping impermeable bed. He gives analytical expressions for the drawdown distribution around wells that completely penetrate a water-bearing layer. Glover (1966) has developed useful theory for the nonsteady state while the water table is lowering around pumped wells (Sec. 2.6). Fundamental theory of two-dimensional and three-dimensional flow for a wide range of conditions is presented by Zangar (1953).

Muskat (1937a) develops theory for steady-state seepage toward a single well, small groups of wells, infinite sets of wells in one-, two-, and three-line arrays, and for staggered wells in line arrays. Muskat's monumental work provides a source for many of the refinements in seepage theory made in recent years.

Leonards (1962) presents theoretical solutions to a variety of drawdown conditions, including flow to two partly penetrating slots midway between two line sources, flow to single wells (fully and partly penetrating), and flow to arrays of wells.

Harr (1962) presents a treatment of seepage theory that includes a chapter on flow toward wells.

Todd (1959) has a chapter on groundwater and well hydraulics that reviews theories of flow toward wells for steady and nonequilibrium conditions; he describes well flow near streams and presents theory of flow to multiple-well systems.

The U.S. Army et al. (1971b), in a comprehensive review of groundwater flow and well theories, make available a large number of charts, diagrams, and illustrative problems to help those who wish to design dewatering systems.

All of the theoretical solutions for groundwater flow and flow toward wells and clusters of wells depend on Darcy's law for flow in porous media. According to Darcy's law ($q = kiA$) the rate of flow of water into dewatering systems depends on the following factors:

1. The permeability and thickness of water-bearing formations.
2. The depth to which the water level must be lowered below the natural

groundwater level and the corresponding hydraulic gradients producing flow toward the excavation.

3. The cross-sectional areas of water-bearing strata perpendicular to the direction of flow.

Since soil permeability k is the most variable of the three factors listed, and has the greatest influence on the amount of water to be pumped and the amount of power required, special efforts should be made to determine the coefficients of permeability of the water-bearing formations. Large-scale field pumping tests (Sec. 2.6) are one of the most dependable methods of evaluating the average permeabilities of large earth masses.

Powers (1981) says that because the quantity of water pumped is critical to the design and the cost of executing a dewatering program, "appropriate safety factors should be used." His recent book contains a vast amount of information that provides "practical guidelines for the design, selection, planning, and execution of effective systems for ground water control."

Three practical methods of estimating the rate of seepage into dewatered excavations in extensive alluvial formations are described in this section with reference to an excavation for a building in California (Fig. 7.8). This figure gives a plan and profile of a foundation excavation that was dewatered for the major portion of a year. The area is characterized by a high water table and highly pervious sands and gravels beneath a thin cover of silts and silty sands. During construction of the foundation the water table was lowered approximately 15 ft by pumping at a rate of 800 to 1,000 gpm. Theoretical estimates of the rate by the methods described in this section agree well with the actual rate.

Estimation of Pumping Rates with Well Formulas, Treating Excavation as a Large Circular Well. If a dewatered excavation is nearly square and can be approximated by a circular hole, the pumping rate can be estimated with a well formula. If seepage is primarily in a confined aquifer, the rate can be estimated by using the artesian formula or two-dimensional planar flow nets (see Fig. 7.10); however, in this example the simple well formula is more appropriate. With reference to the cross section in Figure 7.8c the simple well formula is

$$Q = \frac{\pi k(H^2 - h_0^2)}{2.3 \log_{10} (R/r_0)} \tag{7.1}$$

In Eq. 7.1 H is the thickness of the water-bearing strata at a radial distance R from the center of the well at which distance drawdown and seepage velocities can be assumed to be zero; h_0 is the thickness of the water-bearing strata at the edge of the hole, which has an average radius r_0. The total seepage rate is Q, the average soil permeability, k.

Referring to Eq. 7.1, we see that the rate of discharge from pumped sinks

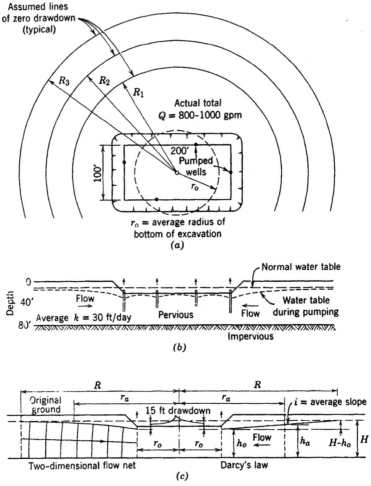

FIG. 7.8   Estimating pumping rates for foundation excavations. (See Tables 7.1, 7.2, and 7.3.) (*a*) Plan of rectangular excavation for a California building. (*b*) Cross section. (*c*) Idealized cross section showing two approximate methods for estimating total pumping requirements.

or wells should (1) increase directly with the soil permeability, (2) increase with the depth of the drawdown, (3) vary inversely with the log of (R/r_0).

In the study of seepage toward wells it is necessary to establish outer boundaries of the fluid systems. Because the distance R cannot be determined exactly before pumping is started, it must be assumed. Theoretical pumping rates for the foundation excavation shown in Figure 7.8*a* for several values of R are summarized in Table 7.1, which shows that the calculated rates vary from a high of 1300 gpm for $R = 200$ ft to 650 gpm for $R = 500$ ft. As previously

TABLE 7.1 Calculated Pumping Rates by Simple Well Formula (Excavation in Fig. 7.8)

Assumed R	Theoretical Q	
ft	cfm	gpm
200	174	1300
300	121	910
400	99	750
500	87	650

noted, the actual pumping rate for this foundation excavation varied from 800 to 1000 gpm.

Estimation of Pumping Rates Using Darcy's Law. Using Darcy's law ($q = kiA$), seepage rates can be calculated at any location in which the effective soil permeability, the area A, perpendicular to the direction of flow, and the hydraulic gradient i are known or can be assigned reasonable values. The total rate Q entering an excavation can be estimated from

$$Q = qL \qquad (7.2)$$

In Eq. 7.2 q is the seepage quantity per linear foot and L is the length of the cross section through which water is flowing. The distance L is the length of an earth dam, for example. For the three-dimensional flow condition represented by the excavation in Figure 7.8 q is the rate of discharge per linear foot at a mean radial distance r_a, and the total rate of flow Q is equal to q times L. The distance L is equal to $2\pi r_a$.

Assume that at some distance R the drawdown is zero. (See the right-hand side of Figure 7.8c.) Also assume that the hydraulic gradient at a radial distance r_a can be approximated at $(H - h_0)/(R - r_0)$. The rate of flow per foot of circumference at r_a is $q = kiA = kh_a(H - h_0)/(R - r_0)$, where h_a is the thickness of the water-bearing strata at r_a. In this simplification h_a is equal to $(H + h_0)/2$.

In this calculation, as with those made by the alternate methods, the distance R at which the drawdown is zero is not known and must be assumed. Rates of pumping obtained for several assumed values of R are summarized in Table 7.2. It will be seen by reference to Table 7.1 that the rates are comparable to but slightly lower than those obtained by using the simple well formula.

Estimation of Pumping Rates Using Two-Dimensional Flow Nets. In two-dimensional flow seepage converges two-dimensionally and flow nets are composed of squares, but in three-dimensional flow, as in the present dewatering problem (Fig. 7.8), seepage converges three-dimensionally and head losses near the center tend to concentrate considerably more than for two-dimensional flow. This characteristic difference is illustrated in Figure 7.9.

TABLE 7.2   Calculation of Pumping Rates by Darcy's Law (Excavation in Fig. 7.8)

Assumed R	Radial distance to midpoint of drawdown, r_a	Area A at r_a $= 2\pi r_a h_a$	Average i	iA	$Q = kiA$	
ft	ft	ft^2			cfm	gpm
200	140	55,000	0.125	6800	140	1050
300	190	75,000	0.068	5100	106	790
400	240	94,000	0.047	4400	92	690
500	290	114,000	0.036	4100	85	640

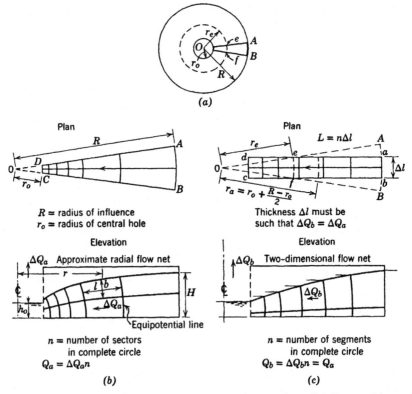

R = radius of influence
r_o = radius of central hole

$r_a = r_o + \dfrac{R - r_o}{2}$

Thickness Δl must be such that $\Delta Q_b = \Delta Q_a$

n = number of sectors in complete circle
$Q_a = \Delta Q_a n$

(b)

n = number of segments in complete circle
$Q_b = \Delta Q_b n = Q_a$

(c)

FIG. 7.9 Use of two-dimensional flow nets to solve certain radial flow problems. (a) Small-scale plan. (b) Circular segment of soil contributing seepage to pumped central hole. (c) Rectangular segment contributing same seepage as circular sector in (b).

Figure 7.9a shows a small-scale plan of the fluid system assumed to exist around the dewatered foundation shown in Figure 7.8. The average radius of the excavation is r_0; the radius of influence is R. The soil that is contributing seepage at a total rate Q toward the sink is a hollow cylinder with an outside diameter $2R$ and a thickness H. Each of n equal sectors similar to AOB (Fig. 7.9a) contributes seepage at a rate ΔQ. *The total quantity Q equal to $n \Delta Q$ must be pumped from the central hole or sink to lower the water level in the hole to height h_o.*

Sector AOB (Fig. 7.9a) is enlarged in Figure 7.9b. The system of flow lines and equipotential lines in Figure 7.9b reveals a typical pattern for radial flow toward a well or other sink. It is shown that flow lines converge both horizontally (upper flow net) and vertically (lower flow net) and that the figures making up the flow net are not squares. Taylor (1948) shows that the characteristic requirement of a radial flow net is that the b/l ratio for each figure (Fig. 7.9b) be inversely proportional to the radius r and that the relationship rb/l be equal for all the figures. In a conventional flow net for two-dimensional parallel flow the ratio b/l is constant and is usually made 1.0.

By trial and error, flow nets can be developed for radial flow to satisfy the requirement that rb/l be equal for all figures; however, the procedure is tedious. Electrical analogy models and sand models have been constructed for this type of flow problem. An approximate analytical method for applying two-dimensional flow nets to three-dimensional flow problems is described in the following paragraphs.

For any circular segment ($ABCD$, Fig. 7.9b) in which a steady rate ΔQ is flowing toward a well there is an *equivalent rectangular segment* (such as *abcd* in the plan view in Figure 7.9c) through which the same quantity ΔQ would flow. The two-dimensional flow net for segment *abcd*, which has parallel instead of radiating sides, is given in the lower diagram of Figure 7.9c. This flow net is composed of square figures and does not exhibit the large head losses near the hole so characteristic of the flow net for radial flow (Fig. 7.9b).

The thickness Δl of the rectangular segment *abcd* that will pass ΔQ with the same head differential $H - h_0$ existing on the circular segment (Fig. 7.9b) can be determined by electrical analogy or sand models if the necessary precautions are taken to minimize errors. Let it be assumed that the distance Δl in Fig. 7.9c has been correctly determined so that the same quantity ΔQ flows in rectangular segment *abcd* as in circular segment $ABCD$ in Fig. 7.9b. The quantity ΔQ must flow across vertical plane *ef* in the plan in Figure 7.9c (and in Fig. 7.9a). A circular arc drawn through points e and f (Fig. 7.9c) is almost identical in length to a chord drawn through these points with a length equal to Δl. The number of sectors n in 360° must be equal to $2\pi r_e/\Delta l$, and the total quantity Q entering the hole at the center must therefore be equal to

$$Q = \Delta Q \frac{2\pi r_e}{\Delta l} = \frac{\Delta Q}{\Delta l} 2\pi r_e \qquad (7.3)$$

As previously noted, the quantity Q is also equal to ΔQ times the number of sectors in $360° = n \, \Delta Q$.

In Eq. 7.3 $\Delta Q/\Delta l$ is the rate of discharge per foot of radial distance and is identical to the unit quantity q determined from the shape factor n_f/n_d for a two-dimensional flow net and Eq. 3.18 (Chapter 3):

$$q = kh \frac{n_f}{n_d} \qquad (3.18)$$

The total quantity Q now becomes

$$Q = kh \frac{n_f}{n_d} L = kh \frac{n_f}{n_d} 2\pi r_e \qquad (7.4)$$

The radius, r_e, of a circle that passes through points e and f (Fig. 7.9a and Fig. 7.9c) is called the *effective radius*. To apply two-dimensional flow nets and Eq. 7.4 to three-dimensional seepage problems it is necessary to establish reasonable values for the *effective radius* r_e. Because of the converging nature of radial flow, the effective radius r_e must always be somewhat less than the *average radius* r_a. Therefore, if the average radius r_a is used in determining the distance $L = 2\pi r$, the calculated pumping rates will be somewhat high. This procedure is believed to be reasonable for most estimates of dewatering rates; however, if two-dimensional flow nets are used for estimating the capacities of water supply wells, r_e can be approximated as $r_0 + 0.30 (R - r_0)$.

To apply two-dimensional flow nets to the foundation represented in Figure 7.8 flow nets were drawn for several values of R; a typical flow net is shown in Figure 7.8c. Calculated pumping rates, using the average radius r_a in lieu of the effective radius r_e as described above, are summarized in Table 7.3. It is seen that the pumping rates agree substantially with the rates obtained with Darcy's law (Table 7.2).

A comparison of Tables 7.1, 7.2, and 7.3 shows reasonably close agreement

TABLE 7.3   Calculation of Pumping Rates with Flow Nets (Excavation in Fig. 7.8)

Assumed R	Radial distance to midpoint of flow net, r_a	Mean length $L = 2\pi r_a$	Shape factor	Total $Q = khL(n_f/n_d)$	
ft	ft	ft	n_f/n_d	cfm	gpm
200	140	880	0.53	147	1100
300	190	1200	0.27	101	760
400	240	1500	0.19	89	670
500	290	1800	0.15	84	630

among the three methods described and good agreement with the actual rate of 800 to 1000 gpm. It is possible that the soil permeability used in these determinations was somewhat low, for a higher permeability would have resulted in close agreement being obtained at larger values of R, which normally would be expected.

If the foundations to be dewatered are relatively long and narrow or irregular in shape or the soil permeabilities and water sources vary materially in various segments of the soil surrounding the excavation, the last two methods (Darcy's law and the flow net) can be used for estimating pumping rates. In these determinations the periphery can be divided into several parts, the quantity entering the excavation in each part calculated by using appropriate values, and the total obtained by adding the parts. Thus, if a high water table is produced by a river near one side of a long excavation and the water table is much lower at the far side, the two sides should be calculated separately, using a higher gradient for the river side than for the far side. Variations in the thickness and permeability of water-bearing formations surrounding excavations can also be treated in this manner. If dependable soil and groundwater data are available and reasonable assumptions are made, Darcy's law and the flow net can be adapted to a wide variety of irregular excavations.

Use of the flow net to estimate quantities of water to be removed from irregular flow systems by deep pumped wells is illustrated with reference to Figure 7.10. Here, buried gravel strata of high permeabilities have direct access to an adjacent river and will be the primary source of seepage from the river to a powerhouse excavation. An existing concrete dam at the site raises the

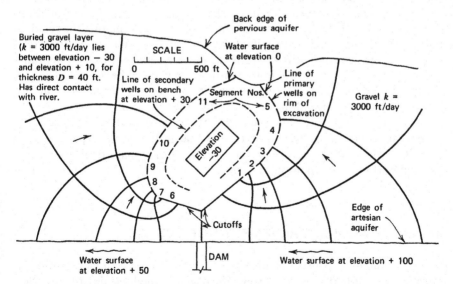

FIG. 7.10 Plan flow net for estimating pumping rate to dewater a deep excavation for a large power house. Artesian flow in buried gravel aquifer.

river to elevation +100 behind the dam (upstream from the dam), whereas the "tailwater" level (downstream from the dam) is elevation +50. It is assumed that two rows of wells will be needed to lower groundwater sufficiently to permit the powerhouse to be constructed "in the dry." For purposes of estimating quantities of seepage to be pumped it is assumed that the average water level along the lines of "primary wells" will be held at elevation 0. It is also assumed that upstream and downstream seepage are separated by an intruding back boundary. If the buried gravel layer is 40 ft thick ($D = 40$ ft) and the effective permeability k is 3000 ft/day (1.1 cm/sec), the estimated total seepage rates into the upstream wells and the downstream wells are as follows:

Upstream. $h = 100$ ft; $D = 40$ ft; $k = 3,000$ ft/day; from the flow net $n_f/n_d = 5/3$

Then $Q = kh(n_f/n_d)D = 3,000(100)(5/3)(40)$
$= 20,000,000$ cu ft/day $= 104,000$ gpm

Downstream. $h = 50$ ft; $D = 40$ ft; $k = 3,000$ ft/day; from the flow net $n_f/n_d = 6/3$

Then $Q = kh(n_f/n_d)D = 3,000(50)(6/3)(40)$
$= 12,000,000$ cu ft/day $= 62,400$ gpm.

For each upstream flow channel $q = 104,000/5 = 21,000$ gpm and for each downstream flow channel $q = 62,400/6 = 10,400$ gpm.

By starting with these quantity estimates and recognizing that the actual number of wells and their capacities will have to be adapted to the actual flows that develop, an initial well system can be designed. Theoretical calculations of head losses around wells in arrays can be helpful in estimating the required spacing of wells and their approximate number and location. Sometimes relatively few large-capacity wells will provide the necessary control; at other times large numbers of wells—often in two or more rows—will be needed. The cost of a system depends not only on the quantity of water to be pumped but also on the difficulty of installing the wells and on their number and size. The quantities of water pumped and the length of time a system must be in continuous operation are often the major factors that influence the type of system selected and its cost. Therefore realistic estimates of seepage quantities are of major importance in estimating the costs of dewatering systems for important projects.

7.2 FOUNDATION IMPROVEMENT BY DRAINAGE

The term *drainage*, as used in this chapter, refers to the artificial removal of water from saturated systems of soil. The drainage methods described in Section 7.1 are used for temporarily controlling groundwater so that construction can be carried on in the dry beneath the normal water level. Because these methods require outside sources of energy, they are seldom used for the permanent stabilization of foundations.

A number of common methods for improving foundations by gravity drainage are discussed in this section.

Surface Drains

When fills are to be constructed over shallow marsh deposits, horizontal drainage blankets can furnish a means of escape for excess water over large areas and thus accelerate consolidation of foundations (Fig. 7.11). If isolated springs or seepage areas occur within foundations, localized blankets, trench drains, or finger drains are often installed. Shallow surface drains are widely used in highway construction and sometimes to improve the foundations of dams and other hydraulic structures. In the past, engineers often used a single layer of aggregate which was expected to provide adequate transmissibility to remove all of the water entering and still prevent piping or the movement of adjacent fine soils into or through the drain.

But, as emphasized throughout this book, obtaining good discharge capacity and good filter protection with a single aggregate material is very difficult, if not impossible. Fortunately, the ready availability of good fabric filters in recent years has opened the way to develop highly effective drains at moderate cost. Excellent drainage with blankets of the kind shown in Figure 7.11 can be obtained by using pea gravel, stone chips, railroad ballast, or other economically available *coarse, clean* aggregate enveloped within suitable filter cloths to prevent clogging.

Collector pipes and discharge pipes are essential features of most drainage systems. Even where a drainage blanket is "daylighted," as shown in Figure 7.11, pipes should almost always be provided, as the outside edges of drains can become clogged or partially clogged by impervious soils washing down on the drains.

Other applications of drains using highly permeable aggregates wrapped within fabric filters are described in Sec. 5.7 under the sub-heading "Applica-

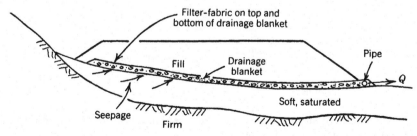

FIG. 7.11 Drainage blanket used for stabilizing shallow foundations.

tions." An example is Figure 5.12c. And the use of these drainage blankets in conjunction with vertical drains is illustrated under "Vertical Drains", later in Sec. 7.2. In all applications where long-term performance is essential to the safety of projects, designers must make sure that any synthetic materials used will have sufficient longevity to function for the expected life of the facility.

In some cases isolated springs which occur in the foundations of fills can be effectively drained by means of small blanket drains, "Finger drains," or other localized drains with outlets (Fig. 7.12b). In other cases a number of springs may occur in a fairly well-defined area. Sometimes a *herringbone* or other pattern of drains can successfully tap a number of sources, as shown in Figure 7.12c.

Just as pipe systems are designed to carry estimated quantities of water, foundation drains should be designed to remove estimated rates of flow. The amount of water that enters these drains can vary from only a few to thousands of gallons per day; hence, the discharge needs vary substantially. The capabilities of several types of narrow drain with the same overall dimensions are illustrated in Figure 7.13. These drains are installed in shallow trenches with about half the material placed below the adjacent ground level and half mounded above the ground level. The drain in Figure 7.13a is built entirely of graded filter aggregate with an in-place coefficient of permeability of 10 ft/

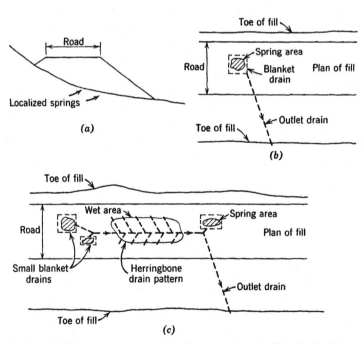

FIG. 7.12 Localized surface drains. (a) Cross section of fill. (b) Draining a single spring. (c) Draining a group of springs.

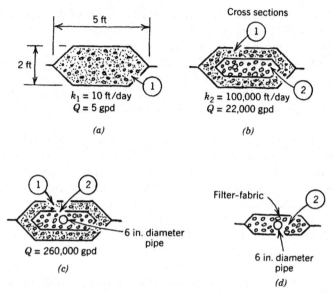

FIG. 7.13 Several types of narrow drain (discharge capacities for $S = 0.01$). (1) is the filter aggregate $k_1 = 10$ ft/day. (2) is the coarse gravel $k_2 = 100,000$ ft/day. (a) All filter aggregate. (b) Coarse gravel enclosed in filter aggregate. (c) 6-in. diameter pipe in gravel. (d) Pipe 6 in. in diameter in gravel enveloped with fabric filter.

day, a class of material often (erroneously) used in highway drains and blankets. According to Darcy's law and on the assumption that the maximum permissible hydraulic gradient in the drain is 0.01, this drain has a discharge capacity of 5 gal/day (0.0035 gpm). Drains of this type can remove only very minor amounts of seepage; however, they are often expected to remove substantial flows.

Improvement can be obtained (Fig. 7.13b) by placing a core of clean washed gravel within a outer ring of filter aggregate. Assuming a permeability of 100,000 ft/day in the core and an allowable hydraulic gradient of 0.01, Darcy's law gives a discharge capacity of 22,000 gpd (15.3 gpm) for this design, more than four thousand times that of the drain constructed with filter material alone (Fig. 7.13a). Drains containing highly permeable cores (Fig. 7.13b) can be built for little more cost than similar drains constructed with a single type of filter aggregate (Fig. 7.13a). Permeable core drains of sound aggregates are suitable for many drainage uses, including foundation drains and outlet drains for dams up to great heights because they can withstand high earth pressures.

This drain can be modified further (Fig. 7.13c) to increase its discharge capacity by the installation of a pipe in the gravel core. If the drain is serving as a collector of seepage, suitable perforations or slots are required to allow the free flow of water into the pipe without the entry of gravel. This drain, which is equipped with a 6 in. diameter pipe, has a discharge capacity of more than 260,000 gpd (180 gpm). Obviously drainage systems containing pipes can be designed to remove extremely large flows.

A still further modification of the line drains in Fig. 7.13 is the substitution of a synthetic fabric filter for the fine aggregate filter surrounding the coarse gravel (Fig. 7.13c), as shown in Fig. 7.13d. If a proper fabric is used (see Sec. 5.7), the discharge capability of this drain will remain at more than 260,000 gpd.

If the drains in Figure 7.13 are required to serve only as conductors to convey water to a discharge point, *without appreciable inflow along the drains,* the enveloping material will need to act only as a *separator* (see Fig. 5.1) to prevent mixing of the adjacent soil with the coarse aggregate in the drain. In such cases the coefficient of permeability of the envelope is not highly important, but if the drains are also serving as *collectors,* with water seeping across the enveloping filter to the interior of the drains, their maximum potential can be assured only if the permeability of the envelope is sufficient to let the water in without excessive head builup across the filter. In order to ensure sufficient inflow capability of drains, the *required minimum permeability* of the enveloping filter can be calculated with Darcy's law, assuming a small hydraulic gradient (and small head loss) across the filter (Fig. 7.13b and c). If a gradient of 0.5 is assumed to be the maximum desirable across the filter, the inflow capacity can be calculated from $q = k(0.5) (aL)$, in which k is the coefficient of permeability of the filter, $i = 0.5$, a is the area of enveloping filter through which water is flowing per unit length of drain, and L is the length of drain through which water is entering across the filter. By rearranging the equation to $k = q/iaL$ the *minimum required permeability* k that will allow quantity q to enter under a hydraulic gradient of 0.5 can be calculated. Table 7.4 lists

TABLE 7.4 **Inflow Capabilities of Filters Enveloping Line Drains of Various Widths.**[a] **For a 100-ft length of drain, $i = 0.5$, and $a = B$; $q = 50 kB$**

k ft/day	Width of drain B (ft)	Inflow capability, q		
		cu ft/day	gpd	gpm
2	5	500	3,740	2.6
	10	1,000	7,490	5.2
	20	2,000	15,000	10.4
10	5	2,500	18,700	13.0
	10	5,000	37,400	26.0
	20	10,000	75,000	52.1
50	5	12,500	94,000	65
	10	25,000	187,000	130
	20	50,000	374,000	260

[a]For designs in Fig. 7.13b and c.

quantities of water than can enter 100 ft of length of drain, assuming that i = 0.5 and that a is equal to the bottom width B of the drains. It is shown that low-permeability filters can severely choke off the flow to drains. The design in Fig. 7.13d eliminates problems of restricted inflows if a good quality, highly permeable filter cloth is used.

When possible, foundations for proposed embankments should be examined during the wettest part of the year, to observe the most unfavorable groundwater conditions. If these conditions are not compensated for in designs, severe damage or deterioration to constructed works can develop.

Vertical Drains

General. Two types of vertical drains are often used for speeding up the consolidation of weak, compressible formations in foundations of fills or for other purposes that require strengthening of the soils. When the drains are filled with columns of sand, they are called "sand drains". When geotextiles are used to form thin, narrow drainage columns, they are often called "wick" drains. As used in this section, the term "vertical drain" refers to types of drains that improve soil formations by allowing excess water to flow upward or downward to a permeable layer (either natural or man-made) that allows the water to flow freely to outer edges of areas being stabilized. The well-known sand drain is the forerunner of the newer "wick" drain.

When structural loads are placed on saturated, compressible soils, the loads for a time produce *excess pore pressures* in the water. These excess pore pressures produce hydraulic gradients in the water that force the excess water out of the soil pores. The soil behaves much like a sponge that is squeezed. In the beginning the new pressure is carried entirely by the water in the pores. When all excess water has been squeezed out, the load is carried on the mineral skeleton. If a soil or sponge has large pores, water will flow quickly from them, allowing new loads to be transmitted to its skeleton in a short time, but if the pores are small, as in fine silts and clays, the transfer of stresses goes on over a long period of time.

The process by which the water content of saturated soils decreases under new loads is called the *process of consolidation.* Terzaghi (1925, 1943) developed his well-known *theory of consolidation,* which permits estimates to be made of the rates of consolidation of saturated foundations under load.

According to the basic theory of consolidation, the time for a given *degree of consolidation* or *percentage of consolidation* varies (1) inversely with the permeability of the soil and (2) directly as the square of the longest drainage path.

Dimensionless *time factors* facilitate computations of rates of consolidation of foundation layers of known thickness when certain data are obtained in laboratory consolidated tests.

The permeabilities of natural soil deposits must be accepted as they are given by nature, but frequently something can be done to shorten the drainage path and thus speed up the consolidation of foundations. Daniel E. Moran obtained a patent in 1926 for his invention of the vertical *sand drain* or *drain well* (U.S. Patent No. 1,598,300), which speeds up the consolidation of foundations. A sand drain system is commonly composed of a group of vertical holes that are backfilled with clean filter sand or pea gravel; although closely spaced, small-diameter drains of various kinds of porous material are being used as substitutes for the conventional sand drain.

Much of the credit for the development of sand drains in the United States can be given to the California Division of Highways, which pioneered them (Porter, 1938). Sand drains have been used widely for stabilizing foundations for highways and to a lesser extent for other specialized foundation problems; for example, they have been employed for preconsolidating mud foundations under sewage treatment tanks and other structures. To some extent they have been used for consolidating the foundations for dams. In the treatment of dam foundations a zone beneath the impervious core must be left untreated; otherwise excessive leakage can be caused by the drain system. Sherard et al. (1963) describe the sand drain installations used for consolidating the foundations of the Boundary Dam in Saskatchewan, Canada, and the Rough River Dam, Louisville District, U.S. Corps of Engineers. Both drain systems not only reduced pore pressures during construction but have the important secondary function of serving as relief wells to control underseepage during the operation of the reservoirs.

Vertical drains have been used for stabilizing soft formations for many kinds of conditions and purposes. Billings (1985) describes a sand drain system used to consolidate the soft clay core of an earth fill dam (San Pablo Dam in Calif.), built by hydraulic fill methods in 1919–1921. Its core was still very soft, and the dam would have been succeptible to slumping in an earthquake more than 60 years later, according to Billings. The treatment included the installation of vertical sand drains in the clay core to bleed pore pressures that could be induced by earthquakes upward to a 30-ft thick sand blanket placed on the upstream slope prior to the placement of a thick earth buttress on the entire upstream slope. Vertical sand vents will bleed the water from the drains upward through the buttress fill and into the reservoir. These procedures will raise the factor of safety during strong earthquakes to allowable levels, says Billings.

The usual method for installing sand drains is to drive a hollow pipe or mandrel with a hinged bottom lid to form a hole by displacing the soil as the mandrel is driven. Then, the mandrel is carefully withdrawn while sand is poured into the cavity left by the mandrel, to form a column of sand to serve as a drain. Sand drains are commonly installed at horizontal spacings of 6 to 10 feet, with a 2 to 3-ft thick layer of sand spread over the entire area to collect the water and conduct it to the sides. Morrison (1982) points out that this method produces large disturbances of the soil around each drain, reducing its permeability, thereby reducing the flow of water to a drain and the efficiency of a system. Because of that and other problems in installing the sand drains, new "wick" drains are virtually eliminating the conventional sand drains. He says that European and Japanese engineers "have known for the past two decades or so—prefabricated wick drains can do the job more efficiently and more inexpensively than sand drains." Since their introduction in the U.S. in 1978, wick drains have claimed as much as 80% of the soil consolidation market (in 1982). In a review of the development of the vertical foundation drains, Morrison points out that about the same time that the sand drains began to be used in the U.S., the director of the Swedish Geotechnical Institute, Walter Kjellman, developed a prefabricated cardboard strip type of drain which became known as the Kjellman Wick. Cardboard proved to be an inferior material because of its poor permeability and low strength when wet. So, when plastics became available for geotechnical purposes, prefab drains using polyethelene or PVC with paper filters came into use in both Europe and Japan. More than 50 types of plastic vertical drains were on the market in 1982, according to Morrison. He emphasizes that modern wick drains offer great advantages over the conventional sand drains. They have many channels to conduct water, good conductivity, and good wet strength. No water is needed during installation, as most plastic wick drains can be put in with a conventional vibratory hammer along with a special fabricated wick drain installation rig—to depths up to 100 feet, which increases their range of suitability. Because of their smaller size, they are installed with much closer spacings than sand drains; however, they are said to be faster to install and more economical than sand drains. Because of their ease of construction, they have opened up new areas for development where sand drains or more costly alternatives had been economically infeasible.

Figure 7.14 shows an interesting application of wick drains. Morrison explains that at a navigation lock and weir project located in a delta area near

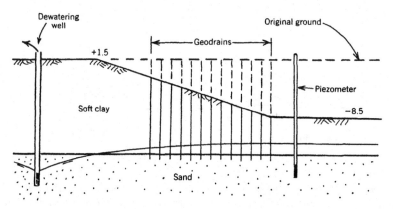

FIG. 7.14 Cross section through excavation using wick drains (and deep pumped wells) to improve strength of soft clay. Reprinted from *Civil Engineering,* (March, 1982), with permission, ASCE.

Basrah, Iraq, very soft clay deposits 15 m or more in thickness had to be stablized before the excavation could be started. Deep dewatering wells were used to lower the water pressure in a sand layer under the clay, reducing uplift pressures beneath the planned excavation area. Also, the presence of an underlying permeable sand layer allowed drainage to take place vertically downward through the wick drains (also called "geodrains") to discharge in the sand. The pumping neutralized uplift in the clay, increasing its effective unit weight from "submerged" to "wet", thereby increasing the rate of consolidation of the clay itself. It was virtually the same as putting on a surcharge, as is done more conventionally. The vertical drains also aided in accelerating the rate of consolidation. Morrison says that about 70% of the total ultimate settlement of the clay took place in about four months, whereas without the wick drains and the pumping from the underlying sand layer (with deep wells), more than three years would have been needed. Thus, the need for a surcharge with upward (conventional) drainage was eliminated, greatly expediting the construction. This example is a notable illustration of the benefits of "bottom drainage", which is the most efficient kind of drainage when it can be used (see also Chap. 9 for other examples of bottom drainage.)

Seim et al. (1981) and Hannon (1982) describe a major highway construction project in California that made use of wick drains to accelerate the consolidation of soft bay muds in the foundations of approach fills for the new Dumbarton Bridge at the south end of the San Francisco Bay. As shown in the simplified cross section in Figure 7.15, fabrics were used to prevent failures of initial fills, sawdust was used for the lower parts of main fills to reduce weight, and wick drains were used under the primary part of the main fill. An eight inch thick aggregate drainage blanket constructed of one inch to No. 4 aggregate allowed free exiting of consolidation water flowing out of the wick drains. The initial work had to be done in open water 3 to 15 feet deep.

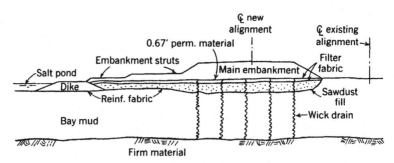

FIG. 7.15 Cross section through approach embankment for new Dumbarton Bridge, using wick drains, fabric, and sawdust fill to aid construction over soft mud foundation. [After Seim, Walsh, and Hannon (1981), "Wicks, Fabrics, and Sawdust Overcome Thick Muck"]. Reprinted from *Civil Engineering,* (July, 1981), with permission, ASCE.

The soft bay muds were 10 to 40 feet deep. A Swedish hydraulically driven wick driving machine was mounted on a backhoe for a test installation prior to the main construction. For the main work the contractor used a two-wick-driving machine using dual mandrels suspended from wide-track 70-ton cranes. Vibratory hammers drove the wick drains through the fill, drainage layer, and fabric to the bottom of the soft mud. Excess pore pressures and settlements were carefully monitored throughout the construction and consolidation periods. As may be seen in Figure 7.16, the actual time for consolidation was about a year, whereas the theoretical time without the drains was around 50 years. This was the first wick drain job in California (Hannon and Walsh, 1982). Because of the success of projects like this one, the wick drains have virtually eliminated the old sand drain.

Long (1986) discusses wick drain systems and comments that careful monitoring of excess pore pressure in soft foundations, and the amounts of settlement occurring as the work progresses are essential if failures are to be avoided. If pressures rise too fast the rate of loading of fill must be reduced until pore pressures have dropped off enough to allow more fill to be placed.

Often, when vertical drains are used, carefully determined slow rates of fill

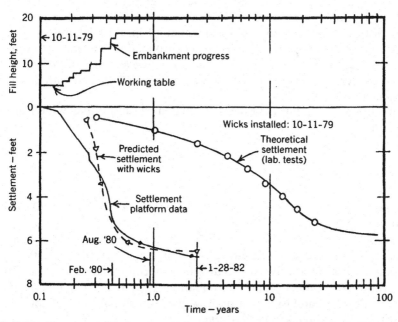

FIG. 7.16 Time-consolidation curves for embankment using wick drains to consolidate soft foundation [After Hannon and Walsh (1982), "Wick Drains, Membrane Reinforcement, and Lightweight Fill for Embankment Construction at Dumbarton," Transportation Research Record 897, Jan., 1982, pp. 37–42]. With permission of Transportation Research Board, National Research Council, Washington, D.C.

placement are specified to avoid overstressing weak foundations. Struts or berms are also employed when necessary to resist side failures. To increase the benefits of accelerated consolidation from vertical drains *surcharge* loads are often placed above the required fill level and any excess material is removed after settlement has leveled off.

Determination of Drain Spacings. Barron (1948) applied the theory of consolidation to drain wells while working on the design of Corps of Engineers' projects which had difficult foundation treatment problems. The solutions developed by Barron include the effect of well smear (remolding of soil adjacent to well) and well resistance (retarding of seepage) on the rate of consolidation with drain wells.

Approximate computations of the probable rates of consolidation with drain wells can be made by using the consolidation drainage path concept, which can be expressed as

$$\frac{t_1}{t_2} = \frac{h_1^2}{h_2^2} \tag{7.5}$$

In this expression t_1 is the time for a clay layer with a maximum drainage path h_1 to reach a given degree of consolidation and t_2 is the time required for a layer of the same soil with a maximum drainage path h_2 to attain the same degree of consolidation.

To illustrate the use of Eq. 7.5, assume that a 1-in.-thick laboratory consolidation sample ($h_1 = 0.5$ in.) reaches 90% consolidation in 60 min. How much time would be required for a 50-ft-thick layer of the same soil to reach the same degree of consolidation without drains ($h_2 = 25$ ft) and with drains at 8-ft spacing ($h_2 = 4$ ft)?

Without vertical drains the time is calculated by substituting numerical values in Eq. 7.5:

$$\frac{1 \text{ hr}}{? \text{ hr}} = \frac{(0.5 \text{ in.})^2}{(300 \text{ in.})^2}$$

The time for the 50-ft layer to reach 90% consolidation is (1 hr) $(300/0.5)^2$, which is 360,000 hr, or 41 years.

If vertical drains are installed, their influence on the rate of consolidation can be approximated by substituting values in Eq. 7.5, recognizing that this equation applies specifically to two-dimensional consolidation and that drain wells produce three-dimensional consolidation. Because Eq. 7.5 does not allow for concentrated head losses near drain wells, estimates of times of consolidation with wells will tend to be *too small*.

Assuming that drain wells are installed at 8-ft centers in this foundation, the maximum drainage path is reduced from 25 to 4 ft, or 48 in. Using Eq. 7.5, the time to reach 90% consolidation is approximately 1 hr $(48/0.5)^2$, which is 9200 hr or 1.05 years. Even though this determination is approximate, it is logical to conclude that without treatment this foundation will continue to settle for 40 or 50 years, whereas with vertical drains at 8-ft centers it will be well consolidated in only a few. More dependable estimates can be made with Barron's solutions, which take into consideration concentrations in head produced by radial flow and the effects of well smear and drain resistance.

Smear and drain resistance have in some instances slowed the rate of consolidation to the point at which the usefulness of sand drain systems has been largely nullified. On one important project theoretical studies, neglecting smear and head losses in the drains, indicated that drains at 10-ft spacing would permit embankment to be placed with the use of struts to a height of 35 ft in about 12 months. Fill placement was started after the installation of a sand drain system which discharged into a sand blanket spread over the entire foundation area. The fill failed disastrously after being placed to a height of only 18 ft with greatly widened struts. Smear and drain resistance are believed to be the primary factors that slowed down the rate of consolidation to a fraction of the expected rate.

If consolidation takes place primarily by radial flow to vertical wells, the time t for a given degree of consolidation U_r can be obtained from the equation

$$t = T_r \frac{d_e^2}{c_v} = \frac{T_r \gamma_0 a_v d_e^2}{k_h(1 + e)} \tag{7.6}$$

Where T_r is the dimensionless *time factor* for radial consolidation, γ_0 is the unit weight of water, a_v is the *coefficient of compressibility* equal to $\Delta e/\Delta p$, d_e is the diameter of the equivalent cylindrical volume of soil being drained by the well, k_h is the horizontal permeability of the soil, e is the void ratio of the soil, and c_v is the *coefficient of consolidation* equal to $k_h (1 + e)/\gamma_0 a_v$.

Hannon (1986) describes wick drain installations and gives the following formula developed from work of Barron, Kjellman and Fellenius for radial drainage to vertical wick drains to estimate the required time for consolidation for a given installation with a specified drain spacing:

$$t = (D^2/8C_h) [\ell n(D/d_e) - 0.75] \, \ell n \, [1/(1 - \overline{U})] \tag{7.7}$$

in which: t = time,

   C_h = horizontal coefficient of consolidation,

   D = zone of influence of a drain,

   d_e = equivalent diameter of a drain,

   $\overline{U}$ = average degree of consolidation

On spacings of three to six feet, wick drain installations have given good performance in many locations.

Vertical drains are most effective for consolidating clays and organic soils that contain horizontal partings of silt or sand that feed water to the vertical drains. These beds can be of great assistance in the consolidation of the overall formation if they can be intercepted by vertical drain wells. Meardi (1957) reports a project that used vertical sand drains for the stabilization of a soft clay foundation containing many horizontal layers of pervious shells. He concludes that the presence of the shell layers greatly increased the benefits of the sand drains in consolidating the foundation.

Deep mud deposits present the soils engineer with some of the most difficult foundation problems. Thick deposits of impermeable muds have reached their present densities and strengths by accumulation and drainage processes that have gone on for decades or centuries. They resist rapid changes. Vertical drains can speed up drainage processes and often make mud deposits more suitable for engineering construction. If economical drain systems combined with flattened side slopes to prevent failures and surcharge loads cannot make poor foundations perform as required, other more drastic choices confront the designer. The following are several alternatives to be considered:

1. Move the project to another location in which foundation conditions are more suitable.
2. Prepare to have the unsatisfactory foundation soil stripped or removed by forced displacement.
3. Design structures that rest on piles or caissons embedded firmly in non-compressible underlying strata.

Head Losses in Drains and Discharge Blankets. When sand drains are used for consolidating weak, compressible foundations, the head losses in drains and discharge blankets can reduce the effectiveness of the systems unless exceptionally clean, permeable materials are used for the drain backfill and discharge blankets. Using Darcy's law, the writer developed equations for estimating the minimum permeabilities needed for the backfill and for blankets, to keep head losses low enough to avoid serious interference with the discharges (see Cedergren, 1967, 1977).

When prefabricated wick drains are to be used, specifications should require suppliers to warranty that the drains to be supplied will have the necessary discharge capacity (transmissibility) to remove estimated flow rates with head losses (in the wick drains) below specified small amounts (such as two or three ft of head loss/100 ft of length of drain).

When discharge blankets for vertical drain systems are constructed with six inches or more of pea gravel or other open-graded aggregates in the size range of 1/4 to 1 inch, covered on top and bottom with suitable filter fabric, the drainage capabilities should be many times greater than needed. But, careful monitoring of pore pressures and settlement rates should always be required during fill placement and consolidation periods to make sure that the systems are performing as expected. When used with adequate care, the new geotextile drainage products can be a big boon to the construction of fills and other engineering facilities on poor foundations.

REFERENCES

Avery, Stuart B., Jr. (1951), "Analysis of Ground-Water Lowering Adjacent to Open Water," American Society of Civil Engineers Separate No. 106, December 1951, pp. 1-16.

Barron, R. A. (1948), "Consolidation of Fine-Grained Soils by Drain Wells," *Transactions,* A.S.C.E., Vol. 113, pp. 718-754.

Billings, H. L. (1985), "Hydraulic Fill Dam Made Earthquake Resistant," *Civil Engineering,* ASCE, June, 1985, pp. 56-59.

Casagrande, Leo (1947), "The Application of Electroosmosis to Practical Problems in Foundations and Earthworks," Department of Scientific and Industrial Research, Building Research Tech. Paper 30, London (H.M. Stationery Office).

Casagrande, L., R. W. Loughney, and M. A. J. Matich (1961), "Electro-Osmotic Stabilization of a High Slope in Loose Saturated Silt," *Proceedings,* 5th International Conference on Soil Mechanics and Foundation Engineering," Paris, July 17-22, 1961, Vol. II, pp. 555-561.

Cedergren, Harry R. (1967), *Seepage, Drainage, and Flow Nets,* John Wiley & Sons, Inc., N. Y., 1st Ed., pp. 291-297; (1977), 2nd Ed., pp. 324-329.

Engineering News-Record (1975), "Slurry-Trench Cutoff Wall Pierces Landslide Debris to Keep Site Dry," November 13, 1975, p. 53.

Esrig, Melvin I., and Steven Majtenyi (1965), "A New Equation for Electro-osmotic Flow and Its Implications for Porous Media," Technical Pub. TP64-1, School of Civil Engineering, Cornell University, Ithaca, N.Y., paper presented at the 44th Annual Meeting of the Highway Research Board, Washington, D.C., January 1965.

Freundlich, H. (1926), "Colloid and Capillary Chemistry," translated from the 3rd German edition by H. S. Hatfield, London (Methuen and Co.).

Glossop, R., and A. W. Skempton (1945-1946), "Particle-Size in Silts and Sands," *Journal of the Institution of Civil Engineers,* London, Vol. No. 25, (2), pp. 81-105.

Glover, R. E. (1966), "Ground-Water Movement," U.S. Department of the Interior, Bureau of Reclamation, Denver, Engineering Monograph 31. 2nd printing.

Gray, D. H., and J. K. Mitchell (1967), "Fundamental Aspects of Electroosmosis in Soils," *Journal,* Soil Mechanics and Foundation Engineering Division, *Proceedings,* A.S.C.E. 93, No. SM6, pp. 209-236.

Hannon, J. and Thomas J. Walsh (1982), "Wick Drains, Membrane Reinforcement, and Lightweight Fill for Embankment Construction at Dumbarton," Transportation Research Record 897, TRB, Jan. 1982, pp. 37-42.

Hannon, J. (1982), "Fabrics Support Embankment Construction Over Bay Mud," Proc. 2nd Int. Conf. on Geotextiles, Las Vegas, Nevada, Aug. 1-6, 1982, pp. 653-658.

Hannon, Joseph (1986), "Wick Drain Selection and Design," *Transportation Research Circular,* TRB, No. 309, Sept. 1986, pp. 6-7.

Hantush, Mahdi S. (1963), "Hydraulics of Gravity Wells in Sloping Sands," *Transactions,* A.S.C.E. Vol. 128, Part I, pp. 1423-42.

Harr, Milton E. (1962), *Groundwater and Seepage,* McGraw-Hill, New York.

Helmholtz, H. (1879), "Studien uber elektrische Grenzschichten," *Wiedemanns Annalen der physik und chemie 7,* Leipzig.

Leonards, G. A. (1962), *Foundation Engineering,* McGraw-Hill, New York, Chapter 3.

Long, Richard P. (1986), "Wick Drains—An Overview," *Transportation Research Circular,* TRB, No. 309, Sept. 1986, pp. 2-6.

Meardi, G. (1957), "Consolidation with Sand Piles of Soft Clays in Levee Foundation," *Proceedings,* 4th International Conference on Soil Mechanics and Foundation Engineering, London, August 12-24, 1957, Vol. II, pp. 334-337.

Morrison, Allen (1982), "The Booming Business in Wick Drains," *Civil Engineering,* ASCE, March, 1982, pp. 47-50.

Muskat, M. (1937), *The Flow of Homogeneous Fluids Through Porous Media,* McGraw-Hill, New York; also lithoprinted by Edwards Brothers, Ann Arbor, Michigan, 1946, p. 149; (1937a), Chapter 9.

Peterson, Dean F., Jr. (1961), "Intercepting Drainage Wells in Artesian Aquifer," *Transactions,* A.S.C.E., Vol. 127, Part III, p. 32-42.

Porter, O. J. (1938), "Studies of Fill Construction over Mud Flats Including a Description of Experimental Construction Using Vertical Sand Drains to Hasten Stabilization," *Proceedings,* Highway Research Board, Vol. 18, Part II, pp. 129-141.

Powers, J. Patrick (1976), "Notes on the Design and Selection of Wellpoint Systems," Short Course and Seminar on Groundwater Analysis and the Design of Dewatering Systems," University of Missouri, Rolla, at St. Louis, January 12-17, 1976.

Powers, J. Patrick (1981), *Construction Dewatering: A Guide to Theory & Practice,* John Wiley & Sons, Inc., N. Y.

Reuss, F. F. (1809), "Sur un nouvel effet de l'electricite galvanique," *Proceedings,* Imperial Russian Naturalist Society, Moscow, Vol. 2, pp. 327-337.

Seim, Charles, T. J. Walsh, and J. B. Hannon (1981), "Wicks, Fabrics, and Sawdust Overcome Thick Muck," *Civil Engineering,* ASCE, July, 1981, pp. 53-56.

Sherard, James L., R. J. Woodward, S. F. Gizienski and W. A. Clevenger (1963), *Earth and Earth-Rock Dams,* Wiley, New York, pp. 110-111; (1963a), pp. 444-445.

Stang, John W., Jr. (1976), unpublished "Dewatering Article."

Taylor, D. W. (1948), *Fundamentals of Soil Mechanics,* Wiley, New York, pp. 192-194.

Terzaghi, K. (1925), Erdbaumechanik auf bodenphysikalischer, Grundlage, Vienna, Deuticke.

Terzaghi, K. (1943), *Theoretical Soil Mechanics,* Wiley, New York, Chapter 13.

Terzaghi, Karl, and Ralph B. Peck (1948), *Soil Mechanics in Engineering Practice,* Wiley, New York, p. 336; (1948a), pp. 337-339.

Todd, David K. (1959), *Ground Water Hydrology,* Wiley, New York.

U.S. Departments of the Army, Navy, and Air Force (1971), TM 5-818, NAVFAC P-

418, and AFM 88-5, Chapter 6, "Dewatering and Groundwater Control for Deep Excavations," p. 17; (1971*a*), pp. 16–17; (1971*b*), pp. 90–170.

Winterkorn, H. F., and W. E. Schmid (1971), "Soil Stabilization Parameters," Technical Report No. AFWL-TR-70-35, Kirtland Air Force Base, New Mexico.

Zangar, Carl N. (1953), "Theory and Problems of Water Percolation," U.S. Department of the Interior, Bureau of Reclamation, Denver, Engineering Monograph 8.

CHAPTER EIGHT

SLOPE STABILIZATION WITH DRAINAGE

8.1 GENERAL CONSIDERATIONS

Natural earth slopes usually stand on the steepest slope that Nature allows, unless they have been flattened by erosion and weathering processes. The maximum steepness and height a given slope can tolerate depends primarily on its shearing strength. Unjointed, unfissured, massive rocks often stand on vertical cliffs of great heights. A notable example is El Capitan, which rises vertically 4,600 feet above Yosemite Valley in central California. Durable rocks that are heavily cracked and jointed can stand to almost unlimited heights on slopes of about 35 degrees (1 ft vertically to 1.4 ft horizontally) in wet climates where rainfall produces seepage with destabilizing effects. In dry areas with hardly any rainfall and no groundwater (except at great depths), such materials could theoretically stand on a slope of about 40 degrees. But, weathered serpentine and other clay-like formations often are unstable on slopes as flat as six degrees to eight degrees. Because many natural slopes are barely stable, with factors of safety little above 1.0, any unusual actions that increase destabilizing forces or reduce strength—such as strong earthquakes, unusually heavy rainfall, broken water lines, loading the upper parts of slopes, or undercutting lower parts—can trigger slipouts or landslides. Improving the stability of earth slopes with drainage (often in combination with other techniques) is one of the most important activities of civil engineers.

Importance of Stable Slopes

Uncontrolled releases of energy cause most of the physical catastrophes that take large numbers of lives and cause heavy property damage. The sudden transfer of potential energy into energy of motion (kinetic energy), when earth masses break loose from mountainsides, can be very destructive. A single large landslide of modern times developed the momentum of 100,000 high-speed freight trains and consumed energy at a rate of at least 20 billion horsepower, over 2,000 times the power output of Grand Coulee Dam.

Stable earth slopes, both natural and man-made, are of great importance to mankind as demonstrated by the examples of slope failures given in Table 8.1.

TABLE 8.1   Common Examples of Slope Failures

Kind of slope	Conditions leading to failure	Type of failure and its consequences
Natural earth slopes above developed land areas (homes, industrial)	Earthquake shocks, heavy rains, snow, freezing and thawing, undercutting at toe, mining excavations	Mud flows, avalanches, landslides; destroying property, burying villages, damming rivers
Natural earth slopes within developed land areas	Undercutting of slopes, heaping fill on unstable slopes, leaky sewers and water lines, lawn sprinkling	Usually slow creep type of failure; breaking water mains, sewers, destroying buildings, roads
Reservoir slopes	Increased soil and rock saturation, raised water table, increased buoyancy, rapid drawdown	Rapid or slow landslides, damaging highways, railways, blocking spillways, leading to overtopping of dams, causing flood damage with serious loss of life
Highway or railway cut or fill slopes	Excessive rain, snow, freezing, thawing, heaping fill on unstable slopes, undercutting, trapping groundwater	Cut slope failures blocking roadways, foundation slipouts removing roadbeds or tracks, property damage, some loss of life
Earth dams and levees, reservoir ridges	High seepage levels, earthquake shocks; poor drainage	Sudden slumps leading to total failure and floods downstream, much loss of life, property damage
Excavations	High groundwater level, insufficient groundwater control, breakdown of dewatering systems	Slope failures or heave of bottoms of excavations; largely delays in construction, equipment loss, property damage

Avalanches, landslides, and sudden slope failures take a large toll of life, frequently wiping out entire villages in a few moments. When the slope of Mt. Toc suddenly dropped into Vaiont Reservoir in October 1963, a wall of water several hundred feet high splashed down the canyon, destroyed several Italian villages, a town of 4600, and drowned more than 3000 persons (see Fig. 8.6).

Much less spectacular than these sudden failures are the countless slow moving slope failures that damage railroads, highways, excavations, subdivision, and other improvements. These slides usually take but few lives but add substantially to the total economic loss cause by unstable slopes, which in the United States alone amounts to many hundreds of million of dollars a year. Slides that took place around the edges of the reservoir created by Grand Coulee Dam (in about 1950) cost property owners and taxpayers at least $20,000,000 (Fig. 8.1).

Water is almost always the agent that contributes to the failure of natural earth slopes and the slopes of dams, levees, highway and railway cuts and fills,

FIG. 8.1 Massive landslide at edge of reservoir of Grand Coulee Dam, caused by rising and spreading saturation from reservoir water; is approaching highway on right (about 1950).

and other types of engineering works; therefore a study of seepage and its control would not be complete without a chapter on the stabilization of slopes by drainage. The importance of drainage in the control of landslides is emphasized in the Highway Research Board's excellent volume on landslides by Baker and Marshall (1958) in the statement that "drainage is without question the most generally applicable corrective treatment for slides." They also point out that drainage methods are the only economical choice for the treatment of very large slides or flows.

In a recent publication, Zaruba (1987) gives good descriptions of major types of slope failures, and recommends improving subsurface drainage in potentially unstable slopes by means of drainage galleries, wells, and horizontal drains (see also Sec. 8.3).

Many natural slopes slide because of water added by man-made activities. An initial small movement in an unstable area may cause the rupture of water mains, sewer pipes, swimming pools, water storage reservoirs, and so on with water from these broken supplies flowing into the soil and increasing the height of saturation. In such cases the added water may be sufficient to set severe landslides in motion. Figure 8.2 shows an expensive home near Sacramento, California, teetering on the edge of a collapsing bank that slid out from under the house after being soaked by water leaking out of broken water pipes (Sacramento *Union,* Jan. 24, 1988, p. A-1; Jan. 27, 1988, p. A-6). Water pipes in the area had been leaking for years, and were gradually being replaced. A spokesman for the water district serving this house said, "We just didn't get in there in time."

Railroads in many parts of the world are constantly fighting slope instability problems, which are often aggravated by wet weather and water. Areas that have remained stable for several "normal" or "dry" years may suddenly collapse during an unusually wet period. Because of the steepness of many of the slopes railroads must traverse, the stability is in a precarious state most of the time. Anything "unusual" can trigger slides. Figure 8.3 shows an area of a major Western railroad in extremely steep terrain. Although the steepness of the slope, combined with the soaking of water into the grade of the track, has caused a slow creep (evidenced by the thick ballast in the foreground), there has been no major failure here.

Excess pore water pressures cause many of the failures of deep-cut slopes in open-pit mines. Gloe et al. (1971) say that among the causes of trouble in very deep open cuts for brown coal in Australia are *piezometric pressures* that build up in the coal and in underlying sand aquifers. To evaluate the effects of the water pressures and forestall failures *piezometers* and *extensometers* have been installed in these cuts to depths of 720 ft (220 m) in areas in which high water pressures and temperatures have created problems.

The need for control of seepage in constructed slopes of dams and other engineering works is widely recognized. Many of the failures in such works have been caused by lack of control over internal water. Of the 206 earth

FIG. 8.2   Water leaking from broken water pipes caused mud slide that left this house near Sacramento, California teetering over the collapsed bank. (Courtesy, The Sacramento *Union*/Bob Moore). (Jan. 1988).

dam failures listed by Sherard et al. (1963) two of every three were caused by uncontrolled seepage. Good drainage is doing much to make many engineering works possible; it is an important factor in the stability of earth slopes.

Factors Influencing Slope Stability

The forces of gravity are constantly pulling downward on all soil and rock, flattening slopes at every opportunity (Table 8.1). Likewise, the pull of gravity on the water in soil and rock produces seepage forces and pore pressures that

FIG. 8.3 Unstable slopes are a major problem for many railroads. A steep slope with a "creep" problem. (Printed with permission of The Western Pacific Railroad Company.)

aid the tearing down processes. Earthquakes produce additional destructive actions, and human activities often lead to landslides.

The degree of stability of a given slope can vary widely, depending on the conditions that exist in it at a given time. Dry or well-drained slopes are the most stable and those with high water levels, the least, as evidenced by the large number of serious landslides that occur during or after heavy rainstorms. Furthermore, the basic strengths and densities of soil and rock formations have a major influence on stability during severe earthquake shocks. Severe landslides occasionally take place in dry, loose, silty soils, but the vast majority occur in *saturated soil during periods of heavy rainfall or during severe earthquakes.*

When subsurface investigations are being scheduled for highways and railways (and any other important facilities) in hilly regions, efforts should be made to have the work done when it can detect the worst wet-weather conditions—groundwater levels, perched water tables, etc.—that can produce slipouts and other water-related problems during construction or after the projects have been completed. If explorations are made in dry summer months (which is often the practice), severe wet-weather conditions can go undetected unless field operations are also made during the wet periods. Observation wells should be installed while the field explorations are being made, and groundwater conditions observed periodically throughout at least one entire wet season. Careful, frequent examinations should also be made for all surface signs of springs, seeps, etc. that can give clues to bad conditions that will develop in wet seasons.

If potentially harmful conditions are not detected either during the explorations, during follow-up examinations, or during construction, serious problems can develop later. Figure 8.4 shows streams of water emerging from the face of a highway cut in northern California the first wet season after a freeway had been built in dry weather. Water is flowing freely out of a porous seam that was not discovered before or during construction. Little damage beyond slight erosion is occurring here, but at another location where a wide embankment had been placed over a similar seam (without adequate drainage), water trapped behind the fill caused large hydrostatic pressures to build up behind the fill, literally "floating" it down the slope (Fig. 8.5), closing an important freeway for many weeks while major repairs were made under difficult conditions.

Sometimes very adverse conditions within slopes are almost impossible to

FIG. 8.4 Streams of groundwater emerging from face of highway cut. (Similar streams, blocked by a fill, caused collapse of the highway: see Fig. 8.5) (Courtesy of the California Division of Highways.)

FIG. 8.5 Groundwater, trapped by fill during a period of unusually heavy rainfall, caused collapse of this highway. (Courtesy of the California Division of Highways.)

detect by routine explorations. I have examined a number of slopes that were still sliding although the slopes were as flat as six or eight degrees. Careful examination of soil samples taken at the depth of the slide movement revealed the presence of grease-like slip planes where slickensides had been caused by the slide movement. The sliding action had actually changed the composition of the material at the slip planes, reducing its coefficient of friction to only a fraction of the original strength. Even though the slopes were very flat, the masses were still unstable and started to move when very small cuts were made at their lower edges. Brandl (1976) tells about slides in Europe where very wet, flat slopes in weathered materials were undergoing progressive failure because of very small residual angles of internal friction at polished slickensided surfaces.

The general degree of stability of earth slopes under a range of physical conditions is shown in Table 8.2, which lists several combinations of conditions in decreasing order of stability.

Table 8.2 is a purely qualitative comparison of stability under the various conditions listed, with an exact position in the scale depending on detailed conditions. By reference to Table 8.2 the most unfavorable conditions should

TABLE 8.2   General Degree of Stability of Slopes

Decreasing order of stability	Conditions
7	Dry slopes, without earthquake
6	Dry dense slopes, with earthquake
5	Saturated, favorable seepage, no earthquake
4	Saturated, favorable seepage, with earthquake
3	Saturated, unfavorable seepage, no earthquake
2	Saturated, unfavorable seepage, *dense* soil, with earthquake
1	Saturated, unfavorable seepage, *loose* soil, with earthquake

be expected in saturated, loose earth deposits subjected to heavy earthquake shocks. This is, indeed, a combination of conditions that is present in many of the severest landslides in nature. Loose, saturated soil formations are often severely remolded by the shaking and shearing actions of earthquakes, and undergo large reductions in strength that lead to major landslides. Just the presence of large amounts of water and high water levels in slopes have caused many slides. Unusually heavy rainfalls that occurred in the Yukon Territory in Alaska between 500 and 1950 radiocarbon years ago triggered some catastrophic landslides. Clague (1981) says they were caused "on steep slopes in pervasively fractured and faulted rocks, with an abundance of talus and glacial sediments available for remobilization as debris flows and debris torrents and *the occurrence of intense rainstorms.*"

The sudden drawdown of water levels in reservoirs has caused many landslides because of the reduction of restraining water pressures against the saturated materials in the reservoir slopes. Fortunately these slides seldom endanger human life, because there is little water left in a reservoir after a major drop in level. But they can be costly, as they can damage facilities around a reservoir and are expensive to clean up. Schuster and Embree (1980) say that the rapid dropping of the water level caused by the sudden failure of Teton Dam in Idaho on June 5, 1976, "produced an extreme example of the 'rapid drawdown' condition in a . . . 17-mile long . . . reservoir." Approximately 3.6 million cu yds of soil and rock slid from the canyon walls after failure of this dam, according to these writers.

Preventing the flow of surface water into potential or existing landslide areas can be helpful in slowing down or preventing slide movement. Methods that are used to reduce surface water inflows into slide areas include the construction of lined ditches above an area to divert surface runoff around the area. Also, the filling in of tension cracks at upper portions of slides, and contouring to eliminate depressions that trap water can reduce surface water inflows. Unloading at the head of a slide area; the building of drained buttresses (see Fig. 8.22) or retaining walls or crib walls, etc. along lower edges;

and installing horizontal drains, vertical wells drained at their bottoms, and drainage tunnels are some of the methods used for stabilizing landslides.

Some of the most devastating earth slides have occurred in saturated sediments whose structure has been disturbed by earthquake shaking. Davis and Karzulovic (1963) describe landslides caused by the disastrous Chilean earthquakes of May 1960. The largest of the slides at Lago Riñihue was evidently a liquefaction type of failure. Severe earthquake shaking caused a collapse of the structure of unconsolidated beds of porous silt and fine sand with the release of pore water that left the overlying materials supported on a soil-water mixture with virtually no strength. Additional shaking started the mass moving toward the river at a rapid rate, mixing river water into the soil at the bottom of the slide mass. The authors estimated that the toe of the original slope moved outward 400 m in less than 5 min.

A rather similar soil collapse occurred during the Alaskan Good Friday Earthquake on March 27, 1964. This earthquake, which had a Richter magnitude of 8.4 to 8.6, struck southern Alaska as one of the most devastating quakes of modern times. Seed (1964) studied the slide at Turnagain with laboratory models that shed light on the mechanisms of earthquake-induced slides in unconsolidated earth. The Turnagain slide was even larger than the large slide at Lago Riñihue, for it involved about 8,000 ft of coastline, and extended in from shore about 600 ft at the east end and 1,200 ft at the west; 5 to 20 ft of sand and gravel lay over about 100 ft of saturated sensitive clay.

Evidently the details of internal movements at Turnagain differed from those at Lago Riñihue; however, the basic cause of both was the structural collapse of saturated loose earth. The extremely flat slopes on which these slides came to rest testifies to the inherent insecurity of slopes supported on saturated, unconsolidated fine-grained soils. A study of slope failures leads to the inevitable conclusion that conditions conducive to liquefaction must be eliminated in all structural embankments and foundations in seismic regions if failures of these works can cause serious losses of life and property.

Sherard et al. (1963a) conclude that the shaking of an earth dam during severe earthquakes may impose the most dangerous conditions to which it will be exposed during its lifetime. These authors (1963b) state that the Sheffield Dam near Santa Barbara, California, which was built in 1918, is the "only known dam to have failed completely as the result of an earthquake." Evidence collected after the failure indicates that poor drainage of lightly compacted fill and improper preparation of the subgrade probably produced the conditions that permitted this dam to collapse under the shaking of an earthquake with an intensity of about 9 on the Rossi-Forel scale.

Many natural earth slopes in youthful valleys are in a delicate state of balance, and the slightest increase in driving forces caused by heavy rains or earthquake shocks or the slightest reduction in resisting forces due to undercutting of slopes or lowering of reservoirs can set large landslides in motion. Once the sliding mechanisms have been started, the resistance to further sliding is greatly lowered. Nasmith (1964), in describing landslides in the Meikle Valley, Alberta, Canada, points out that although the bulk of the soil in an earth slide remained essentially unchanged the remolded soil along rupture surfaces increased substantially in water content and the shear strength dropped more than 50%.

Trying to predict whether a given natural slope or dam can fail during an earthquake is an extremely difficult and complex problem, although it is known that saturated formations in which pore pressures can build up during shakings are most susceptible to failures by liquefaction. Also, as noted in Sec. 8.2, the danger of liquefaction diminishes with increased density. In analyzing the behavior of saturated cohesionless materials in which pore pressures may vary during an earthquake, Seed (1966) makes use of a procedure with the following steps:

1. Determine the initial stresses in the embankment before the earthquake.
2. Determine the characteristics of the motions developed in rock underlying the embankment and its soil foundation during the earthquake.
3. Evaluate the response of the embankment to the base rock excitation and compute the dynamic stresses induced in representative elements of the embankment.
4. By subjecting representative samples of soil to the combinations of pre-earthquake stress conditions, and superimposed dynamic stress applications, determine by test the effects of the earthquake-induced stresses on soil elements in the embankment. These effects will include any evidence of soil liquefaction and the magnitude of the deformations induced by the earthquake loading.
5. From a knowledge of the deformations induced in individual soil elements in the embankment, evaluate the overall deformations and stability of the cross section.

Rocks can lose even more strength than soil, virtually changing from a solid state to a semiliquid state. The tremendous landslide that destroyed Vaiont Reservoir (Fig. 8.6) is an example of a rapid landslide in a geologically youthful and unstable valley, set in motion by minor influences that aided natural earth-leveling processes. Hendron and Patton (1985) say that "The large volume and high velocity of the Vaiont slide combined with the great destruction and loss of life that occurred make it a key precedent land-slide, particularly for slides caused by reservoir filling." Their conclusion after making a very detailed study of the causes of the slide is that "The results of the study also suggest that *the slide could have been stabilized by drainage.*" (My emphasis.)

Although many landslides occur in slopes because of natural influences, human activities such as undercutting, piling earth on unstable slopes, or raising the groundwater level by constructing reservoirs, are important causes of landslides. Unfavorable groundwater and seepage conditions are among the

FIG. 8.6   Landslide at Vaiont Reservoir, Italy, October 1963. (Courtesy, Engineering News-Record.)

most frequent. Water lowers stability and contributes to slope failures in the following ways.

1. By reducing or eliminating cohesive strength.
2. By producing pore water pressures which reduce effective stresses, thereby lowering shear strength.
3. By producing horizontally inclined seepage forces which increase the overturning moments and the possibility of failure.
4. By lubricating failure planes after small initial movements occur.
5. By supplying an excess of fluid that becomes trapped in soil pores during earthquakes or other severe shocks, leading to *liquefaction* failures.

Properties of soils and rocks that influence the ability of slopes to withstand the actions of water and other damaging forces are the following:

1. The shear strength of the basic materials.
2. The plasticity and strength of fillers.
3. The strike and dip of weak bedding planes.
4. The spacing, thickness, and extent of joints.
5. The density of soil and highly weathered rocks which influences their susceptibility to liquefaction under shock.

6. The position, inclination, and extent of nonconformities such as barriers that are impervious to seepage, highly pervious strata, joints, and cracks.

The factors influencing the stability of slopes are so numerous that the true stability of most undisturbed natural slopes can be known only approximately. Some slopes have a high degree of stability, others are barely stable. With the aid of analytical methods embankment slopes for dams and other structures can be designed and built with maximum security against failure under any foreseeable conditions. Natural slopes must be accepted substantially as furnished by nature but can often be improved by adding weight at their toes, removing weight from their heads, installing piling, cables, retaining walls, or cribbing, by electoosmosis (Sec. 7.1), or by draining, with which this chapter deals.

Methods for Studying Slope Stability

The *relative* influence of important factors on the stability of slopes can be compared by a number of methods that have been devised for their analysis. If the strength of earth and rock in slopes can be determined with reasonable dependability, numerical *factors of safety* can be calculated for assumed combinations of conditions, and even though the strength is known only approximately, the *relative* stability can be estimated. The study of the influence of important factors developed in this chapter can improve our understanding of the causes of slope failures and the evaluation of methods for their control and prevention.

Sherard et al. (1963c) divide the various methods for analyzing slope stability into two general categories: *method A—sliding surface methods*, and *method B—the unit stress method*. Under method A are included various shapes of potential shear failure surfaces, including the sliding circle method originated by K. E. Petterson in Sweden in 1916. This method, also called the Swedish method, was refined further by Fellenius (1936), Taylor (1948), Bishop (1955), Bishop and Morgenstern (1960), and others. May (1936) simplified the method by using a planimeter to sum up the forces. Other practical methods such as the *infinite slope method* (Taylor, 1948a) and the *sliding block* or *wedge* method (Sherard et al., 1963d, Taylor, 1948b) have been developed. Chowdhury (1987) presents good descriptions of the methods usually used in analyzing the stability of soil slopes.

Methods of slope stability analysis that determine the forces acting on sliding surfaces express the factor of safety G_s against sliding as

$$G_s = \frac{\text{the sum of the resisting forces or moments}}{\text{the sum of the activating forces or moments}} \quad (8.1)$$

The sliding circle method is applicable to a wide variety of slopes by modifying the shape of the failure plane as necessary to conform to nonuniform conditions. If formations in a slope are cohesive, failure surfaces tend to be deep-seated, as along arc ABC (Fig. 8.7). To determine the *factor of safety* of a slope by this method a number of trial circles are analyzed for stability; the one with the lowest factor of safety, the *critical circle*, represents the failure surface along which the slope is most likely to fail. The numerical factor of safety for the critical circle is the theoretical minimum factor of safety of the slope for the assumed conditions.

A common procedure for calculating the factor of safety for a trial failure circle (Fig. 8.7) is to divide the circular segment into a number of vertical slices. The simplified procedure assumes that the forces acting on the sides of the slices largely balance and can be neglected. For each slice the total weight of soil is determined and plotted as a vertical force through the center of gravity of the slice. For slice 6 (Fig. 8.7) the weight is W_6. The component of W_6 which acts tangential to the failure plane and therefore is the *activating* or *overturning* component is the tangential force T_6. The component of W_6 which is perpendicular or *normal* to the failure plane *and* is contributing to the resisting forces is the normal force N_6. These two components can be determined

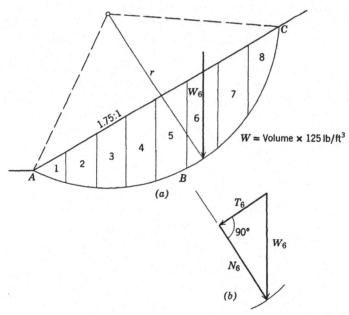

FIG. 8.7 Illustration of simplified Swedish method of analysis of slope stability. (*a*) Cross section and critical circle with slices. (*b*) Simplified force diagram for slice 6, for dry or moist slope.

graphically by plotting W_6 to a convenient scale and constructing a right triangle as shown in Figure 8.7b. The same procedure can be applied to all slices and the factor of safety calculated from the formula

$$G_s = \frac{\Sigma N \tan \phi + cl}{\Sigma T} \tag{8.2}$$

In Eq. 8.2 ϕ is the coefficient of friction of the soil or rock within the slope, c is its unit cohesive strength, l is the arc length over which c is acting, ΣN is the sum of the normal components for all slices, and ΣT is the sum of the tangential components for all slices.

If the formation is cohesionless, c is zero and Eq. 8.2 becomes

$$G_s = \frac{\Sigma N \tan \phi}{\Sigma T} \tag{8.3}$$

In Eq. 8.3 the factor of safety is proportional to the ratio $\Sigma N/\Sigma T$. It can be shown that the steeper the slope, the lower the ratio $\Sigma N/\Sigma T$ and vice versa; likewise, any force that acts detrimentally on a slope lowers the magnitude of $\Sigma N/\Sigma T$ and the stability.

If slopes are made up of cohesionless soils, failure surfaces tend to be shallow and parallel to the slope. The limiting case, an infinitely long slope with uniform conditions throughout, is called the *infinite slope* case (Fig. 8.8). For this condition the factor of safety of the dry slope (*case a*, Sec. 8.2) can be calculated by considering any vertical slice of soil such as *abcd* (Fig. 8.8a). As in the sliding circle method, the weight W of a slice can be plotted to scale at the center of gravity of the slice. If no other forces are acting, W is also the

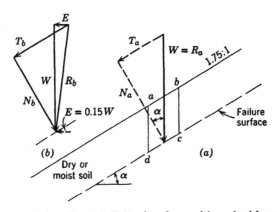

FIG. 8.8   Stress conditions in an infinite dry slope with and without earthquake. (*a*) Cross section and force diagram without earthquake (*case a*). (*b*) Force diagram with earthquake (*case b*.)

resultant body force R_a (the subscript denotes the case). The normal component N_a and the tangential component T_a can be determined graphically by plotting a right triangle, as shown, or N_a and T_a can be determined from the geometry of the slope, for tan $\alpha = T_a/N_a$ and $T_a = N_a$tan α. The factor of safety G_s can be expressed as

$$G_s = \frac{N_a \tan \phi}{T_a} \tag{8.4}$$

and

$$G_s = \frac{\tan \phi}{\tan \alpha} \tag{8.5}$$

If the coefficient of friction ϕ is unknown, the *relative* degree of stability under various conditions can be determined for any given slope from Eq. 8.4 because for any given value of ϕ

$$G_s \sim \frac{N}{T} \tag{8.6}$$

For deep-seated failure surfaces in cohesive formations Eq. 8.6 is not applicable; however, it is useful for studying the effects of important conditions on the stability of cohesionless slopes. To make a study we can select a vertical slice such as *abcd* in Figure 8.8*a* and analyze the various systems of forces acting on that slice.

Many engineering computations are simplified by substituting for *systems of forces* a single *resultant.* In soil mechanics the resultant force acting on a body of soil is called the *resultant body force* or the *body force* (Sec. 3.5). It is the resultant of all significant forces acting on an element of soil. In Figure 8.8*a* the body force is equal to W, the downward weight of soil in element *abcd.* If additional forces are acting, they are combined to obtain the correct resultant, which is used in obtaining normal N and tangential T components. The numerical values of the ratio N/T are an index of relative stability. This is the basis for the studies described in Section 8.2.

8.2 INFLUENCE ON IMPORTANT CONDITIONS ON SLOPE STABILITY

Table 8.1 lists some common examples of slope failures and Table 8.2, the general degree of stability of slopes for a range of conditions. In this section the relative influence of poor versus good drainage, earthquake forces versus no earthquake forces, low versus high density, and various combinations of these conditions are compared on the basis of N/T ratios as indexes of stability. The conditions analyzed (see Table 8.3) vary in stability from *case a*, a dry slope without earthquake forces, to *case f*, a slope of saturated loose soil subjected to earthquake forces. The forces acting, the resultant body forces, and the N and T components for a number of illustrative cases are given in Figures 8.8 to 8.11. To facilitate a comparison of these conditions the N and T components and the N/T ratios are summarized in Figure 8.12. The conditions are described in subsequent paragraphs.

Effect of Earthquake Shocks on Dry Slopes

The forces acting in an infinite dry slope without earthquake are shown in Figure 8.8a, with earthquake in Figure 8.8b. For the dry slope without earthquake (*case a*) the only force acting is the weight of the soil W, which is equal to the dry or moist unit weight of the soil times the volume (the area of the segment times a 1-ft thickness). In this case (Fig. 8.8a) the weight W is the only significant force acting; hence it is also the resultant body force R_a. The ratio N_a/T_a = 1.79, a relatively large number, indicates a relatively high degree of stability.

To study the dry slope with earthquake (*case b*) a horizontal component for a 0.15g earthquake acceleration is assumed to be acting at the base of the segment in the conventional manner (Fig. 8.8b). The forces acting on the soil element *abcd* of Figure 8.8a, as shown in Figure 8.8b, are the soil weight W and the horizontal earthquake force $E = 0.15W$. The resultant body force R_b (Fig. 8.8b) produces a higher T component and a lower N component than R_a in Figure 8.8a. The ratio N/T plotted in Figure 8.12b for *case b* is about 70% of N/T for the slope without earthquake; hence an earthquake force of 0.15W lowers the stability of this slope by 30%.

Effect of Direction of Seepage in Saturated Slopes

Many natural earth slopes become saturated during periods of heavy precipitation in which the water table rises to the ground surface and water flows essentially parallel to the direction of the slope. For this condition (*case c*) soil element *abcd* in the infinite slope has the submerged weight W_o and the seepage force F acting, as shown in Figure 8.9a. A portion of the flow net shows flow lines parallel to the slope and equipotentials perpendicular to the slope. Using either the hydraulic gradient method or the boundary pressure method (Sec. 3.5) the seepage force F can be determined from the flow net. The resultant body force with seepage can then be determined, as shown in Figure 8.9b, by plotting W_o and F. (See also Sec. 3.5 for determination of resultant body forces with seepage.) The resultant R_c (Fig. 8.9b) produces tangential component T_c, which is greater than T_a or T_b, and normal component N_c, which is smaller than N_a or N_b; consequently the ratio N/T for this case (Fig. 8.12b) is lower than for dry slope without earthquake (*case a*) and dry slope with 0.15 g earthquake acceleration (*case b*). It is less than 50% of N/T for the dry slope with-

Study of Conditions Influencing the Stability of Slopes

	Conditions			Transient pore pressure			
Case	Dry or moist	Saturated, with seepage	Earthquake	pos.	neg.	Occurrence	Significance
a	x					Naturally dry, or well drained, little infiltration	Highly favorable condition, the goal of good drainage design
b	x		x			Whenever an area is struck by a significant earthquake	Seldom a cause of serious failures
c		H*				Normal, uncontrolled, unfavorable seepage in slopes saturated by rainfall, etc.	Seepage generally parallel to slopes produces excess pore pressures, lowers stability
d		V†				Favorable, vertical seepage forced by drainage under slope	Vertically downward seepage eliminates excess pore pressures, consumes energy harmlessly
e		H*	x			Saturated slopes subjected to severe earthquake, no volume change in soil	A common design assumption for permanent structures in seismic regions
f		H*	x	x		Saturated, *loose* soils or weathered rocks subjected to severe earthquake; liquefaction tendencies	These conditions conducive to complete collapse of slopes; must be prevented in important works
g		H*	x		x	Saturated, *dense* soils or rock formations subjected to severe earthquake; expansion tendencies	Dense, strong soils and rocks tend to resist earthquake damage

*H = seepage approximately horizontal, or parallel to slope
†V = seepage essentially vertical (downward)

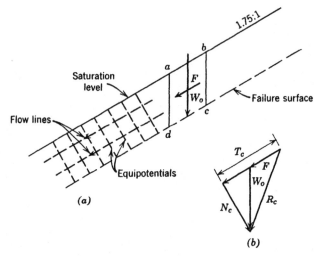

FIG. 8.9 Stress conditions in infinite saturated slope with seepage parallel to slope (*case c*). (*a*) Diagram of slope. (*b*) Force diagram.

out earthquake and about 60% of N/T for the dry slope with earthquake. It is thus seen that horizontally moving seepage can greatly lower the stability of slopes.

Slopes can be highly saturated yet essentially free of excess pore pressures and damaging seepage forces. If the slope under study is underlaid with a highly pervious gravel layer, as shown in Figure 8.10a, the flow net consists of vertical flow lines and horizontal equipotentials. Under this seepage condition the energy of the free water in the soil is consumed harmlessly as it flows vertically downward to the gravel, thence outward to the toe of the slope. For this seepage condition (*case d*) the forces acting on element *abcd* (Fig. 8.10a) are its submerged weight W_o and a downward seepage force F_v. The resultant body force R_d is the sum of W_o and F_v, a vertical force producing tangential component T_d and normal component N_d (Fig. 8.10b). The ratio N/T for this case is identical with that for the dry slope (Fig. 8.12b), although the forces N_d and T_d are slightly larger than for the dry slope because the total weight is slightly greater for the wet slope than for the dry slope (Fig. 8.12a).

Cases c and d lead to a fundamental principle: *Water seeping in a generally horizontal direction destabilizes slopes, whereas water seeping vertically downward produces no destabilizing forces and no pore pressures.* The benefits of drains beneath constructed embankments subjected to heavy rains are described by Terzaghi (1943), and the use of inclined or horizontal drainage layers to force seepage into vertical patterns and thus improve the stability of retaining walls is described by Terzaghi and Peck (1948) (see Sec. 10.2).

Many constructed slopes of dams and other structures controlling seepage are built with drains that keep important soil zones free of water or produce favorable seepage patterns. On occasion naturally existing drainage layers add

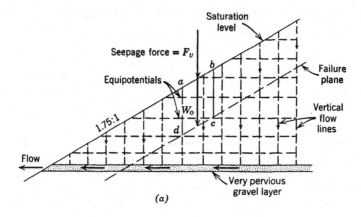

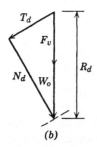

FIG. 8.10 Stress conditions in infinite saturated slope with vertical seepage (*case d*). (*a*) Cross section, flow net, and forces. (*b*) Force diagram.

to the stabilization of natural slopes. Many slopes no doubt remain stable *because* of favorable drainage; for example, a natural earth cofferdam for a deep excavation in an extension to the Bonneville powerhouse in Oregon. Natural, slightly cemented, stratified sandy gravels provided a cofferdam that kept the Columbia River out of the deep excavation for many months. Within the natural beds of earth, which were excavated on a slope of about 1.7:1, were intermittent strata of pervious gravel (Fig. 8.14). Several tiers of wellpoints were installed in the slope, and shallow ditches at the base of the slope fed any seepage that bypassed the wellpoints to sumps, where it was collected and pumped back into the Columbia River. Although the wellpoints removed considerable water, a large amount emerged on the slope through the pervious gravel lenses. The highly pervious gravel layers evidently dominated the seepage pattern in the natural cofferdam, forcing flow into a predominately vertical direction. The flow net in Figure 8.14 shows the surmised seepage pattern in this cofferdam, ignoring the influence of the wellpoints. Possibly this slope would have been stable without the natural horizontal drains, but they evidently provided a substantial increase in stability and helped to keep this important excavation slope entirely stable throughout a long dewatering period.

Many other examples that verify the importance of direction of seepage on the stability of slopes could be cited. In regions in which rainfall is high and water tables are near the surface, landslides and slope problems are generally more severe than in regions in which water tables are low. A notable example is the difference between mine tailings dams in wet regions with high water tables (e.g., Appalachia), in which slope stability problems are generally severe, and in dry desert regions in which water tables tend to be much lower.

Obviously a continuous horizontal drainage blanket cannot be placed under a *natural* slope, but other measures such as intermittent horizontal pipe drains can be used to improve seepage conditions and stability (Fig. 8.24). Fortunately the designer of *constructed* slopes can specify whatever kind of drainage he wishes. Whenever possible, man-made structures should be constructed with drainage facilities that will not permit unstable conditions to develop. The seepage analysis methods described in this book can help designers develop structures with full control over adverse seepage.

Effect of Earthquake Shocks on Saturated Slopes

No Volume Change (case e). "Standard" methods of analyzing the stability of slopes under earthquake shocks apply a horizontal force equal to the weight of soil times an arbitrary percentage of the acceleration of gravity g. On this basis a horizontal earthquake force of $0.15g$ was combined with the dry soil weight to study the influence of an earthquake shock on a dry slope (Fig. 8.8b). The earthquake force E is equal to W times the assumed earthquake acceleration. In this example $E = 0.15W$. The normal analysis assumes that the soil neither expands nor contracts under the applied earthquake force. If cracks and pores in earth formations are not filled with water, volume changes under earthquake shocks are relatively unimportant; however, if these spaces are filled with water, volume changes can be extremely important, as will be shown subsequently in this section.

The influence of an earthquake shock on a saturated slope *without volume change* is illustrated by Figure 8.11a (*case e*). The force system (Fig. 8.11a, right diagram) includes three forces: (*a*) the submerged soil weight W_o, (*b*) the seepage force F, and (*c*) a horizontal earthquake force E. The resultant body force produced by these three forces R_e (Fig. 8.11a, left diagram) produces normal component N_e and tangential component T_e. For direct comparison with the other cases under study these components are also shown in the summary in Figure 8.12a and the N/T ratio is given in Figure 8.12b. It will be seen that this case is the most severe of those studied up to this point, the relative stability (N/T ratio) being about 25% of that of the dry slope without earthquake. The slope would have to be flattened considerably to withstand this force system.

Section 8.1 points out that many severe landslides occur in loose saturated earth during earthquakes. A hypothetical force system that can exist under these conditions is shown in Figure 8.11b (*case f*). The mechanisms responsible

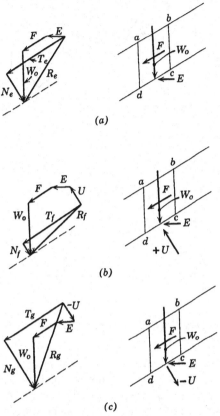

FIG. 8.11   Forces acting in a saturated slope during an earthquake. (*a*) No pore pressures caused by volume change (*case e*). (*b*) Positive pore pressures caused by disturbance of *loose* structure (*case f*). (*c*) Negative pore pressures caused by dilation of *dense* structure (*case g*).

for these adverse conditions are described in the next paragraphs before discussion of cases *f* and *g*.

Critical Density Concept. Earthquake oscillations in the earth's crust produce horizontal and vertical accelerations whose sums must add up to little more than zero. The "standard" method of assuming an acceleration of gravity of some arbitrary amount is believed to be justified only on the grounds that it produces more conservative designs than would be obtained without it. Many earth slopes that have withstood the effects of earthquakes have tumbled down during small undercuttings or loadings; therefore it appears that of greater importance than an earthquake component are the *alterations in soil or rock structure and transient pore pressures that develop as a result of the shaking.*

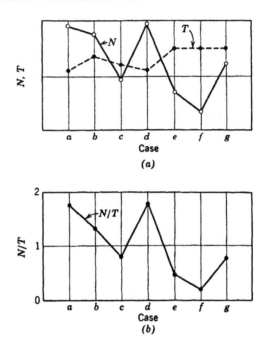

FIG. 8.12   Summary of N and T values and N/T ratios for cases studied.

A basic concept of soil mechanics known as *critical density* was developed by A. Casagrande (1936) to explain important differences in the developed strengths of certain soils in a loose and a dense state. According to this concept, granular soils that are less dense than a certain critical state tend to become more dense under shearing actions or severe shocks, whereas granular soils that are more dense than this critical state tend to expand and become less dense during shearing actions or shocks. The jarring or shearing of loose soils tends to rearrange the particles and allows them to slide and slip into a somewhat denser state, but densely arranged particles tend to ride up or mount over one another, which tends to cause an increase in volume and a reduction in density. At one time it was thought that each soil had a particular "critical density" below which volume reductions could not occur under shear. It is now known that "critical density" varies with the confining pressure and with the state of stress; hence a sand or gravel compacted to a given density may be highly stable in a small dam but unstable if compacted to the same density in a high dam.

If the pore spaces in granular soils are filled with water, shearing actions in soils looser than some critical state tend to produce liquefaction failures because volume reduction tendencies suddenly leave the soil pores overflowing with water. Unless the excess water can escape instantly, it causes a sudden reduction in *effective stresses* and strength and permits serious failures of slopes that previously appeared perfectly stable. This kind of action causes many of the major landslides that accompany earthquakes. It contributed to the serious Turnagain slide in Alaska during the 1964 Good Friday Earthquake. In contrast, the expanding tendencies in soils denser than the critical state create negative pressures in the pore water while additional water is being sucked into the enlarging pores. As a consequence the *effective stresses* and the strength of dense granular soils may increase during shear.

Marsal (1961) investigated damages to various kinds of structures and buildings that occurred during the earthquake in Veracruz, Mexico on August 26, 1959, and concluded that failures were caused "by partial liquefaction of the sand and silt." These materials are nonplastic, in a rather loose state and quite uniform in grain sizes. Marsal concluded that in large masses these soils can undergo rapid displacement and behave as a liquid. He concluded that the phenomena observed in the Mexican earthquake were very similar to those taking place in the Valdivia region as a result of the severe earthquake that struck southern Chile during May 21 and 22, 1960. The greatest protection against serious failures of major dams and other important permanent works during earthquakes can be obtained by the following measures:

1. Utilize effective drainage to keep saturation as low as possible in zones in which liquefaction could lead to failure.
2. As far as possible, eliminate loose, potentially unstable materials from foundations.
3. Require thorough compaction of all embankment zones that can become saturated.

Continuing in this study, stress conditions that can develop in loose and dense saturated formations subjected to severe shock are illustrated in the following discussions of cases f and g.

Loose Formations Under Severe Shock (case f). The right-hand diagram in Figure 8.11b shows a potential system of forces acting on soil element $abcd$ for the assumption that the slope is composed of loose soil or highly weathered rock that tends to consolidate under severe earthquake shaking. In addition to the three forces that were acting in *case e* (Fig. 8.11a), a positive pore pressure U is added to the force system. The magnitude of U depends on the degree to which the soil structure is damaged by the shock. Although U can only be estimated, it is known that transient pore pressures can set off catastrophic slope failures. In this example U is assumed to be about 40% of the submerged soil weight W_o. With a force U of this amount included in the force system, the resultant R_f (left diagram, Fig. 8.11b) produces a very damaging N/T ratio, the relative degree of stability being only 11% of that of the dry slope (Fig. 8.12b). The slope would have to be extremely flat to resist this system of forces.

In highly pervious rocks and gravels it is likely that transient pore pressures, if they develop at all, can dissipate so rapidly that they would have little influence on slope stability. But in typical rock or gravel fill commonly placed in modern earthwork construction it is possible that under severe shock such pressures can be significant. As previously noted, major dams and other works whose failures could cause serious losses of life and property should be made as resistant as possible to damage under severe shock. The three measures listed above are fundamental to the security of important works in areas that can be severely shaken by earthquakes.

Dense Formations Under Severe Shock (case g). In Figure 8.11c, right diagram, the force system for soil element *abcd* includes the three forces of *case e* together with a *negative* transient pore pressure $-U$. The resultant body force R_g (left diagram, Fig. 8.11c) for this system of forces produces normal and tangential components N_g and T_g which are more favorable than for either *case e* or *case f*. This indicates that if slopes in dense granular soils or jointed hard rock fragments are shaken, negative transient pore pressures may temporarily increase the resistance to failure. This factor should not be counted on as we cannot be sure that this force will always be available.

Earthquake and Seepage Forces Compared

By referring to the summary in Figure 8.12b it can be seen that the *dry slope with earthquake (case b)* has a substantially higher degree of stability than the *saturated slope without earthquake (case c).* It therefore appears that under some conditions seepage forces may be more detrimental than rather severe earthquake forces. The *relative* influence of earthquake and seepage forces on slopes *without volume change* is illustrated in Figure 8.13. A portion of an infinite slope with the water table at the surface and flow lines parallel to the slope is shown in Figure 8.13a. The equipotential lines, as for *case c* (Fig. 8.9a) are perpendicular to the slope.

For the slope shown in Figure 8.13a the force system with seepage but without earthquake (Fig. 8.13b) includes the submerged weight W_o and the seepage force F. These forces produce resultant body force R_1, which can be resolved into normal component N_1 and tangential component T_1. If this 1.4:1 slope is assumed to be composed of cohesionless material, the degree of stability is proportional to the ratio N_1/T_1, as in the cases previously examined. Next let it be assumed that this slope when fully drained is subjected to an earthquake shock. If the degree of stability that existed in the saturated slope without earthquake is to be maintained, the ratio N_2/T_2 (Fig. 8.13c) must be equal to the ratio N_1/T_1, determined in Figure 8.13b. The earthquake force E, which must be combined with the total soil weight W to produce N and T components in the same ratio as N_1/T_1, can be obtained as shown in Figure 8.13c, for triangle *abc* is similar to triangle *aef*.

The force diagram of Figure 8.13b is replotted (Fig. 8.13c) to determine the line of action of resultant R_1. The total moist soil weight W is plotted next between points *a* and *g* (Fig. 8.13c). The line of action of resultant R_1 is then extended to some point *d* and horizontal *gc* is drawn to close triangle *agc*. The

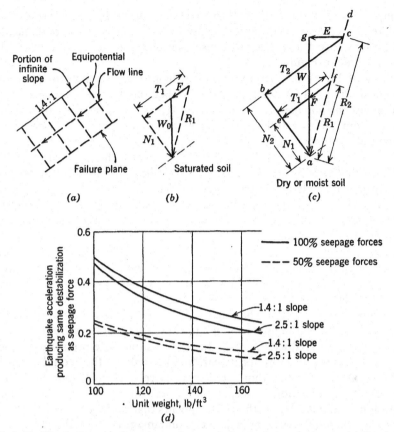

FIG. 8.13 Equivalent earthquake and seepage forces in earth slopes. (*a*) Portion of slope and flow net. (*b*) Force diagram with seepage. (*c*) Earthquake force E equivalent to seepage force F. (*d*) Summary chart.

distance gc is the earthquake force E that was sought. This follows because any combination of forces whose resultant body forces lie on line ad will have N and T components in a constant ratio because the similar triangles make $N_1/T_1 = N_2/T_2$, etc.

If force E (Fig. 8.13*c*) is assumed to be equal to W times a percentage of the acceleration of gravity, the acceleration that produces E is equal to E/W 100%. For this example an acceleration of 0.3 g produces force E, which combines with the total weight W to give the same degree of stability as the seepage force F and the submerged weight W_o.

To facilitate a comparison of seepage forces with equivalent earthquake accelerations several curves are plotted in Figure 8.13*d*. Earthquake accelerations that produce the same destabilizing influence as seepage forces for slopes of 1.4:1 and 2.5:1 are given as solid lines. For comparative purposes earthquake accelerations with *half* the destabilizing influence of the seepage forces are plotted as dashed lines. Because seepage forces are more damaging to lightweight formations than to dense, the unit soil weight is used as the horizontal scale in Figure 8.13*d*.

A study of Figure 8.13*d* reveals that in lightweight formations (densities from 100 to 120 lb/cu ft) steady seepage parallel to the slope is about as detrimental as an earthquake acceleration of 0.4*g*, whereas in heavy formations (densities from 140 to 150 lb/cu ft) seepage is as damaging as an earthquake acceleration of about 0.25*g*.

The relationships given in Figure 8.13 indicate that the physical forces induced in earth bodies by earthquakes of fairly strong intensity (but short duration) may have much less influence than seepage forces on stability. This study is presented as further evidence that the *transient pore water pressures* and *remolding actions of earthquakes* are major side effects that cause much of the damage of earthquakes.

If natural earth slopes are examined, it is evident that a great many have been standing for decades, possibly for centuries, with static factors of safety of not much more than unity. Why do many of these slopes, which remain standing during earthquakes, fail under minor undercutting or other changes? One explanation may be that shaking actions in relatively dense jointed rocks may temporarily open the joints enough to cause transient *negative* pore water pressures or at least reduce the magnitude of positive pore pressures. If such actions do in reality develop during earthquake shaking, they could compensate for a portion of the earthquake forces. By referring to Figure 8.13*d* it is seen that a 50% reduction in seepage forces is equivalent to an earthquake acceleration of about 20%. It therefore appears that slopes composed of fractured rocky materials may be relatively undamaged by earthquakes even though they lack large factors of safety under static conditions. The effect of dilation tendencies on the stability of slopes in dense formations is described under case *g*.

Designing Structures to Resist Earthquake Damage

A review of the cases studied in this section and a study of important landslides and slope failures lead to the following conclusions (see also Sec. 6.4):

1. Good drainage is one of the most effective means of improving slope stability during earthquakes.
2. No amount of slope flattening or "beefing up" of the widths of earth dams or other structures containing important zones of loose saturated soil can prevent liquefaction under severe shocks.
3. The provision of ample zones of dense, strong, well-drained materials within the cross sections of earth dams and other important structures involving water and the elimination of loose, weak, saturated materials

from their foundations is essential for important works whose failure could cause heavy property damage and loss of life.

These basic concepts can ensure maximum protection against failures caused by (1) seepage; (2) earthquake shaking; (3) combined forces of water and earthquakes.

8.3 DRAINAGE METHODS FOR STABILIZING SLOPES

Builders of earth dams, reservoirs, highway fills, and other engineering works have control over the steepness of constructed slopes, internal drainage facilities, and the properties of the materials used. Natural slopes must be used substantially as furnished by nature and such measures taken that are believed necessary to prevent failures. Natural earth formations are always variable to a greater or lesser degree; hence a slope that may on the surface appear to be stable may be inherently unstable and vice versa. Geologic and soil surveys in connection with the slopes of important engineering works should always be made by highly experienced persons.

Slope stability can usually be improved by flattening, by removing earth near the head, or by placing material near the toe. Heavy rock or gravel *buttress* fills placed near the bottoms of slopes can improve stability by serving as drains and by adding weight. Retaining walls and other restraining devices are widely used for improving slope stability. Paved interceptor ditches are often placed around the heads of slide areas, surface soils are compacted, cracks and depressions are filled, and surfaces are contoured to reduce the flow of water into the soil mass.

Control over the groundwater and seepage conditions deep within slopes by drainage is one of the best methods for improving the stability of earth slopes (Fig. 8.14). Several methods that are used for improving groundwater conditions within slopes, singly or in combination, are the following: (1) stabilization-drainage trenches, (2) drainage tunnels or galleries, (3) well systems, (4) horizontal drains. These methods are described in the following paragraphs. In addition, the design of drained toe buttresses is discussed.

Stabilization-Drainage Trenches

Slides frequently damage highway and railway fills and other side-hill embankments in hilly terrain. Failures of fill foundations can often be prevented by the installation of stabilization-drainage trenches (Fig. 8.15). The stabilization or "stab" trench, as it was called by California Division of Highway's engineers who originated it, improves foundation stability in three ways.

1. By replacing weak foundation soils with strong, well-compacted backfill keyed into firm substrata.

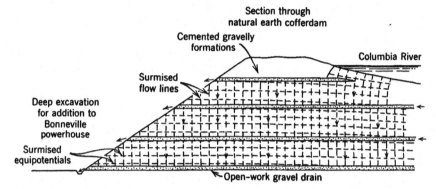

FIG. 8.14 A natural slope with built-in drainage layers. Note simplified, diagrammatic section.

2. By lowering hydrostatic pressures by intercepting groundwater with pervious drainage blankets that discharge through pipe outlets.

3. By serving as large exploration trenches, to forewarn builders of unsatisfactory foundation conditions and allow corrective measures to be taken.

If side-hill embankments for highways, railroads, or other facilities are allowed to block natural groundwater flows, the trapped water can often have devastating effects (e.g., see Figs. 8.4 and 8.5). A properly located and sized stabilization-drainage trench of the kind shown in Figure 8.15 can eliminate such dangers if it is correctly designed. A key to the success of a stabilization

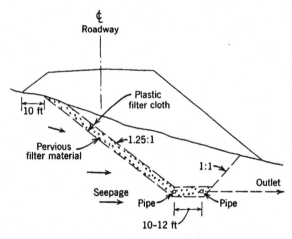

FIG. 8.15 Cross section through a stabilization-drainage trench.

trench is its internal drain, which must have ample capacity to remove all incoming water. Estimates of probable seepage quantities needing to be removed can be made with Darcy's law. Also flow nets can be used to estimate required permeabilities in the drains as shown in Figure 5.9. But very high discharge capabilities in such drains can be assured by using highly permeable crushed rock, 1/2 in. to 1 in. in size, railroad ballast, or other comparable material enveloped within an appropriate fabric filter.

In the past the tendency has been to use well-graded blends of sand and gravel for the filter layer in stabilization trenches of the kind shown in Figure 8.15. If groundwater flows were small, these drains generally improved conditions, but when groundwater flows were large reports are that these drains produced little or no improvement in seepage and groundwater conditions. Figure 8.16 is a view below a major railroad line in which a fill had been constructed over a stabilization trench, using a drain of a sand and gravel blend with a fines content (passing No. 200 sieve) of 5 to 8%.

This embankment survived several "normal" winters of moderate amounts of rainfall, but a few years after construction an extremely large amount of rain fell on the area, causing many landslides and slipouts along both highways and railways. As shown in this photograph, the foundation and the fill had

FIG. 8.16 Embankment for a major railroad has nearly collapsed (second telephone pole has moved 2 to 3 ft to right). New drains of railroad ballast protected with filter fabric successfully stabilized the area. Original construction had a subsurface drain that was totally ineffective because of its low permeability. Light color is a sandy fill material placed in part of subsidence area. (Printed with permission of The Western Pacific Railroad Company.)

slid several feet downhill (the second telephone pole has moved about 3 ft to the right); 2 to 3 feet of new ballast were required to restore the track to grade; furthermore, the concrete culvert has started to fail. Although the amount of movement may not appear to be large, an alarming condition does, in fact, exist. Because of its low permeability, the initial drain was completely ineffective. The unusually heavy rains of that year put this relatively new relocated portion of the line to a very severe test, and caused a number of other slipouts and complete failures. The author was invited to examine a number of the unstable areas and offer suggestions for making repairs and improving the stability of many weak spots. Since railroads have liberal amounts of clean, coarse ballast materials available at all times for maintaining the grades of their lines, my general suggestion was that at locations where improved drainage was needed, to construct drains using railroad ballast for aggregate. And to prevent the clogging of the ballast with adjacent fines, I recommended that the new drains be completely enclosed within suitable plastic filter fabric. This was done at a number of places, including the one shown in Figure 8.16. At this place, a trench was dug behind the fill to the maximum feasible depth (in dry weather), fabric was used to line the entire trench, it was backfilled with railroad ballast, and the top was covered with impermeable soil to keep surface water out. A collector pipe at the bottom of the trench with a gravity outlet drilled under the embankment provided positive drainage to prevent the buildup of hydrostatic pressures under and behind the fill. This drain effectively lowered the ground water level and prevented the very probable loss of the embankment and possible service interruption.

Drainage Galleries

One of the oldest methods of stabilizing troublesome hillsides or landslides, the abutments of dams, and other slopes is the drainage tunnel or gallery. Horizontal tunnels comparable to mining tunnels are dug into the base of a hill or slope in the hope of intercepting groundwater sources deeply enough to relieve hydrostatic pressures and improve the stability of the slope. If tunnels intercept water-bearing joints, cracks, or strata, they can be highly effective in lowering groundwater. They are sometimes used in conjunction with vertical drain wells to serve as gravity outlets for seepage that is intercepted by the wells. Drainage tunnels are perhaps the most costly method of draining slopes and have seldom been used in recent years because of the high installation cost. In some cases they are employed for the protection of expensive buildings or other improvements when limited access or right-of-way prevent the use of other methods or when tunnels are determined to be the best method.

Sharp (1970) discusses the use of galleries for draining slopes and points out that their presence aids in the inspection of conditions that influence the selection of the most effective and economical solution to a given drainage problem. Galleries facilitate the drilling of additional holes (from the galleries) in locations in which the greatest benefit can be obtained at least cost.

In some cases unlined railroad tunnels have served as massive galleries that effectively drained the formations and improved the stability of a large earth slope.

Well Systems

Well systems are used for improving groundwater conditions in troublesome highway slopes, hillsides, dam abutments, and other situations. When installed in the foundations of dams and levees, they are called *relief wells* (Chapter 6). In some cases drainage wells for slope stabilization have been given no outlet for seepage, except by flow from their tops, with the intention of relieving artesian pressures much as they do when acting as relief wells in the foundations of dams and levees. When wells are drilled in earth slopes, they are most effective when freely drained at the bottom. For the temporary lowering of groundwater during the construction of engineering projects wells often are pumped (Chapter 7). Sand-filled wells, or vertical drains constructed with prefabricated synthetic fabric filters around highly permeable plastic cores are often used for stabilizing weak, saturated, compressible earth foundations (Sec. 7.2).

For slope stabilization well systems are sometimes used in conjunction with tunnels that provide gravity discharge of the water removed by the wells. The tunnels may be excavated by mining methods or by belling out the bottoms of closely spaced wells, using miners equipment as necessary for making the tunnels continuous. Pervious filter aggregate is placed in the wells with pipes for increasing the discharge capacity. The tunnels are designed to prevent collapse and loss of adjacent material by piping. If they cannot be economically *daylighted,* low-level pipe outlets are installed in bored holes or jacked into the tunnel from the outside. In some cases pipe outlets are drilled or jacked from within the tunnel or from within belled out drain wells. An oil well instrument can be used to check the alignment of outlet pipes.

Well systems offer the advantage of being flexible because additional wells can be installed at intermediate points if initial spacing proves inadequate to control a seepage and groundwater condition.

A typical well drainage system that successfully stabilized a troublesome earth slope is described by Palmer, Thompson, and Yeomans (1950) in relation to the U.S. Naval Station in Seattle, Washington. Landslides had occurred in a hill composed of discontinuous strata of sand and silt in essentially impervious blue clay. Eight slides were set in motion by excessive hydrostatic pressures that were revealed by borings put down to determine soil and water conditions. The layout of the system of vertical wells and *individual outlets* is shown in Figure 8.17. A cross section of a typical well and outlet with details appears in Figure 8.18. The decision to use drainage for the improvement of this slope proved wise, for the system successfully stabilized this slope.

Vertical drain wells are used for bypassing impervious layers of rock or soil that impede the natural drainage in slopes. One, described by Baker and Mar-

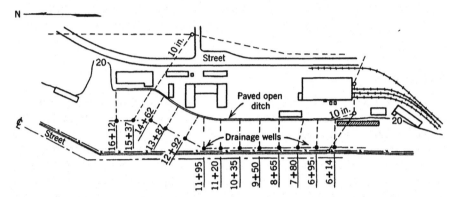

FIG. 8.17 Slide area stabilized with well system. (After Palmer, Thompson, and Yeomans, 1950).

shall (1958a), was installed by the Virginia Department of Highways on U.S. Route 220. A group of eight vertical sand drains allowed vertical seepage into pervious alluvial sand and gravel at the base of the 80-ft holes. The drains bypassed a horizontal clay layer that had made the slope unstable by impeding natural drainage.

To be successful vertical drains should have sufficient discharge capacity to remove the water that reaches them. Seepage studies based on Darcy's law and the flow net (Chapter 3) often permit reasonable estimates to be made of

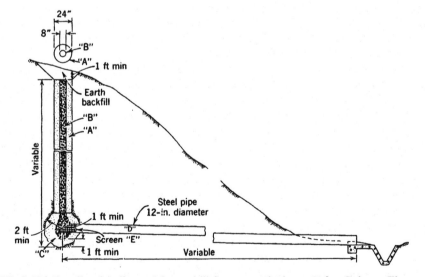

FIG. 8.18 Details of wells used for stabilizing an earth slope. (After Palmer, Thompson, and Yeomans, 1950).

probable seepage rates. If back-filled wells cannot be designed with sufficient capacity to remove the estimated seepage economically, perforated or slotted pipes surrounded with filter materials should be installed. In such cases it is important that the filter materials have sufficient permeability to permit seepage to enter the pipes freely. If the permeability of the filter material is known, its capability to conduct seepage to the pipe can be estimated by Darcy's law or the flow net. Figure 8.19a shows shape factors that can be used to estimate rates of seepage into pipes through filter materials of known permeabilities or to determine the *required* permeabilities of filter materials that will permit given rates of seepage to enter vertical pipes in drain wells surrounded with filter sand.

To illustrate the use of Figure 8.19a assume that wells 18 in. in diameter which contain perforated pipes 6 in. in diameter are surrounded with filter sand with a horizontal permeability of 10 ft/day. What is the maximum capability of the sand backfill to conduct water horizontally to a 30-ft length of pipe with an average head of 3 ft at the outside edge of the sand? Assume that the pipe is completely drained at the bottom by a pump or outlet pipe. In Figure 8.19a for d_p/d_w = 6 in./18 in. = 0.33, a shape factor n_f/n_d = 5.5 is obtained. For the 30-ft length of drain L, the potential rate of infiltration through the sand backfill Q is obtained from

$$Q = kh\left(\frac{n_f}{n_d}\right)L \tag{8.7}$$

Substituting numerical values,

$$Q = 10 \text{ ft/day(3 ft)(5.5)(30 ft)} = 5000 \text{ cu ft/day} = 26 \text{ gpm}$$

When pipes are used in drain wells, the capacity for removal of water is usually very large and the required well interval is determined largely by the closeness of spacing needed for interception of water-bearing strata or trapped bodies of water. But, if pipes are not used, the limiting factor may be the head losses within the wells and the discharge capacities of the individual wells.

The discharge capacities of gravel or sand-filled wells that discharge freely at the bottom can be estimated by Darcy's law, $Q = kiA$. Assuming a hydraulic gradient of 1.0 within wells, the chart in Figure 8.19b gives discharge capacities of gravel or sand-filled wells draining freely at the bottom. It will be seen, for example, that wells 18 in. in diameter backfilled with graded filter aggregate (k = 10 ft/day) can remove only 0.1 gpm, but the same diameter wells backfilled with pea gravel (k = 3000 ft/day) can remove about 25 gpm. This chart is useful not only for designing well systems but it also gives designers an appreciation for the general capabilities of filled wells to remove water. Too much should not be expected of wells filled with fine filter aggregates, for their capacities are definitely limited.

All types of drain should be "designed" by using the appropriate methods

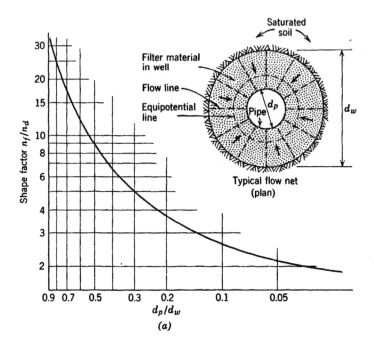

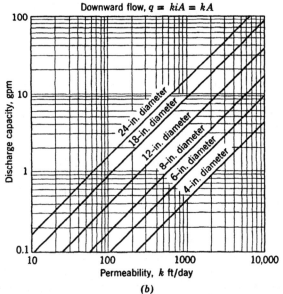

FIG. 8.19 Discharge capacities of drain wells. (*a*) Horizontal flow within filter material surrounding vertical pipes, fully drained at bottom. (*b*) Vertical flow in sand or gravel-filled wells, $i = 1.0$.

outlined in this book to have *sufficient discharge capacity* and to provide *adequate filter protection* to prevent the adjacent soil from piping or clogging the system (Secs. 5.1 and 5.2). Obviously both requirements need thorough consideration in the development of drainage systems, for insufficient capacity for water removal renders drains ineffective and insufficient filter protection can lead to piping of soil with damage or destruction of the facilities being protected or to clogging of the system.

Horizontal Drains

About 1939 the California Division of Highways introduced the *horizontal drain* for stabilization of troublesome highway slopes. A horizontal drain is simply a small-diameter well drilled almost horizontally into a hillside or fill foundation to remove groundwater and seepage. Many of the early installations were made with water-drive drills known as "hydraugers," and some engineers still speak of horizontal drains by that name. Other types of drill are now being used for the installation of these drains, appropriately named *horizontal drain*.

Although horizontal drains are often installed as part of the planned stabilization of cut slopes in wet, unstable ground, they are also installed for the correction of slides that have started to develop in new or old slopes of highways, railways, and other works. Frequently horizontal drains are installed at several levels on benches in cut slopes to lower the water table and prevent slides from taking place while the excavations are being deepened. In shallow cuts horizontal drains are usually installed near road level.

Collector pipe manifolds or paved ditches are provided to take the water away to locations where it can be discharged without danger of reentering the slope and causing further instability. Perforated, asphalt-dipped black iron pipe 2 in. in diameter or slotted PVC pipe 1½ in. in diameter is installed in the full length of a drilled hole. The outer 10 ft. or so is left unperforated, and clay is tamped firmly around the outer few feet of the drain so that all of the water can be safely collected and removed. A pipe plug is screwed into the outer end of each drain, which is fitted with a "Y," so that the discharge of individual drains can be observed and measured. The entire flow of groups of horizontal drains can be measured periodically and the water level in observation wells recorded to determine the behavior and effectiveness of drain systems.

In most horizontal drain installations the initial drains serve as exploratory wells and borings, pointing to areas in which additional drains should be drilled. Finished installations often have drains spaced at intervals of 10 to 50 ft or more. When access to suitable drilling locations is difficult, several drains are fanned out from the most suitable points. Their lengths can vary from about 50 up to 300 ft or more.

Although horizontal drains are used primarily for stabilizing slopes, they sometimes drain and improve the stability of fill foundations. In emergencies a number of horizontal drains can be quickly installed into troublesome slopes or foundations to obtain immediate relief from adverse seepage and groundwater conditions. Under conditions favorable to their use they have stopped many incipient failures when early signs of instability were detected by an alert individual.

An example of a rather complex system of horizontal drains, combined with other measures, is a troublesome landslide near the small town of Towle, California on the west slopes of the Sierras (Smith and Cedergren, 1963). Figure 8.20 shows the plan of the system that was used in the construction of a portion of the Trans-Sierra Freeway.

In the winter of 1957-1958 a serious landslide undermined part of a transcontinental railroad and blocked the two-lane highway shown at the bottom of the plan in Figure 8.20. Waste excavation required to get the railroad back into service was dumped at the head of the slide area between the railroad and the highway. Simultaneously with the dumping of this material, muck poured out over the highway as the loading disturbed the precarious equilibrium of the slide. After the dumping ceased the slide stopped moving, and, subsequently, plans were developed for the construction of a four-lane freeway across this unstable area.

A series of exploratory drill holes (R-1, R-2, R-3, etc.) indicated that firm, undisturbed weathered rock existed at a depth of 40 to 50 ft in the slide area and that stabilization would not be out of the question. After a thorough study

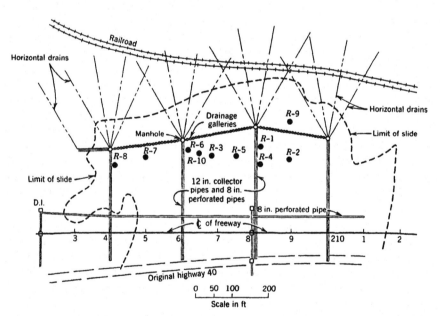

FIG. 8.20 Plan showing stabilization details and boring locations at Towle slide. (Smith and Cedergren, 1963.)

of several alternatives, plans were developed for stabilizing the slide with four transverse stabilization trenches, using horizontal drains to improve the stability of the hillside. Permeable aggregate blankets and pipe drains in the transverse trenches were to be installed to collect and remove underseepage.

The badly disturbed soil within the slide was so weak that extensive predrainage was required to permit the stabilization trenches to be excavated without serious slope failures. Predrainage was obtained with a 700 ft long row of vertical drainage wells installed near the upper ends of the transverse trenches before they were excavated. After these wells were drilled they were interconnected at the bottom by miners who dug out the soil between the wells. After they had been pumped for a number of weeks groundwater conditions improved sufficiently to permit the stabilization trenches to be excavated without serious failures. Horizontal drains were drilled into the base of the hill under the railroad from the bottom of these trenches.

This example is an illustration of a *combined* drainage system that incorporates vertical wells connected by a drainage gallery, horizontal drains, and stabilization trenches.

Horizontal drain installations have been known to produce unbelievably large temporary flows. One of the early systems installed by the California Division of Highways in a slope at Questa Grade (about 1952) produced a combined flow of 3,000,000 gal in 24 days. A single drain in this installation flowed at a rate of 1,000,000 gal/day during the first day and 15,000 gal a day for two years (Fig. 8.21). Four other drains in this system each produced temporary flows of 200,000 gal/day for short periods. In contrast to this example many successful horizontal drains produce flows of 5 gpm or less which indicates that large flows are not always necessary for these systems to stabilize troublesome slopes effectively.

Numerous countries throughout the world have made use of horizontal drains for stabilizing wet slopes. Teixeira and Kanij (1970) describe an unstable area of about 200,000 sq m (50 acres) on the Serro do Mar in the eastern part of the state of Saõ Paulo, Brazil. Two adjacent slides are involved in this area on the Via Anchieta, an important highway connecting Saõ Paulo with Santos. Large pore pressures in the highly micaceous schist, weathered to great depths, caused the slides. Approximately 200 horizontal drains, each about 100 m (328 ft) long, were driven into the soil and weathered rock. According to these authors, these drains improved the drainage sufficiently to raise the factor of safety to a satisfactory level.

Most of the early horizontal drains were installed with perforated or slotted iron pipe, usually asphalt-coated. This type of casing provided no protection from inflowing fine sand, silt, or other erodible fine-grained formations which quickly washed out through or plugged the drains. To overcome these problems a slotted 1½-in. inside diameter PVC pipe has been substituted in recent years for iron pipe.

Horizontal drains are sometimes used for the drainage of dam abutments, narrow ridges around reservoirs, and other earth slopes of water-impounding

FIG. 8.21 "Grand Daddy" of horizontal drains that produced initial flow of 1,000,000 gpd. (Courtesy of the California Division of Highways; photo by G. E. Ebenhack.)

structures. Extreme care is required in all such installations to prevent the use of unfiltered holes that can cause piping failures.

Lutton and Banks (1970) tell about a proposal to use horizontal drains to stabilize Panama Canal slopes in clay shale. After a comprehensive study of the East Culebra and West Culebra slides, making use of slope indicator tubes to define the present depth of sliding, the report recommended the installation of horizontal drains to improve internal drainage, and the sealing of surface cracks and improvement in surface drainage to prevent surface water runoff from entering the cracks.

Toe Buttresses

Many types of improvement involve cutting into the toe of an existing slope. This is often done to provide space for warehouses, garages, homes, parking lots, and roads. If a slope is barely stable, the undercutting can result in sloughing or raveling of material or serious slipouts. To reduce the danger related to undercut slopes, buttress fills are often placed at the toe after the cut has been made. The most desirable time to construct a toe buttress is *before* a slide or slip-out has had a chance to take place. If trouble has developed, it may be necessary to wait for dry weather so that the slope can be undercut (and slide material removed) enough to allow a buttress of sufficient thickness to be placed so that stability can be substantially improved. To be most effective a toe buttress should be as heavy as possible and it should not trap seepage from the formations being stabilized. These objectives are obtained best with *drained buttresses.*

Figure 8.22 shows two methods of constructing drained toe buttresses. In Figure 8.22*a* the fill is constructed of quarry rock, boulders, crushed clean rock, or other highly permeable granular material. In order to safeguard the fill and hold the soil in place, a synthetic fabric filter (or fine aggregate filter) is placed between the soil and the coarse fill material. A shallow trench drain with a suitable pipe conducts water away by gravity drainage and allows the saturation level to be held below the adjacent surface, thus providing an additional degree of stability and eliminating a water nuisance at the toe.

If good quality rock or boulders are not economically available, an effective toe buttress can still be built (see Fig. 8.22*b*). In this design a thin layer of clean crushed rock (1/2 in. or larger sizes) is placed against the slope and on the bottom to provide efficient drainage of water out of the slope. To prevent clogging, this drain is completely enclosed in a suitable fabric filter. The trench

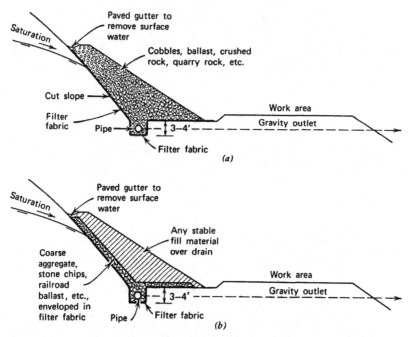

FIG. 8.22 Designs of drained buttresses for stabilizing earth slopes. (*a*) Fill is composed of cobbles, ballast, crushed rock, quarry rock, and the like. (*b*) Fill can be any stable earthen material. *Precaution:* No filter fabric should be placed on a slope unless the coefficient of friction is high enough to preclude the formation of a plane of weakness.

drain is also constructed with the stone, and its bottom and sides are covered with the filter cloth. When the design in Figure 8.22b is used, any well-compacted stable earth fill can be placed over the drain to obtain a fully drained buttress which mobilizes the total wet weight of the fill rather than only the submerged unit weight that would exist without drainage. An effective drain under a buttress fill can at least double the benefits of a buttress in resisting movement of a slope.

Figure 8.23 is a photo of a slide on a Forest Service road in northern California that started out as a small slip after a narrow logging road was cut along this slope, which was barely stable before the cut was made. Each wet season for three years, the slide became progressively larger and higher, until as shown here it extends about 200 feet above the roadway. It was stabilized by removing loose material from the lower part of the slide, and constructing a drained buttress of the general type shown in Figure 8.22b. Without stabili-

FIG. 8.23 A troublesome slide in a Forest Service road in northern California. A drained buttress, installed after this photo was taken, prevented further movement.

zation measures, if any additional efforts had been made to keep the road open to logging trucks, the slide would probably have progressed hundreds of feet further to the top of the slope.

8.4 INFLUENCE OF SOIL AND GEOLOGICAL DETAILS ON DRAINAGE

Minor soil and geological details in earth slopes can have a major influence on the drainage patterns that develop naturally and on the effectiveness of artificial drainage systems. Impervious barriers near the outer extremities of slopes can retard natural drainage, raise the general saturation level, or cause pools of water to build up within slopes. When this occurs, the stability of slopes can often be greatly improved by drainage.

As a general rule, slender drains that contact only minute portions of the total volumes of earth in slopes are most effective if they tap internal pools of water or intercept permeable joints, seams, or cracks. In homogeneous soil or rocks, drains must be closely spaced if they are to control groundwater effectively. In homogeneous, impervious formations horizontal drains or wells may remove such low rates of flow that they will be quite ineffective; however, in pervious or semipervious formations that develop well-defined seepage patterns toward the drains substantial improvement in stability may be obtained. The flow nets in Figure 8.24 show that closely spaced horizontal drains in homogeneous, semipervious formations can induce vertical seepage patterns and reduce excess pore pressures to negligible levels. In such cases stability can be greatly improved even though drainage rates are comparatively low (Fig. 8.10). This principle is stated under case *d*, Sec. 8.2.

> Water seeping in a generally horizontal direction destabilizes slopes, whereas water seeping vertically downward produces no destabilizing forces and no pore pressures.

The importance of this principle is discussed in a number of other places in this book. In his study of the benefits of horizontal drains in earth slopes Nonveiller (1970) correctly concludes that the greatest effect of horizontal drains is achieved when the flow in a hillside is directed vertically downward "because it then cancels uplift."

8.5 GENERAL CONCLUSIONS

High groundwater levels in slopes, both natural and man-made, substantially reduce stability and frequently cause slope failures. Well-drained slopes are inherently more stable than the poorly drained and much more resistant to

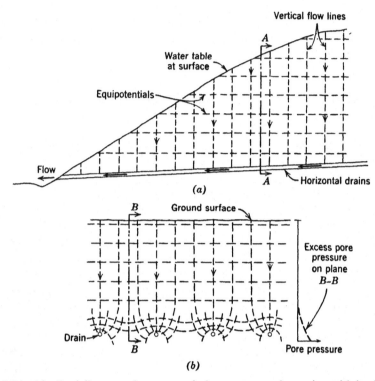

FIG. 8.24 Idealized flow nets for seepage in homogeneous formation with horizontal drains. (*a*) Cross section and flow net. (*b*) Longitudinal flow net at section *A-A*.

failures during severe earthquakes. The construction of appropriate drainage systems based on the analysis of seepage conditions within slopes is one of the best means of protecting people and property from serious slope failures.

A number of drainage methods are available for the stabilization of slopes. Before a new slope is designed or a method is selected for improving an existing slope, a thorough study should be made of soils and geological conditions at the site. Analytical methods of analysis when used in conjunction with thorough field investigations by competent engineers and geologists can help designers and builders to obtain structures that will perform as required with the least risk of failures.

Minor soils and geological details can have a great influence on drainage patterns that develop within slopes and on the effectiveness of drainage systems. Tunnels, vertical well systems, and horizontal drains are most effective when they tap internal pools of water or natural joints, cracks, or seams that serve as secondary feeder drains; however, successful systems can be designed to improve nearly any unstable slope.

As in all other applications of engineering to soil and rock, the effectiveness of designs should be verified by observation of completed works. Piezometer systems should in general be installed in all important slopes stabilized by drainage to verify the effectiveness of the systems and to provide a measure of the inherent stability of the slopes under seasonal fluctuations in groundwater levels. In this manner failures due to unknown groundwater conditions can usually be avoided.

REFERENCES

Baker, Robert F., and Harry C. Marshall (1958), "Landslides and Engineering Practice," Highway Research Board Special Report 29, Washington, D.C., Chapter 8, p. 169; (1958a), p. 179.

Bishop, A. W. (1955), "The Use of the Slip Circle in the Stability Analysis of Slopes," *Geotechnique*, Vol. V, No. 1, March 1955, pp. 7-17.

Bishop, A. W., and Norbert Morgenstern (1960), "Stability Coefficients for Earth Slopes," *Geotechnique*, Vol. X, No. 4, December 1960, pp. 129-150.

Brandl, H. (1976), "Stabilization of High Cuts in Slide Area of Weathered Soils," *Proceedings*, 6th European Conf. on S. M. & Found. Engrg., Vienna, Austria, March 22-24, 1976, Vol. 1, pp. 19-28.

Casagrande, A. (1936), "Characteristics of Cohesionless Soils Affecting the Stability of Slopes and Earth Fills," Journal, Boston Society of Civil Engineers, Vol. 23, pp. 13-32; also reprinted in "Contributions to Soil Mechanics, 1925-1940," Boston Society of Civil Engineers, 1940, p. 257.

Chowdhury, R. N. (1987), "Stability of Soil Slopes," *Ground Engineers' Reference Book*, Edited by F. G. Bell, Butterworths, London, Chapter 11, pp. 11/1 to 11/16.

Clague, John J. (1981), "Landslides at the South End of Kluane Lake, Yukon Territory," Canadian Journal of Earth Sciences, Vol. 18, No. 5, May, 1981, pp. 959-971.

Davis, Stanley N., and Juan K. Karzulovic (1963), "Landslides at Lago Rinihue, Chile," Bulletin. Seismological Society of America, Vol. 53, No. 6, pp. 1403-1414, December 1963.

Fellenius, W. (1936), "Calculation of the Stability of Earth Dams," *Transactions*, 2nd Congress on Large Dams, Vol. IV, pp. 445-462, Washington, D.C.

Gloe, C. S., P. J. James, and C. M. Barton (1971), "Geotechnical Investigations for Slope Stability Studies in Brown Coal Open Cuts," *Proceedings*, 1st Australian-New Zealand Conference on Geomechanics, Melbourne, August 1971, Vol. 1, pp. 329-336.

Hendron, A. J., and F. D. Patton (1985), Technical Report GL-85-5, Waterways Experiment Station Geotechnical Laboratory, Vicksburg, Miss., "The Vaiont Slide, a Geotechnical Analysis Based on New Geologic Observations of the Failure Surface, Vol. I, Main Text, Vol. II, Appendixes A through G," June, 1985.

Lutton, Richard J., and Don C. Banks (1970), "Study of Clay Shale Slopes Along the Panama Canal—Report 1. East Culebra and West Culebra Slides and Model Slope," Technical Report S-70-9, U.S. Army Engineers Waterways Experiment Station, Vicksburg, Miss., Nov. 1970, 326 p.

Marsal, R. J. (1961), "Behavior of a Sandy Uniform Soil During the Jaltipan Earthquake, Mexico," *Proceedings,* 5th International Conference on Soil Mechanics and Foundation Engineering, Paris, July 17-22, 1961, Vol. III, pp. 229-233.

May, D. W. (1936), "Application of the Planimeter to the Swedish Method of Analyzing the Stability of Earth Slopes," *Transactions,* 2nd Congress on Large Dams, Vol. IV, pp. 540-543, Washington, D.C.

Nasmith, Hugh (1964), "Landslides and Pleistocene Deposits in the Meikle River Valley of Northern Alberta," Vol. I, Number 3, *Canadian Geotechnical Journal,* pp. 155-166.

Nonveiller, E. (1970), "Stabilization of Landslides by Means of Horizontal Borings," *European Civil Engineering,* No. 5, pp. 221-228.

Palmer, L. A., J. B. Thompson, and C. M. Yeomans (1950), "The Control of a Landslide by Subsurface Drainage," *Proceedings,* Highway Research Board, Vol. 30, pp. 503-508.

Schuster, Robert L., and Glenn F. Embree (1980), "Landslides Caused by Rapid Draining of Teton Reservoir, Idaho," Proc., 18th Annual Engineering Geology and Soils Engineering Symposium, Boise, Idaho, April 2-4, 1980, pp. 1-14.

Seed, H. B. (1964), "The Turnagain Slide, Anchorage, Alaska," a paper presented at the 1964 National Meeting, Association of Engineering Geologists, Sacramento, Calif., October 28-November 1,1964.

Seed, H. B. (1966), "A Method for Earthquake Resistant Design of Earth Dams," *Journal,* Soil Mechanics and Foundations Division, A.S.C.E., Vol. 92, No. SM1, Proc. Paper 4616, January 1966, pp. 13-41.

Sharp, J. C. (1970), "Drainage Characteristics of Subsurface Galleries," *Proceedings,* 2nd Congress of the International Society of Rock Mechanics, Belgrade, Theme 6, No. 10, 8 pp.

Sherard, James L., Richard J. Woodward, Stanley F. Gizienski, and W. A. Clevenger (1963), *Earth and Earth-Rock Dams,* Wiley, New York, pp. 116-23; (1963a), p. 407; (1963b), p. 159; (1963c), p. 326; (1963d), pp. 354-358.

Smith, T. W., and Harry R. Cedergren (1963), "Cut Slope Design and Landslides," ASTM Special Technical Pub. No. 322, Philadelphia, pp. 135-158.

Taylor, D. W. (1948), *Fundamentals of Soil Mechanics,* Wiley, New York, pp. 432-438; (1948a), pp. 426-427; (1948b), pp. 543-544.

Teixeira, A. H., and M. A. Kanij (1970), "Stability of Slopes in the Serro do Mar Area 500 m Above the Road Connecting Sao Paulo and Santos," *Proceedings,* IV Congress of Brazil Society of Soil Mechanics and Foundation Engineering, Rio de Janeiro, August 1970, Vol. 1, No. 1, pp. IV-33-IV-53.

Terzaghi, K. (1943), *Theoretical Soil Mechanics,* Wiley, New York, p. 254.

Terzaghi, K., and Ralph B. Peck (1948), *Soil Mechanics in Engineering Practice,* Wiley, New York, p. 322.

Zaruba, Q (1987), "Landslides and Other Mass Movements," *Ground Engineers' Reference Book,* Edited by F. G. Bell, Butterworths, London, Chapter 10, pp. 10/1 to 10/14.

CHAPTER NINE

ROADS, AIRFIELDS, AND OTHER SURFACE FACILITIES

9.1 IMPORTANCE OF PROTECTING SURFACE FACILITIES FROM WATER

The Size of the Problem

Throughout the developed parts of the world, flat-lying surface facilities of many kinds are being seriously damaged by water, because they are not well drained (internally). These facilities, which include roads, airfield pavements, city streets, parking lots, playing fields, stadiums, racetracks, and railroad roadbeds, are all exposed to substantial inflows of water, but do not drain readily unless provided with systems of the kind described in this chapter and in other recent publications (see Cedergren, 1974, 1987a, 1988).

Even though roadbuilders have known for centuries that water destroys pavements, hardly any pavements built in the past several decades are well drained. Most modern designers erroneously believe that if they make pavements strong, there is no need for fast internal drainage. This belief evolved during two major experimental road tests: that by the Western Association of State Highway Officials (WASHO) in Idaho in 1954, and that by the American Association of State Highway Officials (AASHO) in Illinois in 1958–1960. Though hundreds of combinations of pavement and base were tested, not a single test pavement was well drained. The members of the Task Force overseeing these tests were thinking only in terms of strength, not at all of drainage as a viable option that could greatly extend pavement life at little or no added original cost, thereby saving billions of dollars a year for those paying for pavement systems.

So, even though these tests proved conclusively that free water in structural sections accelerates damage rates hundreds of times over the damage-rates with no free water present, good drainage was ignored as a better way to design pavements. The thinking of designers at that time is typified by the views of one high-ranking member of the Task Force, who later said to a world-wide gathering of pavement designers: "The pertinent question should be, What is underneath the pavement?—not what falls on top of it." (See Proceedings, International Conference on the Structural Design of Asphalt Pavements, University of Michigan, Ann Arbor, Aug. 20-24, 1962, p. 20). Unfortunately, that designer's belief that fast removal of the water that soaks into pavements through their tops is unimportant has pervaded the pavement design profession and dominates the thinking of most designers even to this day. It is responsible for the premature failures of thousands of miles of pavements everywhere and losses of billions of dollars a year that could have been avoided by good drainage practices. Fortunately, a few designers are adopting the good drainage concepts and are designing truly well drained pavements.

More recent investigations I have made of problems with water in pavements for the Federal Highway Administration (FHWA) (1973), and for the U.S. Army Corps of Engineers Construction Engineering Research Laboratory (CERL) (1974), together with my own personal observations, prove that well drained pavements can outlast conventional poorly drained ones at least three or four times. Figure 9.1 gives a graph with my estimate of the normal life expectancies of the existing undrained pavement systems and the life ex-

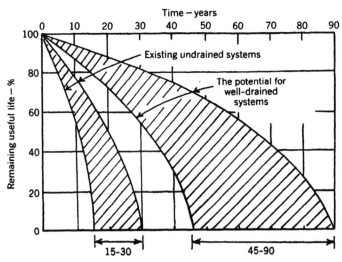

FIG. 9.1 Chart comparing potential life expectancies of well-drained pavements with the conventional "undrained" pavements.

pectancies of pavements designed with good internal drainage systems described in this chapter.

The development of realistic practices for protecting roads, airfields, and other surface facilities from the damaging actions of water is one of the biggest and most demanding jobs facing engineers. Overcoming the archaic belief that drainage is not necessary is the biggest obstacle to progress (see Cedergren, 1982, 1987).

Most of the world's pavements are so leaky that far more water soaks in than can drain away into the subsoil. When most subgrade soils are compacted sufficiently to support heavy vehicle loads, their permeabilities are cut down to a level that allows only miniscule amounts of water to drain downward. Even sandy soils that might appear to the layman to be porous and permeable become nearly watertight when thoroughly compacted. This is true in the San Joaquin Valley of California, where freeways break up from excess water adjacent to vineyards on clean-looking sandy soils. In New England I made percolation tests on a sandy soil subgrade for a major airbase which was said by local engineers to be porous and "free draining." A typical percolation test in a hole drilled under a taxiway pavement (which was cracking up from excess water) gave a coefficient of permeability of 0.002 ft/day (7×10^{-7} cm/sec) (Cedergren and Godfrey, 1974). This level of permeability, which is not at all unusual, would allow about 0.024 in. (0.6mm) of water to soak in per day, or about 1 in. of rainfall in 40 or 50 days. Most subgrades have even lower permeabilities than this.

Compounding the problems caused by impermeable subgrades is the fact that most pavements have extremely impermeable shoulders that serve as "dams" to trap water for long periods of time. The result is major damage and shortened life. The slow drainage of water out of most pavements, which often shows up as "bleeding" stains across shoulders, has led to the use of the terms "box," "trench," and "bathtub" to express the condition existing in most of the pavements throughout the world. Getting water out rapidly with the improved drainage systems described in this chapter could save billions of dollars in the long run for those who pay for pavements.

Even with the abundance of proof that designing undrained pavements is a very foolish practice, it is continuing to the present time. A U.S. Congressman called for a probe to find out why portions of Interstate 75 in Sarasota County, Florida were breaking up in less than five years after construction (see *Engineering News-Record,* Jan. 16, 1986, p. 15). Florida Transportation Secretary Thomas E. Dawdy said the problem could cost more than $10 million in repairs of the $90 million 29-mile segment of I-75. He said improper design— then standard in the state—allowing water to penetrate road joints and become trapped between the structural concrete slab and its lean-concrete base was the cause. "This was the bathtub design used in building Interstates. It is the technique emanating from members of a national task force in the 1950's and 1960's who thought this design was the greatest thing," said Dawdy, according to the *EN-R* article.

Consequences of Internal Flooding

All types of surface that must support the traffic of vehicles, aircraft, and even animals and people must always have sufficient strength to carry the required loads. Excess water in the base course of a highway or in the subgrade of an athletic field can take these facilities out of service. Good drainage capable of preventing prolonged internal flooding is one of the most effective means of ensuring long, trouble-free service of facilities of the kinds discussed in this chapter.

Even though it is known that surface water gets into most pavements, most pavement designers *in fact* visualize that their pavements will behave as well-drained systems. Modern design theories presume a saturated subgrade but start with the assumption that a section will behave as an elastic or semielastic multilayered body in which pore pressures are small enough to be neglected. These theories are not analyzing some of the primary actions that destroy pavements. The well-known phenomenon, *mud pumping,* occurs only when concrete pavements are in a flooded state (Fig. 9.2), and the pounding of traffic on flexible types of pavement that are internally flooded forces mud into incipient cracks, causing the loss of cohesive strength and general deterioration.

Even though paved areas are designed to carry loads on saturated sub-

FIG. 9.2 Water and mud squirting up through joints in a concrete pavement under impact of a truck. "Improved" design with PCC on CTB, yet is deteriorating after only 12 years of traffic; on 50-ft. high fill; no groundwater. (From *Drainage of Highway and Airfield Pavements,* Wiley, New York, 1974, p. 47); updated printing by Robert E. Krieger Pub. Co., Malabar, Florida, 1987).

grades, extensive internal flooding should be prevented. Figure 9.3 is a simplified illustration of pressure distributions ordinarily assumed within roadbeds (Fig 9.3a) and the drastic change that is caused by flooding (Fig. 9.3b). *Lateral spreading of loads as ordinarily assumed (Fig. 9.3a) can occur only in well-drained base layers. All theoretical concepts of stress distributions in layered systems depend on intergranular stresses being developed and the assumption that supporting layers are well drained. When structural sections are completely filled with incompressible water, the applied pressures are transmitted downward to the subgrade with little or no reduction in intensity (Fig. 9.3b). Under this condition, normally assumed stress distributions are completely nullified.* Furthermore, serious damage can be caused by excess pore pressures, erosion, and other actions of free water. Cement-treated bases have totally disintegrated under the pulsating pore pressures of heavy traffic impacts, and both untreated and treated bases have been severely eroded under the violent water action that occurs when pavements are subjected to traffic during the time needed for the entrapped water to slowly work its way out. The presence of free water in structural sections allows *frost action* to take place in susceptible materials, and lengthy exposure to free water is a major factor associated with "D" cracking and "blow-up" of PCC pavements. In short, extensive internal flooding can be blamed for about 90% or more of the serious pavement problems. In my view all pavements of importance should be well drained to cut down on damages from traffic impacts and damages from non-load bearing environmental destructive actions in pavements (see Cedergren, 1988).

A number of detailed studies (Barenberg and Thompson, 1970; FHWA, 1973; U.S. Corps of Engineers, CERL, 1974; etc.) have demonstrated that a major percentage of the damage to most pavements can be blamed on the excess water that enters structural sections (primarily from surface water) and remains for long periods of time. Having seen the devastating effects of free water in major highway and airfield pavements from coast to coast, I am con-

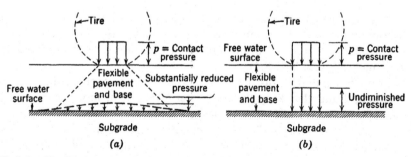

FIG. 9.3 Idealized concept of pressure conditions in well-drained and flooded roadbeds. (*a*) Normal design assumption is that the subgrade is saturated but the structural section is well drained. (*b*) Idealized condition existing in flooded structural sections.

vinced more than ever that the practice of designing important pavements without adequate drainage should be discontinued. To reduce water damage to a minimum fast-draining bases should be required as a standard feature of design for many types of surface facility, including some non-paved areas.

An international symposium held in Paris, France, in March, 1983, on "Drainage and Erodibility of the Base Course Under Concrete Pavement" concluded that "failure of concrete pavements originated from the formation of water between pavement and base course," which supports views expressed in preceding paragraphs.

9.2 PHYSICAL FACTORS COMPOUNDING THE PROBLEM OF DRAINING PAVEMENTS AND OTHER SURFACE FACILITIES

Many Sources of Water

Water can enter subgrades and layers supporting traffic from a number of sources:

1. By flowing downward through porous or cracked surfaces and unsealed construction joints.
2. By flowing laterally into the edges from saturated medians and shoulders.
3. By seeping upward into the structural section from high groundwater and springs.
4. By being pulled by capillarity from the underlying water table.
5. By accumulating as water vapor resulting from fluctuations in temperature and other atmospheric conditions.

The first three sources of water represent seepage under the influence of the forces of gravity, the first *two* from water seeping downward or laterally into pavements, medians, or shoulders; the *third* from hydrostatic pressures that build up because of high surrounding water tables.

Seepage from high groundwater and springs can usually be controlled by the use of longitudinal or transverse trench drains, deep foundation trench drains, open-graded blanket drains protected with filters, or by combinations of these methods (see Sec. 9.4).

The flow of capillary moisture into subgrades can be controlled or reduced by installing drains that keep the free water level 5 ft or more below the top of the subgrade. Also, a layer of coarse, clean, open-graded aggregate in a blanket drain serves as a capillarity barrier or "break" to stop the rise of water into structural sections from saturated subgrades. Neither capillary water from high water tables nor water vapor can be kept from contributing to the moisture content of a subgrade (without special sealing systems); consequently, even in highly arid climates, pavements are designed on the premise that subgrades will become saturated.

Although water from high groundwater, shoulder ponding, and the other miscellaneous sources already noted can contribute to the water in structural sections, the downward flow through porous, jointed, or cracked wearing courses is by far the most bountiful supply in perhaps 90% of the areas in continental United States and in most of the developed countries throughout the world. By far the greatest drainage need for pavements and other surface facilities is that of removing water that enters through their tops, although other sources can be important in limited areas. Ironically, "standard" practice calls for drains for control of groundwater and similar sources but almost *never* for the control of surface water infiltration.

Unfavorable Geometry

Wide flat service areas of many kinds are inherently difficult to maintain in a well-drained condition because of two compounding factors.

1. The areas exposed to surface and groundwater infiltration are large, as are the hydraulic gradients that cause the inflow.
2. The discharge areas and hydraulic gradients available to remove water are small.

As a consequence of these geometrically compounding factors, roads, airfields, parking lots, city streets, playfields, and the like often deteriorate prematurely or are periodically made inoperable unless effective measures have been taken in their design and construction to ensure the rapid removal of the water that enters. Hydraulic principles should be used in designing drainage facilities that will economically accommodate all the water that reaches them.

Problems associated with the drainage of large flat areas subject to intermittent flooding from surface water or groundwater are well known to the farmer. When his fields become saturated by heavy rains or flooding, weeks or months may be required for the land to dry sufficiently by drainage and evaporation before it can be worked. Yet this situation can be greatly improved by the provision of ditches or tile drains spaced at moderate intervals throughout the fields. By narrowing down the widths of individual areas drainage time can be considerably reduced. This simple but fundamental principle is true not only of agricultural lands but also of most of the facilities discussed in this chapter.

The general relationships existing in flat layers subjected to surface or groundwater infiltration can be examined with Darcy's law. The flat, rectangular layer in Figure 9.4, with a thickness h and a total width $2b$, rests on a completely impermeable base. If a uniform rate of infiltration q enters the entire area, a quantity $Q = qb$ flows to each side. According to Darcy's law ($Q = kiA$), the quantity Q is proportional to the hydraulic gradients that are

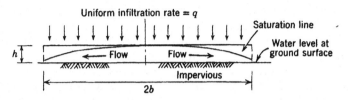

FIG. 9.4 Cross section of base studied with flow nets in Fig. 9.5.

available to induce flow and to the cross-sectional areas through which water is being discharged to the sides. Because the effective hydraulic gradient i is proportional to h/b and the effective discharge area is proportional to thickness h,

$$Q \sim k\,\frac{h}{b} \qquad h \sim k\,\frac{h^2}{b}$$

but the infiltration rate $q = Q/b$, hence

$$q \sim k\left(\frac{h}{b}\right)^2 \tag{9.1}$$

and

$$\frac{q}{k} \sim \left(\frac{h}{b}\right)^2 \tag{9.2}$$

Thus one should expect wide shallow layers of soil or other moderately permeable materials to be extremely slow draining, for the amount of water that can be accommodated varies as the *square* of the ratio of h to b.

Barber (1962) points out that subsurface drains were required in clean permeable sand at John F. Kennedy (Idlewild) Airport because the natural transmission rate of this wide thin deposit was considerably smaller than the annual rainfall.

Equations 9.1 and 9.2 show *qualitative* relationships between the shapes of flat layers and their ability to discharge water to the sides. These relationships can be developed *quantitatively* by drawing flow nets (Fig. 9.5) for a range of values of h and b and determining the shape factors for each of the flow nets. Doing this, and using Eq. 3.18 (Chap. 3) produced the following equation: $q/k = (h/b)^2$, which demonstrates that Eq. 9.2 is a *quantitatively* correct relationship.

Basic relationships developed with Darcy's law and flow nets emphasize the difficulties of securing good subdrainage of multilane highways, airfield pavements, large parking lots, and other large flat areas. The relatively simple studies described here demonstrate that highly permeable drainage layers are

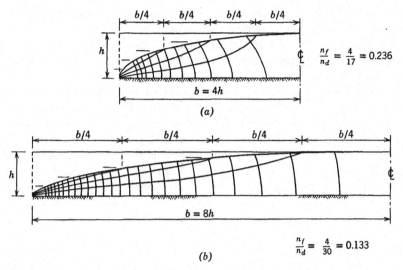

FIG. 9.5 Typical flow nets for infiltration study of base on impermeable foundation (cross section as shown in Fig. 9.4).

nearly always required to prevent prolonged internal flooding of the facilities discussed in this chapter.

9.3 THE BASIC SOLUTIONS

Although it may be possible theoretically to design and build pavements that are stout enough to carry their required loads on flooded roadbeds, it is more economical to provide good drainage systems capable of preventing prolonged flooding. Two basic methods are available for the protection of pavements and other shallow surface facilities from the damaging actions of water: (1) those that keep the water out and (2) those that quickly remove water that has entered.

Methods That Keep Water Out

Careful attention to the crowning of paved areas, the provision of minimum longitudinal slopes, and close construction control to avoid local depressions contribute to fast surface runoff and a minimum of standing of water on pavements. Nearly every designer of roads, airfields, parking lots, and other surface facilities has had the hope that the surfaces of these facilities can be kept watertight and that this will eliminate problems with water (except in areas of springs and high groundwater). Natural elements, such as temperature changes, thermal shrinkage, freezing and thawing, and the like, are constantly working to open cracks and joints in essentially every paved area, and although every effort should be made to try to minimize the amounts of water that can enter pavements *keeping water out* should not be depended on to prevent water damage. *Good subsurface drainage is the best way to control water in structural sections,* for even regularly maintained pavements can absorb water at high rates.

Figure 9.6 illustrates this problem. An inflow test is being made (in 1972) of a supposedly "sealed" construction joint in a PCC pavement at a major airport (constructed in 1969). This sophisticated joint has opened enough in three years to allow a gallon of water to enter 5 ft of joint in a minute. Even though the permeability of the concrete itself is virtually zero, this pavement with joints every 25 ft is accepting water at an equivalent rainfall rate of 0.75 in./hr. This amount may appear to be surprisingly large but should not be unexpected in the light of similar experience throughout the world. Water flowing out of edge joints after only small amounts of rain have fallen is a common sight.

Asphaltic pavement can have effective permeabilities from virtually zero up to several hundred feet a day (0.1 cm/sec). Barber and Sawyer (1952) reported coefficients of permeability of 0.16 to 0.84 ft/day (5.6 $\times$ 10^{-5} to 2.9 $\times$ 10^{-4} cm/sec) for various samples of bituminous concrete compacted in the laboratory under a static load of 3000 psi and then heated to 140°F for 24 hr and 0.0002 ft/day (7 $\times$ 10^{-8} cm/sec) for a traffic-compacted sample taken from

FIG. 9.6 Making inflow test of joint in 3-year old PCC pavement (in 1972). (From *Drainage of Highway and Airfield Pavements,* Wiley, New York, 1974, p. 69); updated printing by Robert E. Krieger Pub. Co., Malabar, Florida, 1987).

a roadway. In 1965 I tested a moderately compacted, dense-graded bituminous pavement slab with an area of 85 sq in. and obtained a coefficient of permeability of 100 ft/day (0.03 cm/sec).

If the permeability k of a pavement is known, the potential rate of infiltration during rains can be estimated with Darcy's law $q = kiA$. By using a hydraulic gradient of unity

$$q = k(1.0)A = kA \qquad (9.3)$$

in which A is the area of pavement through which the rate q enters. For unit area the potential infiltration rate can therefore equal the permeability. Thus a pavement with a permeability coefficient of 40 ft/day is capable of absorbing 40 cu ft of water through each square foot per day. Obviously extremely permeable underlayers would be required to remove rates of infiltration of this order of magnitude.

After a thorough study of the problems caused by surface water infiltering into pavements the Federal Highway Administration (1973) issued a report that says:

Since the present 'State-of-the-Art' of constructing roads does not guarantee watertightness of pavements for more than brief periods of time, subsurface drainage systems are needed for modern rigid and flexible highway pavements.

Since all of the specific locations where water can enter structural sections cannot be predicted in advance, *subsurface drainage systems are needed for the full width of pavements that may be subjected to significant numbers of heavy wheel impacts while the sections contain excess water.*

Drainage Methods

In contrast to dams and other hydraulic structures, which have many of their seepage problems created by the water they control, roads, airfields, and similar engineering improvements are subjected to damage from natural groundwater and precipitation. The design of drainage systems for projects included in this chapter requires a thorough study of climatic conditions and subsurface soil and groundwater conditions.

If there is evidence of subsurface seepage, borings may be needed to delineate these areas and to aid in estimating probable seepage rates. Depending on the conditions and the magnitude of the work, field pumping tests (Sec. 2.6) may be needed for evaluation of in-place soil permeabilities.

In regions in which heavy rainfall can be expected it is good practice to install lightweight casing in borings and to record the depths to the water table during at least one rainy season. Severe groundwater conditions that go undetected can cause serious failures after roads are completed (Fig. 8.5).

Designing drainage systems for roads, airfields, and comparable works depends on locating all sources of water and designing systems capable of intercepting and removing it to prevent its accumulation in the supporting layers.

A method available for estimating quantities of surface water infiltering pavements and needing to be removed by drains is the procedure given by the Federal Highway Administration (1973). The FHWA report recommends that the *surface infiltration rate* be predicted on a *design precipitation rate*. The *surface infiltration rate* is one-third to two-thirds of the design precipitation rate which is the 1 hr/1 yr frequency precipitation rate at a specific location. The fraction used depends on the percentage that the designer anticipates may enter a pavement after it has been in service a number of years. In the U.S. the design precipitation rate varies from 0.2 in./hr in some western areas up to 2.4 in./hr in some southeastern areas.

In determining the amount of water to be removed by drains, the FHWA report says:

> The *design inflow rate* to be used in designing a subsurface drainage layer should be the sum of surface infiltration and all other significant inflows (groundwater, seepage, etc.) that is converted to an equivalent rate in inches/hour.

In rare cases subgrades may be sufficiently permeable to provide some beneficial downward drainage. If pavement structural sections are to be protected from harmful accumulations of free water, the *outflow capabilities* of subsurface drains must be at least equal to the total inflows from all sources less any dependable outflows (Cedergren, 1974, 1987).

Having estimated all of the inflows that neet to be removed from a flatlying surface facility (applying reasonable factors of safety), the *transmissibility* requirements of the drain can then be estimated from the following relationship developed in Chapter 5:

$$\frac{Q}{i} = kA \qquad (5.9)$$

In Eq. 5.9, Q is the quantity of water flowing in the drain, i is the hydraulic gradient *in the drain;* and the product of k (the permeability of the drain) and A (the cross-sectional area of the drain perpendicular to the direction of flow) is the minimum required *transmissibility*. This fundamental relationship should be used for designing drains for all of the facilities in this chapter.

9.4   PAVEMENT DRAINAGE SYSTEMS

Components of Subsurface Drainage Systems For Highways

A properly designed subsurface drainage system for a highway should have the following basic components (FHWA, 1973; Cedergren, 1974, 1987), as shown in Figure 9.7:

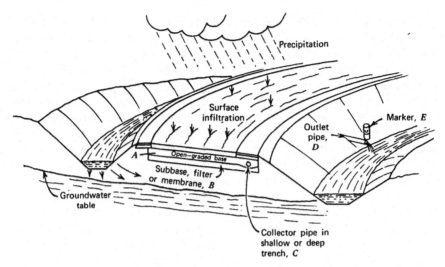

FIG. 9.7   Basic components of subsurface drainage systems (FHWA, 1973).

A. A full-width *open-graded base drainage layer* under the entire pavement being protected, with sufficient transmissibility and permeability to provide protection from all sources of water.

B. A subbase of other suitable aggregate or plastic "filter" or "separator" under the open-graded layer to keep subgrade fines out.

C. Collector pipes along lower edges.

D. Outlet pipes as needed to ensure effective discharge to prevent trapping of water.

E. Outlet markers and posts to protect the pipes and make the outlets easy to find by maintenance personnel.

The open-graded layer is a key feature of a drainage system; hence the performance record of such materials is of great interest to pavement designers. For countless decades, road builders have known of the good structural properties of well-interlocked, angular aggregates. Outstanding road builders who made use of such materials are Thomas Telford, Pierre M. Tresaguet, and John L. McAdam. When coarse open-graded bases are well drained, they provide excellent support for heavy-duty pavements. Richardson and Liddle (1973) tell about U.S. Forest Service roads in the Pacific Northwest that have supported logging trucks weighing 100,000 to 200,000 lb on a layer of emulsified asphalt mix that contains no fines. No shoving or rutting occurs, according to these authors, even when logging trucks used these roads *before the emulsion has had time to cure.* They say that leaving the fines out allows the use of borderline aggregates because there is less breakdown under heavy wheel loads. Winterkorn (1967) and Garbe (1974) also present evidence supporting the conclusion that open-graded aggregate with a high percentage of fractured faces is one of the best materials available for pavement construction. Fortunately it is also one of the best drainage materials available.

Craven (see *EN-R,* 1979), Drake (1979), Haughton (1986), and Lorin (1986) have made notable contributions in pavement drainage by using systems making use of high-permeability drainage layers in their structural sections. Forsyth et al. (1987) describe positive steps being taken by the California Department of Transportation (CALTRANS) to upgrade pavement drainage design standards in their state. After several years of field studies, and a laboratory simulation of faulting in 1981, Caltrans researchers concluded that the State's standard practice (since 1952) of using undrained cement-treated bases under concrete wearing courses slowed erosion and the incidence of faulting, but did not eliminate it. Research under Mr. Forsyth's direction eventually led to today's standard Caltrans designs for new roads, which use a highly permeable base directly under the PCC wearing course, and edge drains using slotted pipe enclosed in filter fabrics to prevent intrusion of soil fines.

When sound drainage principles are used in designing systems for the protection of roads and other facilities described in this chapter, they should function permanently with no serious reduction in efficiency. In the design of base drainage layers, or *blanket drains* for roads and other surface facilities, the criteria given in subsequent paragraphs will aid in ensuring sufficiently high *permeabilities* and *transmissibilities* to meet the needs of these facilities.

Criteria for Designing Blanket Drains. Two important criteria to be checked in the design of a blanket drain for a highway, airfield, or other important pavement are (1) the *transmissibility* must be sufficient to remove all water and (2) the *permeability* must be sufficient to drain it before serious damage can occur (most important in cold climates). When an analysis is being made of these factors, the seepage analysis should be made along the direction of the flow of particles of water. A strip of pavement and drain 1 ft wide can be analyzed conveniently. Flow in a drainage layer will always be *downslope* or crosswise to a level line (contour or equipotential line). This is illustrated in Figure 9.8 in which a blanket drain has a longitudinal slope g and a cross slope s. A flow path of water particles will make an angle α with the longitudinal slope, with $\tan^{-1}\alpha = s/g$: for example, if the cross slope is zero, $\tan^{-1}\alpha = 0$ and flow will be longitudinal with no cross component; if the longitudinal slope is zero, $\tan^{-1}\alpha = \infty$ and $\alpha = 90°$ and flow will be purely lateral with no longitudinal component; if the two slopes are equal, $\tan^{-1}\alpha = 1.0$, and flow will be diagonally at an angle of $45°$ with the longitudinal and cross slopes.

In preparing to make an analysis of flow in a drainage blanket, a flow path is easily located because the distances W and X (Fig. 9.8) are proportional to slopes s and g, respectively. If a cross width W is known, the longitudinal distance $X = W(g/s)$. In order for line AC in Figure 9.8 to be a flow path a line intersecting AC at $90°$ must be horizontal (flow lines must intersect level lines at $90°$ in homogeneous isotropic media). In Figure 9.8 line BE must be

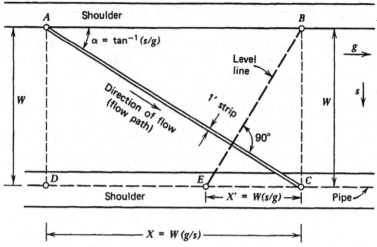

FIG. 9.8 Locating a flow path in a blanket drain with a longitudinal slope g and a cross slope s.

a level line if line AC is a flow line as presumed. From the similar triangles ADC and ECB it is easily demonstrated that line BE must indeed be a level line.

Proof. From similar triangles ADC and, ECB, $W:X::X':W$, and, because $X' = W(s/g)$ the drop in elevation between points E and $C = X'g = W(s/g)g = Ws$; but the drop in elevation between points B and C must equal the cross width times the cross slope $= Ws$. Thus it has been shown that the drop in elevation between points B and C is the same as between points E and C. Because the differences are measured from a common point C, points B and E must be at the same elevation; hence line BE must be a level line (contour). Because line AC is at right angles with line EB, line AC must be a flow path.

Example. Assume that $W = 40$ ft, $g = 0.05$, and $s = 0.02$. Then $X' = W(s/g) = 40(0.02/0.05) = 16$ ft, and the drop in head between points E and $C = X'g = 16(0.05) = 0.8$ ft; but the drop in elevation between points B and $C = Ws = 40(0.02) = 0.8$ ft. Therefore line EB must be a level line.

To ensure that a blanket drain will be able to remove all water that enters a structural section the minimum *transmissibilities* can be estimated by Darcy's law in the form

$$kA = \frac{Q}{i} \qquad (5.9)$$

In Eq. 5.9 Q is the quantity of water that needs to be removed by a drainage blanket, i is the allowable hydraulic gradient (usually its slope), k is the coefficient of permeability of the blanket, and A is its cross-sectional area normal to the direction of flow. Once the required transmissibility is known, any suitable combination of k and A that provides at least that transmissibility (with a reasonable factor of safety) will remove the inflowing water. The transmissibility chart in Fig. 5.7 facilitates the selection of suitable values for k and A for any amount of transmissibility needed.

Assume, for example, that a strip of drainage blanket 1 ft wide must remove 200 cu ft/day under a hydraulic gradient of i of 0.02. Then $Q/i = 200/0.02 = 10,000$ cu ft/day $= kA$. In Fig. 5.7 it is shown that a transmissibility of 10,000 cu ft/day can be obtained by using 0.5 ft with $k = 20,000$ ft/day, 1.0 ft with $k = 10,000$ ft/day, and so on.

The second criterion is that the permeability must be sufficient to allow water to drain out before damage can occur. If freezing in cold climates is to be prevented, any water that enters a drainage system should be able to drain out in a relatively short time. Suggested criteria are 1 hr for highway pavements and 2 hr for the thicker pavements on most airfields. According to Darcy's law, seepage velocity v_s is expressed as

$$v_s = \frac{ki}{n_e} \qquad (3.1)$$

and the minimum *required* permeability is

$$k = \frac{v_s n_e}{i} \qquad (9.4)$$

In Eq. 9.4 v_s is the seepage velocity in the drain, n_e is its effective porosity, i is the hydraulic gradient *in the direction of flow,* and k is the minimum required permeability.

Assume, for example, that water must flow a distance of 100 ft on a slope of 0.015 in not more than 1 hr in a drainage layer with an effective porosity of 0.3. The required $v_s = 100$ ft in 1 hr $= 2400$ ft/day, and the minimum required permeability $k = v_s n_e/i = 2400 \, (0.3)/0.015 = 48,000$ ft/day. This is the permeability needed to allow water to flow out of the blanket drain in 1 hr.

In freezing climates not only must water flow rapidly out of structural sections and drains but the exit regions of drains must be protected from freezing. Ring (1974) suggests insulating pipes as a possible remedy to the problem of ice-blocked pipe outlets.

Transverse Interceptor Drain For Highways

Seepage that enters roadbeds in hilly terrain can flow downgrade into fills and cause subsidences and slipouts if not intercepted. To keep groundwater out of fills, transverse interceptor drains should be constructed at the downgrade ends of water-bearing cuts, as shown in Figure 9.9. Shallow transverse interceptor drains are also constructed at frequent intervals in hilly terrain to remove water from shallow blanket drains installed for the removal of surface water and groundwater.

Some highways that have been constructed with underlying so-called "pervious" bases and interceptor drains for the removal of groundwater have been severely damaged by excessive seepage because their permeabilities were too low. One road in hilly terrain in a relatively high rainfall area was built on a "pervious" drainage layer with transverse interceptor drains at the lower ends of a number of cuts. At a typical trouble spot the pavement was damaged throughout the length of a wet cut, and seepage emerged from the pavement near the lower end of the cut for many weeks after the rainy season. The transverse drain remained dry throughout the year. An investigation disclosed that the "pervious" drainage layer contained 5 to 6% of plastic material finer than a number 200 sieve and had permeability coefficients of only 1 to 2 ft/day. An interesting study by Darcy's law emphasizes the complete inadequacy of this base to conduct water to the transverse drain. If groundwater enters the roadbed in a typical cut as shown in Figure 9.9 how much time would be required for seepage to travel a distance $L = 300$ ft (an average distance) to reach the drain at the lower end of the cut?

Darcy's law can be written (Chapter 3)

$$v_s = \frac{ki}{n_e}$$

For a coefficient of permeability of 2 ft/day, a slope of 0.05, and an assumed effective porosity of 0.20, $v_s = 2$ ft/day $(0.05)/0.2 = 0.5$ ft/day, and the time for seepage to travel 300 ft to reach the drain $= L/v_s = 300/0.5 = 600$ days. Obviously, this roadbed is not well trained, even though supposedly it was provided with a drainage system.

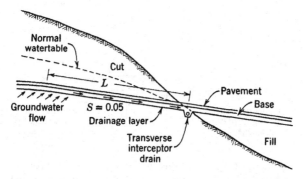

FIG. 9.9 Transverse interceptor drain for highway.

Longitudinal Trench Drains For Highways

In mountainous or hilly terrain water frequently flows downhill with a free surface somewhat parallel to the ground surface. If deep excavations are required for the construction of highways, serious slipouts and slides may damage the roads (Fig. 8.5) unless stabilization trenches, horizontal drains, or other special drainage measures are provided.

In regions of high groundwater, either in hilly and mountainous country or on flat terrain, the groundwater must be intercepted before it can enter roadbeds. Longitudinal trench drains (Fig. 9.10) can often be used for effective control of subsurface water. In a sloping terrain a trench may be excavated along the uphill side of the roadbed, as in Fig. 9.10a. Actual locations and depths of drains depend largely on hydrostatic pressures and soil permeabilities and should be determined with the aid of soil and groundwater surveys. In areas in which the ground is level or nearly so drains may be required along both sides (Fig. 9.10b) and for wide roadbeds along the center also (Fig 9.10c).

The construction of trench drains (Fig. 9.10) requires the use of perforated,

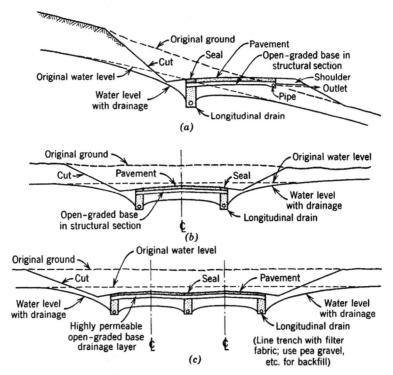

FIG. 9.10 Longitudinal highway drains. (a) Side hill construction. (b) Narrow road in flat terrain. (c) Wide road in flat terrain. In all cases free flow of surface water from structural section into trench drain must be assured.

jointed, slotted, or porous pipe placed near the bottoms of trenches and surrounded with pea gravel or other selected pervious filter aggregate (Sec. 5.2). The main backfill should be specially selected pervious filter aggregate designed to allow unrestricted flow of water to the pipes. The minimum required permeability of the trench backfill material can be determined with Darcy's law, using a downward gradient of 1.0 and a factor of 10 to 20 to allow for contamination. Assume, for example, that a flow of 200 cu ft/day will enter a line drain with a width of 1.5 ft. The minimum $k = 20 (200)/1.5 = 2670$ ft/day. Pea gravel, stone chips, or comparable materials are needed to ensure this level of permeability. When a drain is excavated in erodible materials, synthetic filter fabrics should be used to line the sides and bottom of the trench to prevent soil from entering the coarse backfill in the drain (see Sec. 5.7).

The location of perforations and open joints in pipes should be adapted to conditions at individual locations. Perforations or slots of sufficient size and number should always be provided to allow unobstructed flow to pipes.

If a drainage pipe is completely surrounded with specially selected coarse filter aggregate (Sec. 5.2), perforations can completely surround the pipe. Tight unjointed sections of pipe should be used to lead the collected water across areas in which the feeding of water into the soil from drains must be prevented.

The required diameters of both corrugated metal and concrete or plastic (smooth) drain pipes for a wide range of discharge quantities can be determined from the nomograph in Figure 9.11, which was developed by the U.S. Army Corps of Engineers.

Most drains should be equipped with pipes because gravel or rock-filled trenches have limited discharge capabilities even when clean aggregates are used. The discharge capabilities of drainage trenches backfilled with clean stone or gravel, as estimated by Darcy's law, are given in Table 9.1.

In contrast with quantities in Table 9.1, a 6-in. diameter smooth pipe can discharge 225 gpm on a slope of 0.01 and 80 gpm on a slope of 0.001.

When large hydrostatic pressures exist in pervious strata beneath roadbeds, closely spaced and relatively deep trench drains may be needed to prevent excess pressures from building up. A flow net solution for this condition is summarized in Figure 9.12b and a typical flow net is reproduced in Figure 9.12c. An excess hydrostatic head h is assumed to exist in a pervious stratum at a depth d below the roadbed, and trench drains to depth z are assumed to be excavated into the foundation soil at a spacing $b = 2d$ (Fig. 9.12a). An excess head h' develops at the bottom of the roadbed. For various ratios of h/d the chart (Fig. 9.12b) relates the ratio h'/d to z/d. It is seen, for example, that if the excess head h is 40% of d ($h/d = 0.4$), drains must be more than half the depth d if the excess uplift head h' is to be reduced to a small amount. If larger amounts of head h existed in the pervious stratum beneath the roadway, closer spacing or deeper drains would be needed for effective control over uplift. Factors of safety of roadbeds against blowouts due to excess uplift pressure should be checked with Eq. 10.4.

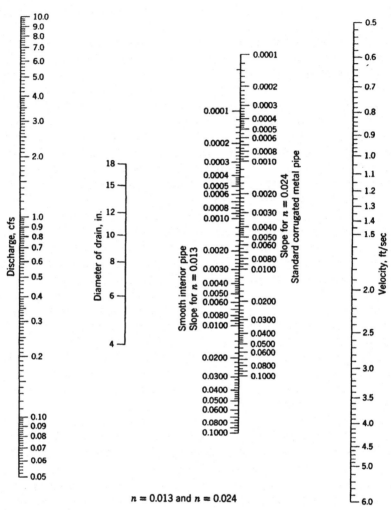

$n = 0.013$ and $n = 0.024$

FIG. 9.11 Drainage nomograph for computing required size of circular drain, flowing full. (U.S. Army, Corps of Engineers.)

TABLE 9.1 Discharge Capacities of 3 × 2 ft Cross Section of Stone-Filled Drains

Size of stone	Permeability, ft/day	Slope	Capacity	
			cu ft/day	gpm
$\frac{3}{4}$ to 1 in.	120,000	0.01	7200	38
$\frac{3}{4}$ to 1 in.	120,000	0.001	720	4
$\frac{3}{8}$ to $\frac{1}{2}$ in.	30,000	0.01	1800	9
$\frac{3}{8}$ to $\frac{1}{2}$ in.	30,000	0.001	180	1
$\frac{1}{4}$ to $\frac{3}{8}$ in.	6,000	0.01	360	2
$\frac{1}{4}$ to $\frac{3}{8}$ in.	6,000	0.001	36	0.2

356

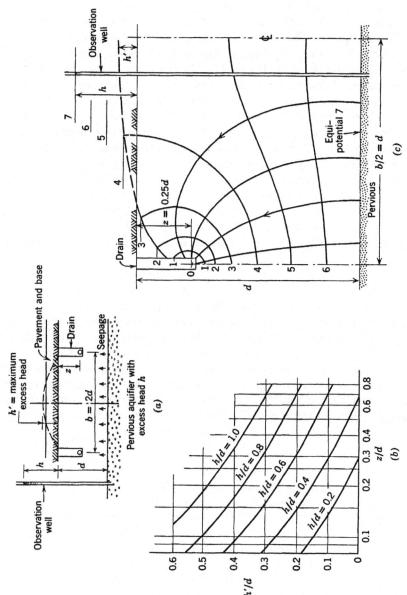

FIG. 9.12 Flow net study of trench drains. (*a*) Cross section. (*b*) Graph giving relationship between depth of trench drains and hydrostatic head beneath impervious pavements. (*c*) Typical flow net for half section.

In many cases in which excess uplift pressures occur beneath roadbeds it is more economical to place graded filter *blanket drains* on the subgrade in conjunction with trench drains or shallow collector drains.

Properly designed blanket drains containing a layer of coarse one-sized crushed rock or gravel (FHWA, 1973; Cedergren, 1974*a*) can protect structural sections from all sources of water, including upward groundwater flows and surface infiltration (Fig. 9.7). They also serve as barriers to the upward rise of capillary water. When blanket drains are used in conjunction with trench drains, as indicated in Figure 9.10, the open-graded base drainage layer (blanket drain) should be in direct contact with highly permeable aggregate filling the trench. Between the trenches a suitable subbase or filter layer should be placed to complete the structural section and also prevent the intrusion of subgrade soil into the open-graded layer (as shown in Fig. 9.7).

Table 9.2 gives permeabilities that were obtained by testing several untreated and asphalt-treated open-graded aggregates. Extremely low heads were used in the tests on the coarser materials to keep turbulence to a minimum. It is a good guide for the selection of sizes of aggregates needed to obtain the high levels of permeability required for drains for roads and other surface facilities.

Open-graded bituminous mixes of aggregates between $\frac{1}{4}$ in. and $\frac{3}{4}$ in. with 2 to 3% asphalt binder will compact readily to form firm nonshifting drainage layers with a high degree of permeability. If open-graded layers are to be placed under flexible pavements, tests should be made to establish their capabilities for withstanding traffic loads at the planned depths. When used in shallow drainage courses, these mixes should be constructed with hydrophobic aggregates that have an affinity for asphalt and will not strip under the actions of water and traffic. If open-graded bituminous mixes are used beneath the normal section, the asphalt binder is required only to prevent shifting during construction; hence stripping after construction of no consequence.

TABLE 9.2 Permeabilities of Untreated and Asphalt-Treated Open-Graded Aggregates

	Permeability, ft/day	
Aggregate size	Untreated	Bound with 2% Asphalt
$1\frac{1}{2}$ to 1 in.	140,000	120,000
$\frac{3}{4}$ to $\frac{3}{8}$ in.	38,000	35,000
No. 4 to No. 8	8,000	6,000

(Lovering and Cedergren, 1963.)

Airfield Pavements

The fundamentals of pavement drainage, as just presented, apply to airfield pavements as well as highways and other roads; however, because airfields are constructed on flat to moderately rolling terrain and have large areas exposed to water infiltration but only flat slopes available for its removal, they are even more difficult to drain than most highways.

As with highways, airfield drainage systems should be adapted to the conditions at individual sites and should provide for removal of water from all potential sources. Subgrade drains and interceptor drains should be provided as required for control of groundwater, springs, and side-hill seepage. Highly permeable base drainage layers (blanket drains), shown in Figure 9.13, should be provided for the removal of surface water that enters through the tops or edges of all heavily traveled pavements. In the design of drains in contact with soil appropriate filter criteria should be checked to make sure that no soil infiltration can occur (Chapter 5) and seepage principles should be used to be certain that sufficient permeability is provided in all elements of drains for rapid removal of water from the systems. Pipes are nearly always needed in trench drains, collector drains, and outlets because of the limited capabilities of aggregates to conduct water.

City Streets, Parking Lots, and Other Paved Areas

Subsurface drainage systems for the pavements included in this subsection should be designed using the same fundamental principles described for highway and airfield pavements. City pavements, even more than highways, need protection against infiltration of standing water and rainfall, as many are subjected to additional infiltration from lawn sprinkling, car washing, and the like throughout the summer months. Also, city pavements are usually constructed at slightly lower elevations than adjacent properties and serve as collectors for storm drainage.

Gravity discharge of seepage is often impossible for city pavements; however, storm drains are nearly always constructed under streets and can be used for the removal of seepage if suitable connections are provided. As a rule, it is not necessary to enlarge storm drains to accept the small additional flows required to accommodate groundwater and surface infiltration, for these flows are usually very small in relation to peak storm flows and lag behind the surface waters in reaching drains. Important street systems should be analyzed by experienced drainage and hydraulic engineers, and subsurface drainage facilities should be designed to meet the needs of individual installations. Principles of hydrology and hydraulics should be used in estimating storm runoff and in designing storm sewers; seepage principles described in this book can be used to estimate probable rates of infiltration from groundwater and through porous pavements.

Many city streets, parking lots, and other paved areas have light traffic

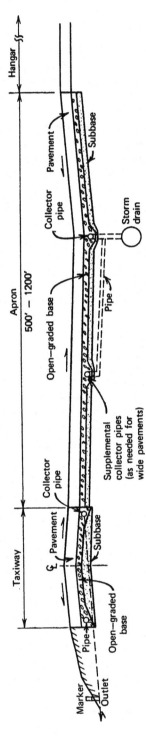

FIG. 9.13 Drainage system for airport parking apron and adjacent taxiway. This kind of system, using a highly permeable drainage layer directly under the primary pavement layer, with collector pipes and outlet pipes, can remove water in an hour or less, virtually eliminating water-related damages; and extend pavement life 300 to 400 percent beyond that of conventional slow-draining pavements. (From *Drainage of Highway and Airfield Pavements*, John Wiley & Sons, Inc., N. Y. 1974, p. 176; Updated printing by Robert E. Krieger Pub. Co., Malabar, Florida, 1987, p. 176).

with comparatively little kneading action to aid in closing cracks and keeping surfaces watertight. As a result, these pavements develop cracks and become relatively permeable to the entry of water. Unless adequate subsurface drainage is provided, water that enters remains for long periods of time and leads to untimely deterioration and costly upkeep.

Figure 9.14 presents drainage features that can protect city pavements permanently from damage due to surface infiltration and groundwater intrusion. The "heart" of the system is a hydraulically designed open-graded drainage layer 4 to 6 in. thick which feeds surface water or groundwater to longitudinal drains containing pipes, thus preventing the pavement and base from remaining flooded for more than short periods of time.

Economics of Drained and Undrained Pavements

When comparisons are being made of the relative costs of drained and undrained pavements, all important costs over the life of a project must be considered. When this kind of comparison is made, drained pavements will win out over undrained pavements because their effective lives (years of service before major repairs or restorations are required) are much greater than those of their undrained counterparts (see Fig. 9.1).

Substantial documented information has been accumulating on the unexpectedly high costs of keeping our nation's systems of undrained pavements in serviceable condition. A General Accounting Office (GAO) report to Congress in July 1970 said that surveys of pavements put down prior to October 1963 indicated some 2,800 miles of these Interstate pavements already needed overlays at an estimated cost of $200 billion. Fifteen years later the "1985 Needs Report" to Congress said that "the percentage of the Interstate pavements needing repair increased from 9 percent in 1981 to 14 percent in 1982." That report also said that failing pavements "will result in over one million miles of major roadways requiring work through the end of the century."

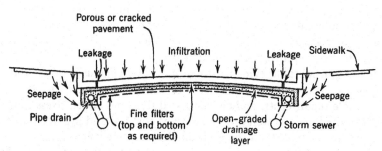

FIG. 9.14 Suggested subdrainage system for city streets subjected to heavy infiltration and seepage.

Though very few of the important pavements built in the past several decades were designed with good internal drainage, all of the well drained pavements I have been privileged to examine have performed excellently. The graph
in Figure 9.1 gives my comparison of the life expectancies of the undrained
pavements with well drained pavements (designed using the principles in this
book). It gives a range of 45 years to 90 years for well drained pavements,
which is three times the life expectancies of the poorly drained ones. As noted
elsewhere, some recently constructed undrained pavements started to break up
in as little as five years. In my view hardly any of the major pavements built
in the past several decades would be needing more than normal maintenance
had they been built with good internal drainage.

To place a dollar value on the losses caused by the practice of designing
undrained pavements in the United States, I used factual information obtained
in studies of problems with water in pavements for the Federal Highway
Administration, FHWA (1973), the U.S. Army Corps of Engineers Construction Engineering Research Laboratory (1974), and a review of major road test
data from experiments by the Western Association of State Highway Officials
(WASHO) and the American Association of State Highway Officials (AASHO).
Evidence proves that during the times pavements remain filled with water, the
damage rates (per traffic impact) can be up to several hundreds of times
greater than when little or no free water is present. Also, calculations made
for seven Case Study sites in the FHWA studies (FHWA, 1973) indicated that
typical state or Interstate highways from coast to coast can remain filled with
water at least 1/4 to 1/3 of the time each year, although in dry desert regions the
time is much less. With this basic information and data released by the FHWA
on costs of restoring major pavements in the U.S. (see *Engineering News-
Record*, Nov. 10, 1977, p. 14), I estimated that the losses caused by designing
undrained pavements in the U.S. are at least $15 billion a year (see *Engineering
News-Record*, June 8, 1978, p. 21) and world-wide they could reach a trillion
dollars over a 30 or 40 year period. The graph in Figure 9.15 gives my estimate
of the split between funds required for restorations and repairs between 1976
and 1990 (the study period used in the FHWA report) and the amounts left
for beneficial construction. It shows that $217 billion is being wasted in that
period of time because of poor drainage. Clearly, it would be difficult to place
an exact dollar value on the money being wasted because of bad drainage, but
there can be no doubt that it is substantial.

There is considerable evidence to support the conclusion that undrained
pavements are expensive and wasteful of natural resources and energy, increase vehicle repair costs, and cause wasted time and added stress to drivers;
and that in the long run well-drained designs cost less and provide longer last-
ing trouble-free service than the undrained. At a time when natural resources
and energy must be conserved it is imperative that design procedures be updated to take advantage of the great benefits of good drainage wherever pavements are exposed to water.

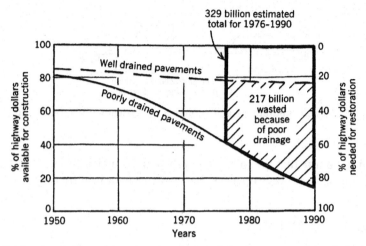

FIG. 9.15 Graph illustrating the magnitude of the waste caused by designing pavements as undrained systems (estimate is the author's). (Used with the permission of *EN-R,* June 8, 1978 issue).

9.5 UNPAVED SERVICE AREAS (PLAYFIELDS, STADIUMS AND ATHLETIC FIELDS)

Unpaved areas that must be used for professional and amateur competitions during a long period extending into rainy seasons often become unusable because of excessive moisture. Although playing fields are crowned slightly to speed runoff, they are often turfed; hence they drain rather slowly. Tile drains at a depth of 4 to 5 ft can speed the removal of water from such areas, as in agricultural drainage. A highly pervious, open-graded drainage layer (Fig. 9.16) is suggested as an alternate means of reducing excessive soil moisture in athletic fields.

Drainage is accomplished by placing a layer of coarse pea gravel or other highly permeable aggregate beneath a 2- to 3-ft cover of select subsoil and topsoil. Figure 9.16 shows that this drainage layer feeds seepage to longitudinal pipe drains embedded in the coarse filter aggregate. Screenings or fine filter aggregate should be placed in a thin layer on the subsoil, and a thin fine filter layer should be placed over the top of the coarse layer to prevent clogging. An alternate is to completely surround the open-graded aggregate with a suitable filter fabric. The system should discharge through pipes to a sewer or to a gravity outfall. If valves are placed in the discharge lines, they can be closed in the summer, and if the subsoil beneath the blanket is sufficiently impervious to prevent large downward flows of water, the field can be subirrigated through the system during dry seasons.

In developing important exhibition fields, agricultural soil experts should be

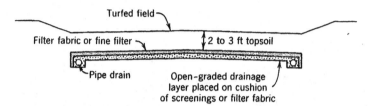

FIG. 9.16   Suggested subdrainage and subirrigating system for playfields and exhibition fields.

consulted to be sure that the soils used in their construction will hold sufficient moisture and have other properties that will sustain a healthy turf.

9.6   RAILROAD ROADBEDS

Railroad rails must be maintained within close tolerances of line and grade. The rails are laid on wood ties which hold them to correct gage and distribute the loads to the ballast. The ballast of crushed stone, crushed furnace slag, prepared gravel, or other hard, durable material provides shallow surface drainage and reduces the pressures on the subgrade. To maintain the rails true to grade additional ballast must be added periodically to compensate for any settlement or yielding of the subgrade that occurs.

After a number of years of maintenance roadbeds often take the general shape shown in Figure 9.17a. The uneven, dished condition shown results in the trapping of water in pockets and unless it is removed contributes to further instability that requires further ballasting and may lead to major instability. Hay (1956) states that, "Drainage is the most important single item in railroad construction and maintenance." He adds that, "Roadbed stability and low maintenance costs can only exist when adequate drainage is provided."

Methods for stabilizing foundations (Chapter 7) and slopes (Chapter 8) apply to railroads as well as highways and other similar constructions. When the condition shown in Figure 9.17a develops, stabilization can be achieved by a number of methods:

1. Grouting the ballast with cement-sand slurries.
2. Driving piles to increase the mechanical stability of the roadbed.
3. Installing vertical drains.
4. Roadbed drainage.

Good drainage always promotes roadbed stability. To remove water from low pockets when the conditions shown in Figure 9.17a develop, drains may be installed in trenches backfilled with filter material. Lateral drains or stubs should be located wherever required to intercept and remove trapped water.

When roadbed conditions on important railroad lines become more than a

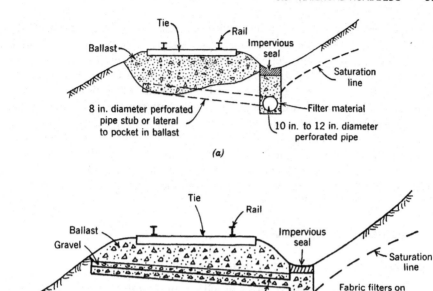

FIG. 9.17 Railroad roadbed drainage (*a*) (After Hay, 1956); (*b*) Use of geotextiles and permeable drainage aggregates to improve roadbed drainage.

matter of draining a few pockets as shown in Figure 9.17*a,* entire sections of track are sometimes lifted or moved to the side while all unsatisfactory ballast materials are removed down to the subgrade, which is graded and recompacted as necessary to provide a good boundary for construction of a drainage system to conduct water to a pipe drain along the lower side. Filter fabric can be placed on the prepared subgrade and highly permeable drainage material placed on the fabric to conduct water to a trench, as shown in Figure 9.17*b.* A fabric can be placed over the open-graded drainage aggregate to prevent its becoming clogged by intrusion of fines from above. Some engineers may place a gravel subbase over the drainage layer before replacing the ballast and track.

Newby (1982) discusses the use of a layer of fabric to prevent the formation of depressions and pockets that can trap water in railroad subgrades. After installing 1,000 miles of corrective treatments for troubled sections of track, he found that the fabric layer between the ballast and subgrade "virtually eliminates the instability problems they were having at many locations." He says, "Tensile strength is important to any geotextile, but separation, filtration, and particularly planar permeability for the rapid release of pore pressures are equal or greater in importance to the stabilization of railroad subgrade." (See Sec. 5.7 for a discussion of factors limiting flow capabilities in the plane of a fabric).

Amsler (1986) tells about an instability problem with several lengths of the double track Geneva-Lausanne of the Swiss Federal Railways. Severe instability problems existed when sections were repaired by conventional methods of building upon existing subgrades. So they looked for a better way and constructed test sections that could be compared with conventional. The new methods, which utilized non-woven geotextile between the graded subgrade and a gravel subbase under the regular ballast supporting the wood ties and the rails, gave much improved performance. Pipe cross drains were placed on the geotextile every 2 m to aid evacuation of water. Whereas deformations of the tracks under the usual treatment (without fabric or drains) reappeared in a year or two the new cross-section remained good, according to Amsler.

9.7 AGRICULTURAL LANDS

Importance of Draining Agricultural Lands

The prevention of high water levels in agricultural lands is widely recognized as essential to the healthy growth of most crops. Although sufficient moisture is needed, an excess robs the roots of air, leads to harmful accumulations of salts, and turns the land sour and unproductive. The artificial application of water to promote crop growth dates back to ancient times, but the realization that irrigation without proper drainage can cause irreparable damage to the land is comparatively recent.

Willardson (1974) says:

> Drainage, natural or artificial, is a fundamental factor in the equation of efficient agricultural production. In humid regions, many of the most productive lands cannot even be cultivated until they are drained. In arid regions under irrigation, control of water tables and salinity is directly dependent on adequate drainage. Drainage is as vital to agricultural production as it is to continents or river basins, or cities, or human beings. In a double sense, drainage underlies modern civilization.

Some investigators believe that numerous ancient civilizations that depended on irrigation faded away when their agricultural lands became useless because of a lack of awareness of the need for drainage. Willardson expresses the belief that the disappearance of many of these civilizations can be attributed to "the thesis of salinized soils and diminishing crop production . . . ," and "this same process of salinizing soils and decreasing production is evident in the modern world and is a concern of modern governments." In relation to the demise of many ancient civilizations that depended on irrigation, he says (Willardson, 1974*a*):

> There is, however, no clear cut evidence that these old irrigation developments were accompanied by corresponding drainage works. This historical record is graphic proof that drainage is a necessary part of irrigation development.

Great reductions in crop productivity have been attributed to poor drainage in countries throughout the civilized world. Major crop reduction attributed to bad drainage has been a serious problem in the western United States, China, and other countries. Yield reductions of 30 to 100% due to lack of drainage have been reported in China; a crop loss of 8% of China's total crop production represents a month's loss of food in a year.

Willardson's (1974*b*) basic conclusion, which is surprisingly similar to that reached by John L. McAdam (1820) on pavement drainage, is that the problem of agricultural drainage ". . . reduces to *matters of recognition of the need for drainage* and the willingness . . . to make the capital investments necessary to provide adequate drainage."

When the water used for irrigation is pumped from beneath the lands being irrigated, the tendency is for the water table to drop progressively year after year, sometimes reaching depths of several hundred feet (as in the San Joaquin Valley in central California, where the depths to water have greatly increased the cost of irrigation). In such areas drainage of the irrigated lands is usually improved by the pumping, but, if water for irrigation is imported, the tendency of the water table to rise may ruin the land for crops unless extensive drainage is carried out. Flat, low-lying lands are inherently slow draining even when the permeabilities of the substrata are high.

The sluggish drainage of many valleys throughout the world is illustrated by Figure 9.18, a plan view of an irrigated valley in an Arizona desert region (shown in more detail in Fig. 2.23). Although the formations under this valley are extremely permeable (1,000 to 3,000 ft/day or 0.3 to 1.0 cm/sec), the average hydraulic gradient is around 0.0005, and the natural drainage capacity of the valley is in the order of 25 acre-ft/day. With irrigation inflows of 2,000

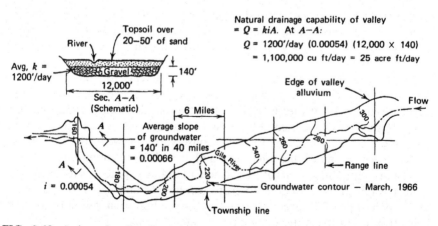

FIG. 9.18 Irrigated valley in Arizona desert. Heavy pumping from deep gravel-packed wells is required for control of groundwater. Irrigation water is imported and may at times exceed 2000 acre-ft/day. (Printed with permission of Los Angeles District, U.S. Corps of Engineers.)

acre-ft/day in summer months an extensive system of gravel-packed wells with capacities of 1000 gpm and greater is required for control of groundwater. Without an aggressive drainage program this valley would be useless for crop production. This is true of a great many of the irrigated lands around the world.

Development of Agricultural Drainage Systems

The development of a successful drainage system for wet agricultural lands depends on a correct diagnosis of the problem. This requires knowledge of the source or sources of the water, the transmissibility (permeability) of the soil formations, and the drainage needs of the crops to be planted on the land. The U.S. Department of Agriculture, Soil Conservation Service (1971), emphasizes that drainage of wet lands is often improved by first giving attention to surface drainage. Flat areas are subject to ponded water caused by (1) uneven land surface with pockets, ridges, etc., which retard or trap surface runoff on slowly draining lands, (2) low-capacity channels that do not remove the water effectively, and (3) outlet conditions that hold the surface water above ground level. Surface contouring and the construction of interceptor ditches and construction or improvement of outlet ditches and drains often help the situation, says the Soil Conservation Service.

In addition to improvement of surface drainage, subsurface drains are often required for agricultural lands. These may be trench drains, mole drains, pumped wells (as in the desert valley just described), or downdraining wells. Criteria for the design of drainage systems may be developed from empirical data obtained by evaluating existing systems with similar soils and from theoretical analysis of the problem with application of known physical laws. No system should be designed until the sources of water are known and basic design criteria have been established. It is often helpful to compare the conditions at a given site needing drainage with other installations. Every completed system should be carefully checked and any deficiencies corrected by supplemental measures.

Various guides for the design of drainage systems for agricultural lands have been published by pipe manufacturers and others (Armco, 1955; American Iron and Steel Institute, 1967; Kirkham and de Zeeuw, 1952) but needs vary so much with local conditions that most agriculturalists rely mainly on local practices developed by the Soil Conservation Service, the U.S. Bureau of Reclamation, and other agencies familiar with local conditions.

The proper management of irrigation and drainage practices to ensure good quality water in the root zone of crops can help to maintain crop production at the highest possible level. Use of irrigation methods that do not oversaturate helps to prevent the waterlogging of irrigated lands.

In a recent study the Committee on Research of the Irrigation and Drainage Division of the American Society of Civil Engineers (1974) reached the following conclusions about the overall problem of salinity control:

"Techniques for minimizing the adverse effects of drainage water are sorely needed to control the degrading of the quality of downstream water supplies." The ASCE committee also says:

> The day is rapidly approaching when some irrigated regions will operate as an essentially closed system. Thus, all (or nearly all) return flows would be collected and recycled or treated. Additional technology regarding treatment processes and the deleterious effects of poor quality water upon plant growth will be extremely important.

REFERENCES

American Iron and Steel Institute (1967), "Handbook of Steel Drainage and Highway Construction Practices," April 1967, pp. 234–235.

American Society of Civil Engineers, ASCE (1974), "Water Management Through Irrigation and Drainage," Committee on Research of the Irrigation and Drainage Division, *Proceedings*, A.S.C.E., *Journal*, Irrigation and Drainage Division, Vol. 100, No. IR2, June 1974, pp. 153–178.

Amsler, P. (1986), "Railway Track Maintenance Using Geotextiles," Proc., 3rd International Conference on Geotextiles, Vienna, Austria, 1986, Vol. IV, pp. 1037–1041.

Armco Drainage and Metal Products, Inc. (1955), "Handbook of Drainage and Construction Products," Middletown, Ohio, Chapter 41.

Barber, E. S., and C. L. Sawyer (1952), "Highway Subdrainage," *Proceedings*, Highway Research Board, Vol. 31, pp. 643–66.

Barber, E. S. (1962), "Discussion of 'Seepage Requirements of Filters and Pervious Bases,' by H. R. Cedergren," *Transactions*, A.S.C.E., Vol. 127, Part I, p. 1110.

Barenberg, Ernest J., and Owen O. Thompson (1970), "Behavior and Performance of Flexible Pavements Evaluated in the University of Illinois Pavement Test Track," Highway Engineering Series No. 36, Illinois Cooperative Highway Research Program Series No. 108, Urbana, Ill., January 1970.

Cedergren, Harry R. (1974), "Drainage of Highway and Airfield Pavements," Wiley, New York; (1974a), pp. 94–96, 104–142, 188–199; (1974b), pp. 215–240; (1987), updated printing by Robert E. Krieger Pub. Co., Malabar, Florida, 1987.

Cedergren, Harry R., and Kneeland A. Godfrey, Jr. (1974), "Water: Key Cause of Pavement Failure?," *Civil Engineering*, A.S.C.E., September 1974, Vol. 44, No. 9, pp. 78–82; (1974a), p. 81.

Cedergren, H, R.(1982), "Overcoming Psychological Hang-ups is Greatest Drainage Challenge," Proc. 2nd Int. Conf. on Geotextiles, Las Vegas, Nevada, Aug. 1–6, 1982, pp. 1–6.

Cedergren, Harry R. (1987a), "Undrained Pavements: A Costly Blunder," *Civil Engineering*, ASCE, in *Forum*, April, 1987, p. 6.

Cedergren, Harry R. (1988), "Why All Important Pavements Should be Well Drained," prepared for presentation at 67th Annual Meeting, Transportation Research Board, Wash, D.C., Jan. 11–14, 1988.

Drake, E. B. (1979), "New Subbase Design Uses Commercially Available Aggregates," *Highway and Heavy Construction*, Vol. 122, No. 10, Oct., 1979, pp. 106–108.

Engineering News-Record (1978), "Poor Pavement Drainage Could Cost $15 Billion Yearly," (By H. R. Cedergren), June 8, 1978, p. 21.

Engineering News-Record (1979), "Open-Graded Base Course Placed at Portland Airport," (describes design by Terence Craven), July 5, 1979, p. 16.

Federal Highway Administration, FHWA (1973), "Guidelines for the Design of Subsurface Drainage Systems for Highway Structural Sections," written by Jorge Arman, Harry Cedergren, and Ken O'Brien of the Cedergren/KOA Joint Venture for the FHWA; printed in 1972; issued in 1973.

Forsyth, Raymond A., Gordon K. Wells, and James H. Woodstrom (1987), "The Road To Drained Pavements," *Civil Engineering,* ASCE, March, 1987, pp. 66–69.

Garbe, Carl W. (1974), (1976), private communications.

Haughton, D. R. (1986), "Open-Graded Aggregate Highway Construction, Pine Pass Project, Canada," Ministry of Transportation and Highways, Geotechnical and Materials Branch, Victoria, B.C., 1986.

Hay, William W. (1956), *American Civil Engineering Practice,* Robert W. Abbett (Ed.), Wiley, New York, Vol. I, Chapter 6, pp. 6–12.

Kirkham, D., and J. W. deZeeuw (1952), "Field Measurements for Tests of Soil Drainage Theory," *Proceedings,* Soil Science Society of America, Vol. 16, pp. 286–293.

Lorin, Roger (1986), private communication, describes Mr. Lorin's introduction of porous concrete drainage layer under PCC wearing courses for major airport pavements in France.

Lovering, W. R., and H. R. Cedergren (1963), "Structural Section Drainage," *Proceedings,* International Conference on the Structural Design of Ashpalt Pavements, University of Michigan, Ann Arbor, Mich., August 20–24, 1962, pp. 773–784.

McAdam, John L. (1820), "Report to the London Board of Agriculture."

Newby, J. E. (1982), "Southern Pacific Transportation Co. Utilization of Geotextiles in Railroad Subgrade," Proc. 2nd Int. Conf. on Geotextiles, Las Vegas, Nevada, Aug. 1–6, 1982, Vol. II, pp. 467–472.

Richardson, Emory S., and William A. Liddle (1973), "Open-Graded Emulsified Asphalt Pavements," Region 10 Office of Federal Highway Projects, Implementation Division, Office of Development, Federal Highway Administration (FHWA), Washington, D.C., June 1973.

Ring, George W. III (1974), "Seasonal Strength of Pavements," *Public Roads,* Vol. 38, No. 2, September 1974.

U.S. Department of Agriculture, Soil Conservation Service (1971), *National Engineering Handbook,* Section 16, "Drainage of Agricultural Land," May 1971.

U.S. Department of the Army, Corps of Engineers, Construction Engineering Research Laboratory (CERL), (1974), "Methodology and Effectiveness of Drainage Systems for Airfield Pavements," Technical Report C-13, November 1974; prepared by Harry R. Cedergren for CERL, Champaign, Ill. Chicago District.

Willardson, Lyman S. (1974), "Drainage for World Crop Production Efficiency," *Proceedings,* A.S.C.E., Specialty Conference *Contribution of Irrigation and Drainage to the World Food Supply,* Biloxi, Miss., August 14–16, 1974, p. 9; (1947a), p. 10; (1974b), p. 17.

Winterkorn, Hans (1967), "Application of Granulometric Principles for Optimization of Strength and Permeability of Granular Drainage Structures," *Highway Research Record* No. 203, Highway Research Board (now Transportation Research Board).

CHAPTER TEN

STRUCTURAL DRAINAGE

10.1 PROBLEMS OF STRUCTURES CAUSED BY WATER

Any well drained structure is inherently safer and more economical than its undrained counterpart. Principles presented in this chapter fortify that statement.

This chapter reviews the control of seepage and groundwater beneath or behind retaining structures, overpour weirs, masonry dams, spillway chutes, basements, drydocks, canal and reservoir linings, and comparable types of engineering works.

Most of the structures reviewed in this chapter have rigid members in contact with earth or rock foundations, which usually are somewhat erodible and often somewhat compressible. The placement of rigid, relatively impermeable structural elements against water-bearing earth leads to two damaging conditions:

1. *Excess uplifting or overturning pressures* caused by trapped water. Concrete and other impermeable elements of structures create major discontinuities where they join the earth. Because they are virtually impermeable, they obstruct natural paths of seepage and may cause water pressures to build up to dangerous levels if not relieved.

2. *Channeling of seepage and piping* caused by the presence of *permeable* discontinuities. Any contact between a rigid structural member and soil or rock is a potential plane of weakness because the slightest separation or opening attracts seepage. If the soil or rock is erodible, piping can develop and unless arrested can lead to serious damage or complete failure.

Another damaging condition in saturated soil is *piping failure caused by heave,* which occurs when the uplift forces due to seepage equal or exceed the downward forces due to the submerged weight of the soil.

Many of the serious failures of dams, reservoirs, and other civil engineering works of our times have been caused by one or more of these damaging actions of water. The objective of *structural drainage* is to control water pressures and seepage forces in the earth adjacent to structures and thus prevent their untimely damage, deterioration, or failure.

In Chapter 3 methods are described for estimating rates of seepage (Sec. 3.4), for determining the forces exerted by seeping water (Sec. 3.5), and for determining the uplift pressures exerted by water under structures (Sec. 3.5). Applying these principles is especially important when designing the types of structures described in this chapter. In Chapter 5 emphasis is placed on the need for adequately designed and constructed filters and drains to ensure long, trouble-free performance of dams and other structures involving water. Filters and drains must be designed to prevent piping (Sec. 5.2), yet have sufficient discharge capacity (Sec. 5.4) to remove quickly any water that reaches them without large buildup of hydrostatic head. These fundamental principles of drain and filter design are of great importance to all the structures described in this chapter. Many of these works have costly structural elements that can be properly and economically protected from the damaging actions of water only by features incorporated in the original construction. If adequate drainage facilities or other seepage control measures are not provided at this time, repairs or corrective measures can be extremely costly.

Strict adherence to sound drainage principles is probably the most important single aspect of the design of all the structures described in this chapter. Almost every serious failure of structures of these kinds has been caused by lack of control of groundwater or seepage.

Throughout this book emphasis is placed on the need for specifying and obtaining drains that provide filter protection and have high water-removing capacities. The mere provision of drains of the proper dimensions in the correct locations does not of itself ensure that structures will be properly drained (Secs. 1.2, 1.3, and 9.4). The *needs* of drainage systems should be determined by the methods described in this book and realistic designs and specifications prepared and strictly followed in the construction.

With the rapid improvements that have been made in synthetic fabric filters and prefabricated plastic drains, many of the problems of obtaining drains with high levels of filter protection and high discharge capabilities have been removed, as brought out in many places in this chapter (also see Chapter 5).

10.2   RETAINING STRUCTURES

As defined in this chapter, a *retaining structure* may be for the purpose of retaining or supporting earth fills and slopes or for holding back water. Structures for the retention of earth may be retaining walls, cribs, steel sheet pile cells, bin walls, or tied sheet piling. Water-retaining structures discussed in this chapter include flood walls and cofferdams but exclude earth dams and levees.

Earth-Retaining Structures

Retaining walls and other types of earth-retaining structure are used to restrict earth slopes within limited rights-of-way or to prevent embankments from spilling down long slopes. They are also used widely to stabilize steep cut and fill slopes, for wing walls for embankments, and for numerous other situations in which substantial differences in the elevation of earth are required within short distances.

Three general types of retaining structure are in widespread use: (1) gravity structures (masonry, crib, or bin walls), (2) cantilever walls, and (3) counterfort walls. The general features of these three types are shown in Figure 10.1. Levinton and Feldstein (1957) describe these types as well as several others and set forth basic principles of design. Regardless of the type, earth pressures act to slide walls forward on their foundations and to overturn them. Walls must be set on foundations with sufficient strength to resist these overturning and sliding forces, and their dimensions should be established with full consideration of the properties of the backfill material and the foundation. Groundwater and infiltering rainfall must always be controlled by adequate drainage. Clay or other highly impervious materials should not be used for backfilling crib- and bin-type walls if drainage is important to a specific wall, as it usually is. Compacted pervious granular materials make stable, well-drained backfill for walls of these types.

Brandl (1987), in a comprehensive review of retaining wall design, says in regard to backfilling, "Drainage measures are absolutely necessary." He suggests using highly permeable backfill material and always observing filter criteria for clayey-silty subsoils. He points out that the recent trend is to use geotextiles instead of reverse filters to prevent piping. Also, he warns that weep holes, pipes or hoses must be provided in places where required to facilitate drainage.

New techniques for building a wide variety of retaining structures with the aid of geotextiles have evolved in the past few years. Fukuoka and Imamura (1982) tell about experimental retaining walls in Japan which demonstrated that fabrics and geogrids (reinforcing pads) can adequately reinforce earth fill and produce long-lasting and economical retaining walls. They emphasize that they are easily constructed, and can be used without piling on relatively soft foundations.

Jones (1982) describes case histories of both full-scale trials and completed retaining walls and bridge abutments, proving that "geotextiles can be used on the most demanding sites to produce economic structures of high aesthetic appeal with no defects or distortion." He says that the critical element to the successful development of geotextile structures is the use of proper construction techniques and "the theoretical reservations (about distortions) are groundless."

For many years it has been customary to place a vertical blanket of "pervi-

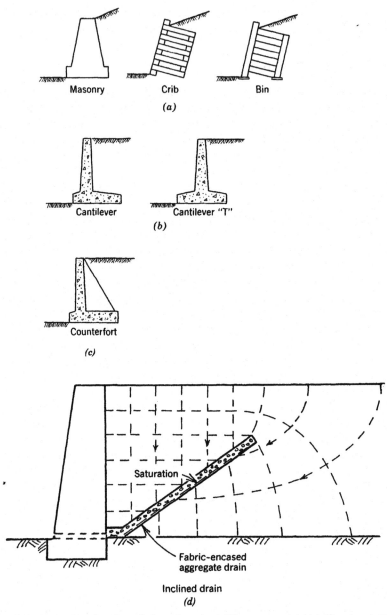

FIG. 10.1   Several types of retaining structures (diagrammatic). (*a*) Gravity walls. (*b*) Cantilever walls. (*c*) Counterfort wall. (*d*) Inclined drain for reducing water pressure behind wall.

ous" sand and gravel behind retaining walls for protection against hydrostatic pressures. Terzaghi (1943) demonstrates that even though the back face of a wall is drained with a vertical blanket appreciable pore pressures can exist in the earth behind the blanket, thus increasing the pressure on the wall. To overcome this deficiency, Terzaghi suggests inclining the drainage blanket to force seepage into a vertically downward pattern, thus eliminating excess hydrostatic pressures in the sliding wedge that enters into the earth pressure computation. One of the best ways to assure that all such drains will be effective is to place a filter fabric on the excavated stable slope in back of a wall, place a few inches of permeable crushed rock ¼ in. to 1 in. in size over the fabric, and cover with another layer of fabric before backfilling over the drain (see Fig. 10.1*d*). A prefabricated composite plastic drain might be another possibility if it had sufficient compressive strength, shearing strength, longevity, and conductivity to meet the requirements of a specific project (see Sec. 5.7).

The desirability of forcing seepage into vertical patterns to improve slope stability is demonstrated in Chapter 8 (Figs. 8.10, 8.14, and 8.24). The principle is stated in Chapter 8: *Water seeping in a generally horizontal direction destabilizes slopes, whereas water seeping vertically downward produces no destabilizing forces and no pore pressures.*

Gilkey (1959) points out that the removal of all possible moisture from the soil behind retaining walls greatly reduces the size of the prism of earth that is actually effective in pressing against the wall and concludes: "Probably the most important rule in the design of retaining walls, therefore, is to provide adequate drainage."

Retaining walls have failed during rainstorms even though the builders thought the walls were protected by "pervious" drainage blankets. An example is a large wall in Portland, Oregon, which toppled over when a drainage layer behind the wall failed to protect it (*Engineering News-Record,* November 6, 1958, p. 23). An investigation of the failure disclosed that "pervious" gravel backfill contained a high percentage of silt and clay and had a much lower permeability than needed for the elimination of hydrostatic pressures behind the wall.

Flood Walls

Flood walls are retaining walls that confine rivers to limited portions of flood plains or river channels. These walls must be capable of withstanding the overturning pressures and sliding forces of the water, as well as the uplift pressures and seepage forces that tend to cause piping of soil out from under them. A typical design, shown in Figure 10.2*a*, is a counterfort wall with a narrow footing on the landward side. The cutoff at the riverward toe is needed for protection against scour and undermining by river currents. Analysis of the flow net in Figure 10.2*a* shows that this design is vulnerable to piping at the landward toe, for an upward hydraulic gradient of more than unity is indicated at point *C*. This undesirable condition can be improved by excavating a

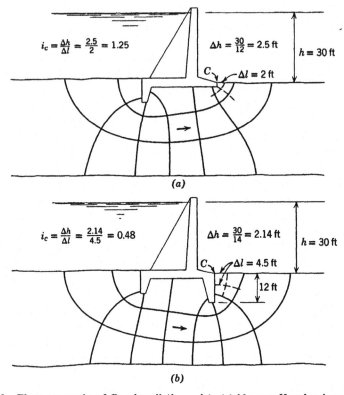

$$i_c = \frac{\Delta h}{\Delta l} = \frac{2.5}{2} = 1.25$$ $$\Delta h = \frac{30}{12} = 2.5 \text{ ft}$$ $$h = 30 \text{ ft}$$ $$C \quad \Delta l = 2 \text{ ft}$$

(a)

$$i_c = \frac{\Delta h}{\Delta l} = \frac{2.14}{4.5} = 0.48$$ $$\Delta h = \frac{30}{14} = 2.14 \text{ ft}$$ $$h = 30 \text{ ft}$$ $$C \quad \Delta l = 4.5 \text{ ft}$$ $$12 \text{ ft}$$

(b)

FIG. 10.2 Flow net study of flood wall ($k_h = k_v$). (a) No cutoff at landward toe. (b) Cutoff 12 ft deep at landward toe.

cutoff into the foundation at the landslide toe, as shown in Figure 10.2b. A cutoff 12 ft deep at this point reduces the hydraulic gradient at C to about 0.5, which makes this design two and one-half times as safe against piping as the design shown in Figure 10.2a. Flood walls can be made safer against piping by the use of *weighted filters* (using either mineral aggregate or fabric filters covered with coarse gravel) at the landward toes or by any measures that effectively reduce seepage quantities and uplift gradients. Flow nets are useful in comparing alternate designs and in developing economical and safe designs.

Seawalls and Crib Revetments

Preventing erosion along the shores of oceans, large lakes, and wide rivers is of major importance in many areas of the world. Heavy waves and strong currents—often made even more destructive by hurricanes and other bad weather conditions—can severely undercut shores if proper prevention measures are not provided. Many shore protection structures make use of large boulders and derrick stone, because of their resistance to being moved about by the forces of water. Many of these structures can be threatened by undermining when built on the erodible soils that often exist in the areas in which protection is needed the most. Fundamental to the safety of all such structures is the prevention of undermining by the use of properly designed filters that hold the soil in place while allowing water to drain out. In some situations gravel filters have been used; however, certain synthetic filter cloths (see Chapter 5) have been employed successfully in many shore-protection structures.

Barrett (1966) discusses the use of synthetic filter cloths as a replacement for graded aggregate filter systems in coastal structures. He states that filter cloths came into use because of the difficulties of constructing sand and gravel filters that would function permanently under the severe conditions existing in coastal structures. A lack of readily available materials with required grading, careless placement, and lack of control of the uniformity of the aggregates and their placement are cited. In contrast, this author continues, woven filter cloths have a uniform filtering ability that is factory controlled. He is discussing woven cloths, not nonwoven fabrics (see also Chapter 5). He says that the "independent tensile strength" of filter cloths prevents the loss of soil from under rubble and other large rock used in many coastal structures.

As evidence of the durability of cloth filters, Barrett cites a rock revetment constructed for erosion protection at Deerfield Beach, Florida, immediately after the March 1962 storm which attached the East Coast of the United States and caused devastating damage. Rocks weighing 500 lb to 2.5 tons were placed directly on the cloth. After four years, during which three hurricanes occurred, the revetment showed no problems at all, according to Mr. Barrett. Although many revetments have been constructed, in which large rocks were placed directly on the filter cloth, Barrett recommends that a layer of gravel or crushed stone be placed immediately on top of the filter cloth. This layer acts as a pad to prevent rupture of the filter cloth by the heavy rocks when movement occurs during a storm or hurricane. Also a layer of clean stone or crushed gravel allows water to be released from the soil over the entire surface area of the structure.

Two types of shore protection structure which make use of bedding layers of gravel or crushed stone and filter cloth to allow free drainage while holding erodible soils in place are shown in Figure 10.3. Figure 10.3a illustrates a crib revetment that uses batter piles with large boulders to provide a highly permeable backfill and a gravel bedding on filter cloth holding the soil in place. A vertical seawall with stone toe protection against undermining is shown in Figure 10.3b. Here large armor stone is laid over 300- to 500-lb stones (or other recommended sizes) on a crushed stone bedding. Between the crushed stone bedding and the underlying soil is a layer of plastic filter cloth. (A breakwater with filter cloth holding erodible foundation soils in place is shown in Figure 5.12a).

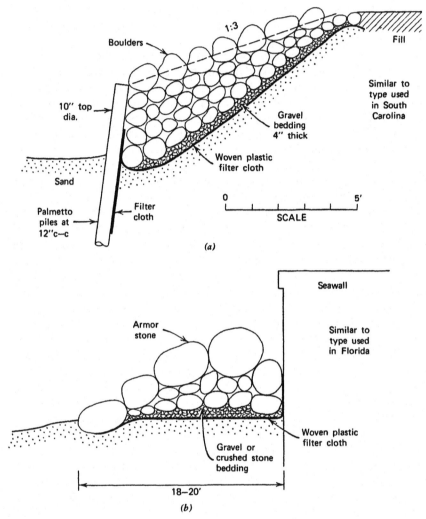

FIG. 10.3 Typical seawalls and crib revetments using filter cloths to hold fine soil in place. (*a*) Crib revetment. (*b*) Seawall. (After typical designs in *Filter Handbook*, printed with permission of Carthage Mills Incorporated, Cincinnati, Ohio.)

Cofferdams

The construction of powerhouses, dams, and other works in river beds often requires the use of temporary structures called *cofferdams*. If the water is shallow (up to 5 ft deep) and velocities are not more than a few feet per second, earth-filled sand bags in two rows may be used to retain clay puddle to form a relatively watertight barrier. A single row of steel sheet piling may be used to somewhat greater depths, but if the water is deep and fast, a more elaborate design is required.

If the working area is plentiful, homogeneous earth fills (Fig. 10.4a) or zoned fills (Fig. 10.4b) may be used; however, space limitations and fast currents often dictate the use of sheet pile cofferdams. For low heads a single row of sheet piling may be driven into overburden materials lying over bedrock, which can be supported with timber braces, as shown in Figure 10.4c, or backed up by earth fill. If the foundation contains layers of sand that tend to pipe or boil inside the construction area, it may be necessary to place weighted filters at the inside toe of the cofferdam (Fig. 10.4c) or elsewhere in the working area.

For large depths of water and fast currents two rows of steel sheet piling may be tied together with tie rods and the space between the piling filled with earth or broken rock (Fig. 10.4d). For large and important excavations *cellular cofferdams* are frequently constructed with interlocking steel sheet piling. Timber cribs filled with rock are often used when bedrock is at the surface or so shallow that piling cannot be driven to an adequate depth (Fig. 10.4e).

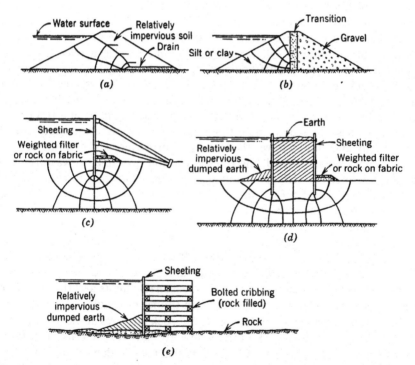

FIG. 10.4 Several types of cofferdam. (a) Earth fill with drain ($k_h = k_v$). (b) Zoned earth fill ($k_h = k_v$). (c) Braced sheeting ($k_h = k_v$). (d) Tied sheeting with earth fill ($k_h = k_v$). (e) Rock-filled timber crib cofferdam.

Cofferdams must be resistant to hydrostatic *overturning forces*, to *piping*, to *seepage forces* causing blowouts, to *sliding* on the foundation, and to *bursting* or *collapse* because of internal soil pressures. Seepage in pervious foundations must be cut off by driving the piling down into these formations or controlled by dumping impervious earth at the water side (Figs. 10.4*d* and 10.4*e*) or by the use of *weighted filters* (Section. 5.1) at the landward toe (Figs. 10.4*c* and 10.4*d*). An alternate to the use of weighted filters constructed entirely of mineral aggregates would be to spread a suitable fabric filter and cover with permeable rock or gravel of sufficient thickness to prevent heave.

Schroeder (1987) emphasizes that for cofferdams to perform satisfactorily, "the foundation supporting the cofferdam . . . must provide an adequate margin of safety against a bearing capacity failure . . . and must be protected against the effects of underseepage. Many cofferdams have been built on deep cohesionless soils and measures for controlling seepage have been particularly expensive and important. Usually a combination of deep wells and wellpoints inside the cofferdam is used"(under such conditions). (Note that "inside" means the dry side or working side of the cofferdam).

If cofferdams are based on rock or dense gravel foundations, their stability is ensured if they have sufficient width and weight to overcome the water pressures acting against the outer side; but if they stand on sand or other yielding earth adequate seepage reduction and drainage must be provided to prevent piping or other failures. Filter materials should be placed over the natural ground before an earth cofferdam is constructed (Fig. 10.4*a*). Filters and drains should be designed on the principles described in Chapter 5 to ensure protection of the soil against piping and to provide sufficient discharge capacity to remove the seepage safely without large head requirements. If a drain is not many times more permeable than the soil it drains, saturation can rise substantially and greatly lower the stability. Furthermore, if material containing large pore spaces is used, the fines in the soil may start to wash through, leading ultimately to collapse if not detected and corrected in time. Usually working space behind cofferdams is at a premium. Correctly designed filters and drains can ensure maximum stability in the minimum amount of space.

Sheeting used for the construction of timber or steel sheet piling cofferdams is never completely watertight, and the backfill and foundations are often highly pervious; hence many cofferdams leak badly. Seepage control by the use of impervious earth at the water side and filters at the land side may be necessary for the reduction of leakage to reasonable limits and the prevention of piping failures. Control over seepage in timber or cellular cofferdams can be obtained by placing impervious earth and rock mixtures in the half adjacent to the river and pervious rock in the landward half. If materials of distinctly different permeabilities are placed in this manner, cofferdams can, to a degree, have the benefits of the zoning used in earth dams.

Cofferdams must be designed for the conditions at individual locations requiring their use. In some cases they must be safe not only against underseepage and internal seepage but against overtopping.

10.3 PARTLY SUBMERGED STRUCTURES

This section describes seepage control measures for several types of structure that lie partly below the water table either permanently or intermittently. It is possible to design any of these structures with sufficient mass and strength to resist full hydrostatic pressures permanently just as a boat is designed to have sufficient strength to resist water pressures acting on it. Permanently located structures, unlike boats, must have sufficient weight to stay permanently in place and not be floated out of their foundations at high water levels. These structures can be designed to be heavy enough to resist full water pressures or they can be designed with drainage to relieve a portion or all of these pressures. Several important types of partly submerged structure are drydocks, highway boat sections, basements, and canals below the saturation level.

Drydocks

Although *drydocks* or *graving docks* are built in limited numbers, they are usually massive structures, often with major drainage problems. Drydocks must be located adjacent to the protected waters of harbors; hence, groundwater levels are high and seepage and groundwater problems are usually severe. As the size of ships requiring maintenance has increased, the dimensions of drydocks have also increased, intensifying already difficult situations. Drydocks for large navy vessels must be large and deep and their foundations must be capable of supporting tremendous loads.

A ship that is to be serviced must be floated into its drydock, the entrance to which is blocked with gates or caissons. Water is then allowed to flow out during low tide and the balance of the water is pumped out to permit repairs or reconstruction of the vessel in the dry. The vital elements of the construction and operation of drydocks that depend on drainage are (1) groundwater control during construction and (2) control over uplift pressures from groundwater when the drydock is dewatered.

Control over groundwater during the construction of drydocks is often a major operation. Among the more common methods used (see also Chapter 7) are sheet pile walls, chemical or cement grouting, wellpoints, and pumped wells. In some cases portions of the work are carried out in pneumatic caissons; in others some of the construction is done under water.

Estimates of the overall rates of pumping required to dewater foundations during the construction of drydocks can be made by using the methods described in Section 7.1.

Although some large drydocks have been constructed with permanent drainage facilities to reduce uplift pressures, most drydocks on earth foundations are capable of withstanding full hydrostatic uplift pressure. Under some conditions artesian pressures can increase uplift above the normally expected maximums. These conditions can be determined only by careful soil and groundwater investigations in advance of the design. Walls should be designed as retaining walls, and floors should have sufficient beam strength to transfer part of the uplift pressures to the heavy side walls which furnish additional mass to overcome uplift. Piles or caissons are used for support under fully loaded conditions and to furnish additional pull to resist uplift. If piles are depended on, they should in no case be allowed greater pull then the submerged weight of the soil into which they are driven.

Abbett, Halmos et al. (1956) classify drydocks according to the degree of drainage provided as (1) full hydrostatic, (2) partially relieved, and (3) fully relieved.

Merriman and Wiggin (1946) describe a drydock that was constructed on piling in an earth foundation (Fig. 10.5a). Sheet piling that surrounds the entire site at the bottom was placed to cut off the flow of water in a sand and gravel bed above a clay subsoil. The massiveness of the construction is readily apparent; the bottom slab has a thickness of 20 ft in the center.

The total effective weight of the walls and floor, together with any downward pull of piling or caissons, is made sufficient to overcome the greatest uplift, which occurs when the drydock is pumped out and the outside water is at its highest level.

Under favorable conditions, drydocks have been built in impervious rock formations with little or no drainage and thin floor sections. In ideal conditions rock is excavated to the rough dimensions of the dock and lined with masonry, with or without drainage. Abbett, Halmos et al. (1956a) describe a drydock (Fig. 10.5b) that was constructed on good rock with a thin lining of concrete.

A notable drydock embodying excellent groundwater control by drainage is operated by the United States Navy at the Puget Sound Naval Shipyard, Bremerton, Washington. Zola and Boothe (1960), in describing this drydock, point out that "Super Carriers" constructed since World War II created the need for this dock, which has inside dimensions of 180 by 1,152 ft and a depth of 61 ft. The water depth at mean lower low level (MLLW) is 42 ft.

Considerable depths of soft sediments were dredged from the site, which was then backfilled to form a working table on which to carry out the construction. Earth fill formed dams on either side of the construction area, and lines of steel sheet piling were driven into the central part of each dam and across the inboard end. The earth fill and cutoff walls are part of the permanent facility. An 11-cell sheet pile cofferdam was constructed along the outboard side of the construction area and braced steel sheet piles formed the cofferdam at the inboard end.

Dewatering for construction of the Bremerton Drydock was accomplished by partly cutting off seepage by the sheet pile walls and drilling 66 deep wells, supplemented by two additional large suction pumps, to lower the water table to elevation 65. The second stage of the dewatering lowered the water table to elevation 51 with 390 wellpoints on 10-ft centers just outside the drydock construction area but inside the deep wells.

A soil boring and testing program indicated that after the initial dewatering

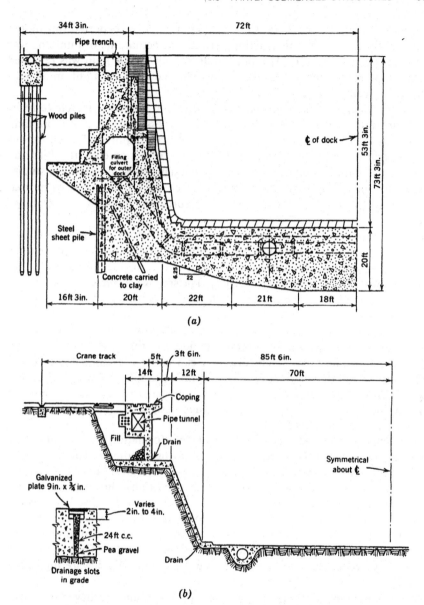

FIG. 10.5 Drydocks constructed in earth and rock foundations. (*a*) Drydock constructed on piling in earth foundation. (Fig. 47*c*, p. 1842, *Civil Engineer's Handbook,* Merriman and Wiggin, Wiley, New York, 1946.) (*b*) Drydock constructed in rock foundation. (Fig. 121, p. 21-145, *American Civil Engineering Practice,* R. W. Abbett, Wiley, New York, 1956.)

a continuous seepage of 20,000 to 30,000 gpm could be expected during the balance of the three-year construction period. The actual rate of pumping was less than expected. Tate (1961) reports that "about 10,000 gpm was removed during initial dewatering stages; this soon stabilized after the addition of well-points to a steady rate of 7,000 to 8,000 gpm."

If this drydock had been designed with sufficient mass of concrete to resist total hydrostatic uplift, a floor slab with a thickness of 43 ft, and more than 85 ft below MLLW at the bottom, would have been required. The enormous construction problems and high costs of this type of design eliminated it from serious consideration and led to the accepted design. Soils and drainage studies were important in developing a *relieved* type of dock that would be permanently protected from uplift and side pressures by an elaborate drainage system.

The pressure relief system for the Bremerton drydock depends on (1) selected granular drainage materials beneath the dock floor and outside the dock walls, (2) granular backfill of adequate permeability outside the walls, and (3) sheet pile cutoff walls to reduce the inflow. The seepage is collected by a system of pipes leading to drainage tunnels in the sidewalls which carry the water to sumps and pumps.

The design of this drydock represents a highly advanced drainage system that was developed by seepage analysis methods. In this type of construction the need for adequate permanent drainage cannot be overemphasized. If the system were to fail during the life of the drydock, serious uplifting and cracking of the floor slab and vital machinery could be expected. Its success is a testimony to the benefits of modern drainage design. The completed project was dedicated in April 1962.

Designers can never have too much advance information about foundation and groundwater conditions at the sites of important projects. This is particularly true of major drydocks. Thorough soil and geological investigations, together with careful analysis of groundwater conditions and seepage control systems are essential to the development of these projects. The behavior of important works should be checked by periodic observation of water pressures in piezometers. The Bremerton Drydock has a permanent system of piezometers outside the walls and under the floor to check on the efficiency of the drainage system and to observe the head differential during dewatering of the drydock.

Regular, permanent monitoring of water pressures under relieved drydocks can forewarn of increasing uplift pressures that could endanger these structures, and allow corrective measures to be taken in time to avoid damage. When the pumping from wellpoints or wells is depended on for removing sufficient water to control uplift, monitoring is particularly important, as the capacities of these devices can fall off considerably from decreases in permeability of the soils surrounding them caused by consolidation under large hydraulic gradients near them, formations of incrustations from salts, iron oxide, iron bacteria, and the like (see Sec. 6.3 for a discussion of problems with relief wells and procedures for increasing well discharges). In 1972 I reviewed very detailed records of piezometers measuring uplift pressures under a relieved drydock that used a permanent system of small-diameter wellpoints connected to suction headers to control uplift pressures under its thin floor. Pressure increases had occurred in the past and had been temporarily relieved by use of acid washes that reduced incrustations around the wellpoints and increased their discharge capacities. But the problem was getting progressively worse. After reviewing all pertinent information and constructing flow nets to help understand the nature of the seepage conditions, I recommended the installation of suction wells using six-inch diameter slotted PVC pipes surrounded with selected filter aggregate. I hoped that these larger-diameter wells would reduce seepage concentrations such as existed around the small-diameter wellpoints, increase discharge rates, and eliminate the clogging tendencies experienced with the wellpoints. My suggestion was carried out, and the uplift pressures dropped to lower levels than had ever existed under this drydock.

Highway Boat Sections

Underpasses and depressed highway sections are frequently excavated below normal groundwater levels. If the soil formations are relatively impermeable and the natural water table is only a few feet above the road grade, it is often possible to control groundwater with trench drains and pipes feeding the water to sumps. Intermittently operative pumps can keep the roadbed drained for nominal cost, but if the water table is high and the soil formations are permeable, drainage and pumping quantities may be too large to be practical. The construction of depressed highways is avoided in areas in which the soil and groundwater conditions are conducive to large, continuous pumping rates, but occasionally underpasses and portions of freeways must be constructed below the general level of the land in areas with extremely adverse groundwater conditions. In such cases it is necessary to build the submerged portions as *unrelieved* or *partially relieved* boat sections. Heavy reinforced floors and walls are designed with sufficient weight and strength to resist uplift and prevent damage caused by full or partial hydrostatic pressure, much as drydocks must be designed to resist these forces. Extensive groundwater control by deep wells, wellpoints, and sheet piling is required during the construction of these projects.

Figure 10.6 shows general soil and groundwater conditions at a location on which a major freeway had to be built as a depressed boat section within a few hundred feet of a river. Beneath river-deposited silts and clays that covered the site to depths of 12 to 15 ft are beds of silt and sand which in turn are under laid by highly pervious sands and gravels that extend beneath the river. During low river stages the water table lies 8 to 10 ft below the ground surface. At high river stages, which are of short duration, uplift pressures in the pervious substrata rise to about ground level. The right half of the cross section in Figure 10.6 shows diagrammatically the design of an *unrelieved* section for

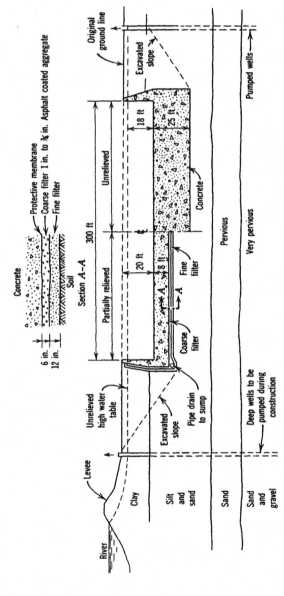

FIG. 10.6 Design of *unrelieved* and *partially relieved* highway boat sections for freeway below the water table.

this portion of freeway, one that is capable of resisting full uplift. This construction would have been massive and costly.

An alternate *partially relieved* design shown also in Figure 10.6 depends on a drainage system to prevent uplift pressures from rising above predetermined levels that exist for moderate periods of time each year. A coarse filter layer beneath the floor and adjacent to the sides of the walls feeds seepage to pipes that carry it to sumps, where it is lifted by permanently installed pumps. Wells with permanent filter protection relieve artesian pressures in the pervious substrata, when needed. Although the drainage system is depended on primarily for high river stages and high groundwater levels, it must function properly without clogging for the life of the project.

When a highway boat section is designed with a coarse filter layer, the permeability requirements of this layer should be determined by using the methods described in this book (Chapter 5). This layer (Sec. *A-A,* Fig. 10.6) should be protected from clogging by placing a membrane over it to keep dirt and mortar out during construction. In addition, a fine filter layer or a suitable filter fabric should be placed under it wherever it rests on fine soils that could clog it.

The decision to design a depressed portion of a highway as a completely unrelieved section (right half, Fig. 10.6) or as a partially relieved section (left half, Fig. 10.6) depends on relative costs and hazards. An unrelieved section cannot be put out of service by power failures, but it requires substantially more concrete than a partially relieved section. On the other hand, a relieved section can become damaged or flooded if its pumping and drainage system fails during periods of high groundwater.

Structural Foundations

Many kinds of structure, varying from single dwellings to massive power plants, are constructed on foundations in which water tables fluctuate within a few feet or less of the ground surface. When concrete floors slabs or mat foundations for houses, garage buildings, warehouses, power plants and the like, are constructed on such foundations, it is customary to install a layer of clean coarse crushed rock or gravel to break capillary pressures and provide drainage to prevent water from wetting the slabs or creating uplift pressures under foundations. Sometimes a watertight membrane is placed between the stone or gravel layer and the concrete as further protection against moisture.

If the water table is at or near the ground surface when the subgrade is being prepared for a concrete slab or other foundation, the placement of the rock or gravel cushion can be difficult. Cases in which the construction equipment mires down in wet sand, silt, or clay under such conditions are common. To overcome these problems filter fabrics can be spread on the wet ground before the aggregate is placed. By this procedure it is often possible to overcome wet foundation conditions. Fast drainage of excess water out of a subgrade up into a coarse aggregate layer (which allows free drainage to the sides) stabilizes the subgrade rapidly and permits the work to be carried out without difficulty. If the water table rises intermittently (or permanently) above the bottom of the drainage layer, collector pipes at the outer edges, or other locations as needed, and gravity or pumped drainage facilities should be provided to prevent this layer from filling with water. (Figure 5.12*b* shows a filter fabric being used to keep foundation soil out of a drainage layer under a road being constructed over a swamp.)

Office buildings, department stores, parking garages, and other structures frequently have basements or walls that are beneath the permanent water table. The weight of these structures is usually more than sufficient to withstand the uplift pressures of the water, and the problems are (1) to make floors strong enough to resist cracking under hydrostatic pressures and (2) to prevent leakage through the walls and floors.

The design of floors and walls below the water table involves estimation of soil and earth pressures and the use of structural theory to develop adequate cross sections. To keep water out, dense impervious concrete is used, construction joints are sealed with water stops, and impervious membranes are placed on the surfaces in contact with the water.

Two suggested methods for keeping basements and walls dry under troublesome groundwater conditions are illustrated in Figure 10.7. The method in Figure 10.7*a* uses (1) seals on the outside surfaces of walls and floor, (2) exterior drains composed of graded filter aggregate or other suitable porous mate-

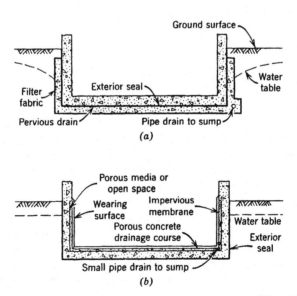

FIG. 10.7 Exterior and interior drainage systems for basements below the water table. (*a*) External drainage system. (*b*) Internal drainage system. *NOTE:* Any filter fabric used should have an adequate life expectancy.

rial feeding water through pipes to sumps where it is periodically removed by pumps.

The second method (Fig. 10.7*b*) uses (1) seals on the outside surfaces of walls and floor, (2) interior drainage layers of porous concrete or other suitable pervious media that feed water leaking through the walls and floor to small sumps, where it is allowed to accumulate and be periodically removed by small pumps, and (3) interior waterproof membranes and linings to keep inside surfaces dry.

The systems shown in Figure 10.7 are suggested for locations in which high groundwater exists, and thorough control over leakage is required. Both systems have double lines of defense; that in Figure 10.7*a* has an exterior drainage course as the first line of defense and an exterior seal as the second; that in Figure 10.7*b* has an exterior seal as the first line of defense and an interior drain as the second.

Ordinarily, exterior drainage that requires permanent pumping is not practical except in relatively impervious soils because pumping costs over a long period of time can be more costly than other methods. Frequently, basements and walls constructed under houses and other structures in hilly terrain can be protected effectively against high groundwater by exterior drains that discharge by gravity (Fig. 10.8). Walls are sealed on the outside with a coating of asphalt or tar or bentonite panels and protected with exterior drains constructed with porous material and tile laid in gravel.

Even though walls and floor slabs are built against relatively impermeable clay or rock below the groundwater level, they must be protected against water pressure unless they are designed to withstand full hydrostatic pressure. To ensure long-time operation drains for floors and walls must not become

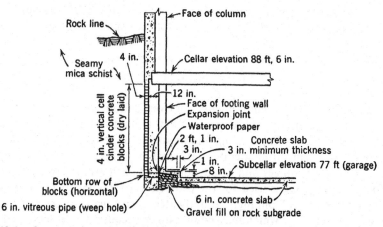

FIG. 10.8 Cross section of wall drainage for seepage in rock (after Feld, 1957). (Section *A-A* of Figure 24, p. 50, Jacob Feld, *American Civil Engineering Practice*, R. W. Abbett, Wiley, New York, 1957, Vol. 3.)

clogged. Feld (1957) points out that a layer of sand placed against saturated clay may become clogged and cease to provide drainage. He reports a condition that caused the failure of concrete basement walls in Hartford, Connecticut and suggests the use of dry-laid porous concrete blocks with hollow cells laid vertically to a thickness of 4 to 8 in. A cross section through a drain of this type designed for the New York Coliseum is shown in Figure 10.8. The bottom course of blocks is laid with the cells horizontal to prevent sealing the open channels. Weep holes feed the seepage to an aggregate drain containing continuous pipe at least 6 in. in diameter installed on a slope of at least 2%. If natural gravity outlets above the water table cannot be reached in reasonable distances, it is necessary to connect the system to a sump from which the seepage is pumped continuously; otherwise the design must allow for full hydrostatic pressure.

An alternate to the porous concrete blocks shown in Figure 10.8 would be a prefabricated synthetic drain with high crushing strength, proven longevity, and adequate discharge capacity (see Chapter 5).

In many older buildings that were built with little or no protection moisture has gradually soaked into the walls from the wet soils on the outside. The dampness can be objectionable from an aesthetic standpoint and because of the physical deterioration it causes. Anderegg (1972) tells of problems caused by capillary water rising from the ground into the masonry walls of many old buildings in Europe. The flow of water can be reversed, according to Anderegg, by the use of electroosmosis induced by a short circuit between wall and soil. He describes practical methods for carrying out this method.

As a large builder of partly submerged structures, the U.S. Navy, Naval Facilities Engineering Command (1982), employs two general approaches to the protection of basements below ground.

1. When the permanent water table is above the top of the basement slab, provide pressure-resistant slab or relieve uplift pressures by underdrainage.
2. When the water table is deep but infiltration of surface water dampens backfill surrounding a basement, dampproof walls and slabs.

Detailed U.S. Navy requirements for foundation waterproofing, dampproofing, and waterstops are given in Navy manuals (see U.S. Dept. of the Navy, Naval Facilities Engineering Command, 1982, NAVFAC DM-7.2). Several types of waterproofing membranes, workmanship requirements, applicability, and the advantages and disadvantages of each are described.

Whenever pumping is used for the protection of basements or other partly submerged structures from high groundwater, care must be taken not to damage adjacent buildings which may be in danger because of the depressed cone of drawdown. Permanently lowering the groundwater level in compressible silts and clays can lead to settlement and cracking of adjacent buildings that rest on these formations.

The design of economical systems and procedures for controlling water against partly submerged structures requires thorough field reconnaissance, explorations, and tests of soil permeability, because the *needs* of individual structures can vary over an almost unlimited range. If the soil permeabilities are high, it may be most economical to design structures that are sealed to keep out the water and heavy enough to resist full uplift and lateral pressures of soil and water. If soil permeabilities are not excessive, drainage may be the more economical choice.

Slurry trenches (see also Section 6.6) are being widely used to control seepage into foundations for many types of below-the-water structures. Zierke (1971) points out that in Germany, after 1959, reinforced concrete slurry walls were used in underground railway construction. In some cases the slurry trenches were unsatisfactory because cracking in adjacent buildings was only partly avoided, but because this form of construction causes *little vibration and noise* it can be used in densely populated areas for underground construction of railways, subways, and the like.

Canals

Adequate control over groundwater and seepage by drainage is an important feature of the design of canal systems. Ordinarily the alignment of canals can be adjusted to avoid deep cuts; however, it is not always possible in the construction of canals of major proportions. Soil surveys and exploratory borings should be made for important projects, and if high groundwater and unstable conditions exist, horizontal drains, vertical wells, and other methods similar to those used for stabilizing slopes for highways and railways (Chapter 8) should be used.

To reduce water losses into pervious soil or rock, canals are often lined with gunite, concrete, impervious asphalt, or other material. Any lining is subject to leakage through cracks, joints, and other imperfections, with the result that pockets of water can develop behind linings, even though the natural water table is low. Groundwater also produces high water levels behind linings. As long as a canal is full of water, the water pressure behind the lining is largely balanced by the water pressure in the canal; but when the canal must be emptied for repairs or for other purposes, unbalanced water pressures can lift it off its foundation, causing serious damage.

Figure 10.9 shows two methods for lowering the hydrostatic head behind canal linings. Both depend on a pervious drainage layer under the lining to feed leakage and groundwater to pipes. One method (Figure 10.9a) continuously discharges seepage through pipes to gravity outlets. This arrangement removes seepage while the canal is full or empty and protects adjacent properties from water infiltration. If this degree of groundwater control is required and the lost water is high in cost, it may be economical to pump the collected seepage back into the canal. If leakage into adjacent land is no problem, outfalls may be omitted, and the pipes can be terminated in sumps from which

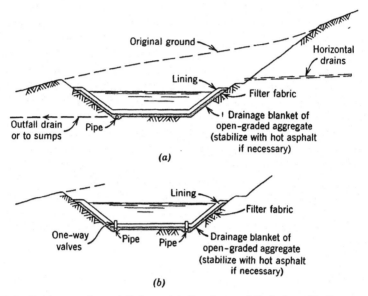

FIG. 10.9   Drainage of canal linings. (*a*) Gravity outfall drains. (*b*) One-way valves into canal.

the leakage can be removed by inserting portable pumps when the canal is emptied.

Another type of control illustrated in Figure 10.9*b* utilizes flap valves or other suitable check valves that relieve pressures by permitting water to flow into the canal as the water level is drawn down. Check valves must be noncorrodible and must not be clogged or held open by silting or other causes. Those used by the U.S. Bureau of Reclamation for the Delta-Mendota Canal in California became obstructed and clogged by minute shellfish that flourish in this canal.

When canals are excavated in water-bearing rock formations that are highly resistant to piping, it may be possible to place highly permeable, open-graded drainage aggregate directly against these formations without danger of clogging the drains. If, however, the formations are erodible, it will be necessary to use a fine aggregate filter or a filter fabric to hold the soil in place, as shown in Figure 10.9. No filter fabric should be used unless it has an adequate life expectancy.

10.4   MASONRY DAMS

Masonry dams (see Figure 10.10) represent a major class of structures throughout the world. Developing measures for controlling seepage through and under

FIG. 10.10 A typical masonry dam. Pine Flat Dam in central California backs up 1,000,000 acre-ft of water as flow of 18,300 cfs pours through the spillway gates. (Courtesy of Sacramento District, U.S. Corps of Engineers.)

these dams has been a major concern of civil engineers for many decades. Some of the basic concepts are outlined in the following paragraphs.

Seepage through Masonry Dams

Carlson (1957) estimates that the permeability of concrete is so low that equilibrium pore pressures (steady seepage) are not likely, if ever, to develop in a moderate number of decades.

In those parts of the concrete in a dam in which equilibrium pore pressures develop, evidence suggests that these pressures probably exist over 100% of its gross area. Differences of opinion have existed among dam designers regarding the true percentage of the area on which the neutral pressure is acting, and the actual uplift caused by the pore water has always been an uncertain factor in the analysis of forces acting in gravity dams. Fortunately its exact value is not extremely important, for the upstream face is made as watertight as possible, and drains are installed a short distance back from the face in an effort to keep uplift pressures small throughout the width of the dam (Fig. 10.11). Even with the best efforts, seepage generally shows up in inspection and drainage galleries a few months after the reservoir is filled and at the downstream face within a year or two. No doubt, localized uplift does develop but its extent is likely to be rather limited and it probably subjects very little

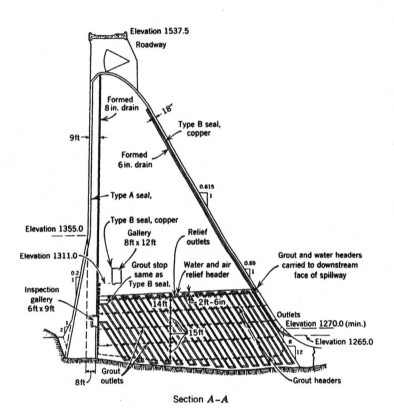

Section *A–A*

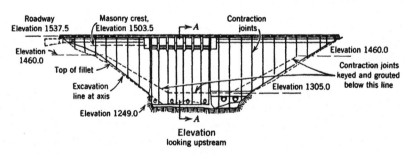

Elevation
looking upstream

FIG. 10.11 Construction details of a typical concrete gravity dam. (After Pearce, 1940. From "Design of Hiwassee Dam," *Civil Engineering,* A.S.C.E., July, 1940, p. 433.)

of the cross section to appreciable uplift. Uncontrolled leakage through gravity dams can be detrimental; however, the excellent performance record of many of these dams throughout the world is evidence of the adequacy of widely used construction practices.

An ASCE subcommittee on uplift in masonry dams (1952) studied the problem of uplift for several years and presented a substantial amount of factual data and an interpretation of the problem. The committee listed two constituent elements in uplift pressure: (1) the *area factor* and (2) the *intensity factor*.

The area factor is the proportion of a horizontal section at the base of a dam which is assumed to be subject to uplift pressure. Its maximum value is 1.0 or 100%. Although smaller values have been used in designing many dams, based on the designers experience and judgment, the committee advised the use of an area factor of 1.0.

The intensity factor is the assumed ratio borne by intensity to an intensity gradient extending from maximum headwater to maximum tailwater.

The committee found it inadvisable to recommend definite, or even minimum, values for the intensity factor and concluded that, "the circumstances peculiar to each project must guide the designer in making the proper assumption."

The ASCE committee concluded that the inspection and maintenance of drainage systems are important and that "if there is not adequate assurance that drains will be inspected and maintained, no credit for their performance should be assumed in design."

Seepage under Masonry Dams

The dam builder has substantial control over the properties of the structure he builds, but he must take the foundation as it is furnished by nature. He can improve on its compressibility and shear strength by consolidation grouting and he can reduce leakage to some degree by cutoff grouting, but he has much less control over its properties than over those of the dam itself.

Nearly all failures of masonry dams are caused by uncontrolled seepage or excessive pore pressures in foundations and abutments. If the rocks are not resistant to erosion, scour can lead to the undermining and failure of rigid dams. Erosion of a soft "conglomerate" probably led to the failure of the St. Francis Dam in California in 1928 (Fig. 1.1). Undetected pervious seams or joints can allow water pressures to build up to dangerously high levels that can destroy dams. Water infiltration into a thin fault (described as a clay seam about 1.5 in. thick) in the left abutment of the Malpaset Dam in France (*Engineering News-Record*, January 21, 1960, p. 26) was believed to have caused the failure of this thin arch dam in December 1959. Londe (1972) reports that, after an extremely thorough 5-year investigation into possible causes of the failure, he concluded that the tightening up of rock joints under compressive pressures produced by the arch, allowed full reservoir head to build up in critical joints, and caused a wedge of the rock to literally blow out from under the arch, leading to its nearly instantaneous failure.

Failures of thin arch dams are sudden; the release of the stored water is almost instantaneous. On the other hand, if an earth or earth-rock dam should start to fail, some time is needed for erosion to develop and for a major flow of water to begin. Some dam engineers feel that even a few hours of advance warning could save many lives in the event of a major failure and that this consideration should enter heavily into the choice of the basic type of dam to be built—particularly above heavily populated areas.

The development of a safe masonry dam requires thorough knowledge of the character of its foundation and abutments, which can usually be developed by thorough geological examinations and explorations. If the rocks in the foundation and abutments are strong enough to resist the stresses created by the dam and the water in the reservoir, the problem is adequate control of seepage forces and pore pressures. This usually can be done by (1) grouting and (2) drainage.

In designing and building filters, grain sizes must be selected to *prevent* the movement of soil particles; however, in the selection of materials to be used in grouts to seal leaky foundations the particles must be small enough to *move freely through pores or cracks.* King and Bush (1963) state that, ". . . . for effective injection to take place, the pore size should be at least three times the effective maximum grain size of the cementing material." Or, ". . . if the diameter of grouting particles is not over $1/5 \times 1/3 = 1/15$ of the particle diameter of the material to be grouted, the grout will pass freely through the void system.

Many of the larger dam builders install devices for measuring the uplift pressures at the contact between the concrete and the foundation rock, along construction joints, and to some extent within the mass concrete. Leaders in this work have been the U.S. Bureau of Reclamation, the U.S. Army Corps of Engineers, and the Tennessee Valley Authority. Keener (1951) describes methods used by the U.S. Bureau of Reclamation and summarizes some of the typical information obtained.

There are two basic purposes for making uplift pressure measurements under masonry dams:

1. To provide a basis for making uplift assumptions in the design of future dams.
2. To obtain regular readings of the uplift in dams so that if considerably more uplift develops than assumed in the design calculations relief measures can be taken to safeguard the dam.

The design of masonry dams that are safe against seepage effects is largely a matter of careful planning:

1. Making thorough, experienced explorations of sites.
2. Designing structures with adequate factors of safety, aided by adequate seepage control.
3. Following up by regular measurements of pore pressures in completed dams, foundations, and abutments.

The actual seepage patterns that develop in rock formations depend on the jointing and crack systems in the rock and seldom can be predicted accurately. Flow-net studies can, however, give an appreciation of the general benefits (or lack of benefits) that can be expected from various possible alternate designs and from localized treatment of the formations by grouting. The results of a study of grout curtains beneath earth dams are given in Figure 6.8. To further illustrate use of flow nets for this purpose Figure 10.12 summarizes a flow-net study of uplift pressure distributions under a masonry dam with a grout curtain penetrating various depths into jointed rock overlying impervious rock. No drains are shown in the foundation. In this study the grouted zone is assumed to have its permeability reduced by 90% to one-tenth of that of the ungrouted rock. It is shown (Fig. 10.12*b*) that even with 100% penetration of this grout curtain the seepage quantity is still more than 60% of that without grouting. The uplift pressures for various depths of penetration of the grout curtain, shown in Figure 10.12*a*, indicate that even with 80% or greater penetration of the curtain into the pervious rock uplift pressures correspond to an intensity of nearly 50%.

10.5 STORAGE RESERVOIRS

General

When populations overdevelop the natural water supplies of areas or overflow into semiarid lands, water must be imported if life is to continue. The long canals and aqueducts that are required are subject to occasional shutdowns during damage, cleaning, or routine repairs; hence to ensure uninterrupted supplies of water and to provide space for regulation of flow, storage reservoirs are an essential part of most water supply systems. Their design requires a high level of engineering and geological knowledge and experience.

Historical. Wegmann (1922) gives the following account of storage reservoirs in India and Ceylon:

> From a remote period of history earthen dams were constructed in India and Ceylon to form reservoirs called "tanks" for irrigation purposes. . . . The earth dams forming these tanks aggregate about 30,000 miles in length.

> In Ceylon there are 30 immense tanks, besides 500 to 700 smaller ones in ruins. Most of these works could be easily restored to service. The great tanks of Ceylon

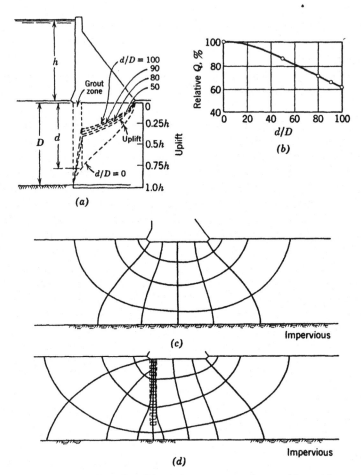

FIG. 10.12 Flow-net study of uplift under masonry dams with partly penetrating grout curtains. (Permeability of grouted zone equals one-tenth that of ungrouted rock; no drains are provided.) (*a*) Cross section and uplift pressures for various depths of grout curtain. (*b*) Relative seepage quantity versus depth of grout zone having permeability one-tenth that of ungrouted rock. (*c*) Typical flow net 1, $d/D = 0$. (*d*) Typical flow net 2, $d/D = 0.8$.

exceed the Indian tanks in extent and grandeur. The tank of Kalaweva, constructed A.D. 459, had a circumference of about 40 miles. It was formed by an embankment 12 miles long, which had a spillway of stone.

Even when compared with modern water developments many of these ancient works represent great feats of construction.

In the second half of the twentieth century the exploding populations of the earth have led to the construction of many storage reservoirs in heavily populated areas. Because reservoirs need to be at higher elevations than the areas served, they are often located in the hills above the communities they serve. Thus every storage reservoir is somewhat of a threat to the security of those living below. Seepage control and drainage are topmost among the methods engineers can use to make reservoirs as safe as possible.

If storage reservoirs are to serve their purpose properly, they must not lose excessive amounts of water by leakage, for water is costly. Horsky (1969) tells about a reservoir that had so much leaky limestone in the foundation that it was completely useless. Figure 6.1 shows a small reservoir in the western U.S. that could not hold water because of undetected open-work gravel under the reservoir and dam.

Many modern storage reservoirs have special impervious linings that reduce water losses and keep the water clean, and drainage systems that control the leakage that does occur. Frequently reservoirs are carved into hillsides or hilltops and thin natural ridges form portions of their circumferences. Thorough soil and geological investigations should be made of the earth formations under and around proposed reservoirs, and seismic activity and fault patterns in the vicinity studied thoroughly to determine their probable influence. Land subsidence from water or oil extractions in the vicinity should also be considered.

Security from Earthquakes, Earth Movements, and Roofing. Known faults and areas in which subsidence is known to be taking place should be avoided. Nevertheless it is possible that nearly every reservoir and nearly every dam will at some time be subjected to a severe earthquake shock or a crustal movement. Fundamental criteria that increase the overall resistance of earth masses to earthquake damage are discussed in Sec. 8.2. These criteria should be considered in the design of storage reservoirs. In addition, it is believed that flexible, self-healing elements should be provided in critical portions of the slopes of storage reservoirs and in the cross sections of all important earth dams. If faulting or other earth movements are considered even remotely possible during the life of a storage reservoir, a layer or zone of cohesionless sand and gravel should be located in the cross section where it can slough into incipient openings, fissures, or faults, or trap migrating fines, thus retarding or preventing piping failures.

Casagrande et al. (1972) discuss the failure of the Baldwin Hills Reservoir in 1963 and conclude that it was caused in foundation strata that were highly sensitive to erosion and crossed by faults. A porous "popcorn" concrete drainage layer (very rigid) had been constructed under the impervious clay lining. These authors say that if the brittle drainage layer had not been used, it might have taken years longer to produce the degree of underground erosion that led to the failure, although eventually the reservoir might still have been destroyed.

Weighted filters on the outside slopes of reservoirs provide a second line of defense by trapping soil particles and preventing piping failures.

Whenever rigid structural elements are placed over semirigid or erodible foundations, seepage that causes any movement of particles from the foundation, tends to cause "roofing" and possible failure of hydraulic structures. Wegmann (1922a) states that, "Gravel will tend to fill up a hole that may be formed in a dam, but clay is apt to arch over an opening which may be enlarged and lead to the rupture of the structure." Obviously arching is equally dangerous if an opening forms under concrete or other rigid structures. Terzaghi and Peck (1948) say:

> . . . the majority of piping failures have occurred several months or even years after the ill-fated dams were put into operation. Hence, it appears that most if not all piping failures of actual dams were caused by subsurface erosion and not by heave. [And also] Erosion tunnels with unsupported roofs are conceivable only in soils with at least a trace of cohesion. The greater the cohesion, the wider are the spaces that can be bridged by the soil.

The preceding comments were made in relation to dams. They support the concept described in this section of requiring *cohesionless, self-healing* elements for the protection of reservoirs against earth movements and piping failures.

Basic Criteria for Reservoirs. Experience suggests that storage reservoirs in critical locations should meet the following requirements:

1. Be safe from overtopping or breaching due to instability of either the upstream or downstream slopes.
2. Be safe from piping along pipe outlets, conduits, or other rigid elements.
3. Be safe from earth movements that might cause offset displacements or tension cracks.
4. Be safe against slumping failures due to earthquakes.
5. Be safe against sudden earth or rock slides into the reservoir.
6. Be designed with seepage control measures to ensure the security of the reservoir and to protect adjacent properties from high groundwater due to leakage from the reservoir.

Drainage Design

Reservoir drainage often is needed for the following purposes:

1. To control seepage and ensure stability of the natural or constructed earth dams surrounding the reservoir.
2. To prevent damage to linings during "rapid drawdown" of the reservoir.
3. To control seepage and prevent nuisances due to raised groundwater levels in developed lands below reservoirs.

Figure 10.13 shows the essential features of drains for reservoir linings. The cross section in Figure 10.13a illustrates a lining of asphalt plank, butyl rubber,

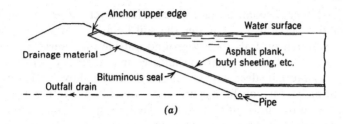

(a)

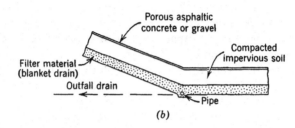

(b)

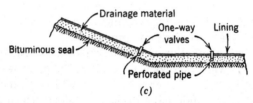

(c)

FIG. 10.13 Methods for relieving hydrostatic pressures beneath reservoir linings. (a) Impervious membrane lining with blanket drain and gravity discharge. (b) Impervious soil lining with blanket drain and gravity discharge. (c) One-way valves to relieve pressure under lining during reservoir drawdown.

or other impermeable material. Under this lining is a drainage layer that should be designed according to the permeability and discharge criteria described in Chapter 5 and subsequently illustrated (Fig. 10.14).

A perforated or slotted pipe drain at the toe of the slope in Figure 10.13a collects seepage and discharges it through outfall pipes. This arrangement will safeguard the lining against blowouts or ruptures during drawdown of the reservoir and will prevent the buildup of saturation in natural ridges and constructed earth dams surrounding reservoirs. Although this arrangement is most suited to small reservoirs with low to moderate heads, it is sometimes used for large reservoirs. As long as the lining is watertight, it provides a high degree of control over seepage; however, large hydraulic gradients are created across the lining, which are conducive to the development of leaks through points of weakness. Maintenance of these linings is therefore very important.

An alternate design shown in Figure 10.13b uses impervious soil for seepage reduction, a porous asphalt or aggregate surface layer for prevention of wave damage, and a pervious layer under the soil lining for drainage. Seepage is discharged through outfalls to low ground outside the reservoir. The design shown in Figure 10.13b is similar to that used in a number of modern storage reservoirs. An alternate design using one-way valves is shown in Figure 10.13c.

If drains under reservoir linings are to provide full protection against uplift pressures, they must, in addition to providing filter protection, have sufficient discharge capacities to remove all seepage reaching them with small buildup of head (Chapter 5). A desirable method to use in designing drainage blankets is to use flow nets or Darcy's law to estimate probable discharge rates (Sec. 5.4). If drainage layers are designed to remove several times the estimated rates, reasonable reserve capacities will be available to take care of unanticipated seepage conditions.

The design in Figure 10.13b depends on the use of a highly impervious clay soil for the lining ($k = 1 \times 10^{-4}$ ft/day or less). Linings with this level of water-tightness should restrict seepage to amounts that can be removed by a single layer of aggregate that will also have adequate discharge capacity (to be verified by the filter and permeability criteria in Chapter 5). Where compacted earth linings permit greater amounts of seepage than can be removed by a single layer, it will be necessary to use dual-layer drains composed of a filter protecting the soil from piping, and an open-graded layer under the filter to conduct the water to discharge pipes.

Designing a blanket drain for the conditions just described is illustrated in Figure 10.14, which shows a rolled earth blanket with a permeability k_b as the seepage-reducing element. All seepage through the earth blanket must be discharged by the underlying drainage blanket which has a thickness H_d and a permeability k_d to a collector pipe located to the right of Section A-A. From Darcy's law or flow nets the total infiltration quantity Q_1 entering the drain can be estimated and several times this quantity used in designing the drainage blanket. Liberal seepage quantities should be included in these determinations, for large discharge capacities of blanket drains are easily obtained with the proper selection of drainage materials.

Two approaches are available in designing drainage blankets for reservoirs:

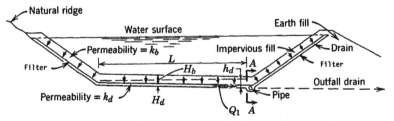

FIG. 10.14 Designing blanket drains for reservoir linings.

1. Assume a thickness H_d and a small allowable hydraulic gradient h_d/L across bottom width L and calculate the required permeability k_d with Darcy's law,

$$k_d = \frac{Q}{iA} = \frac{Q_1}{(h_d/L)H_d} = \frac{Q_1 L}{h_d H_d} \tag{10.1}$$

2. Assume a coefficient of permeability k_d for the drainage blanket and calculate the required thickness H_d from Darcy's law:

$$H_d = A = \frac{Q}{ki} = \frac{Q_1}{k_d(h_d/L)} = \frac{Q_1 L}{k_d h_d} \tag{10.2}$$

With Eqs. 10.1 and 10.2 various combinations of blanket thickness and permeability can be studied and the most suitable and economical selected. Drainage aggregates should always be protected from clogging by the adjacent soil, as described in Chapter 5, and liberal discharge capacities should be allowed in drainage blanket design.

When the stored water is costly and losses must be kept to a minimum, lining systems with drainage features of the kinds just shown are often used. But. if the water in a reservoir is relatively inexpensive, or provisions can be made for collecting the seepage in sumps and pumping it back into the reservoir, some reservoirs have been constructed without any special linings. One such reservoir, that for storing cooling water for a power company's generating plant in Florida, was constructed with a soil-cement facing to prevent wave erosion and compacted local soils for its embankment. After water had stood in it for 18 months, it failed very suddenly on the night of October 30, 1979 (see *Civil Engineering* Magazine, ASCE, Jan. 1981, pp. 46–47). Although the precise failure mechanism could not be determined, it was believed to be caused by piping of fine soils out of the foundation through a very porous seashell formation. Accordingly, the panel of engineers reviewing the failure and making recommendations for its repair, decided that the repair should include a comprehensive drainage system along the downstream toe of the entire dike to protect the toe and downstream slope from future problems with seepage. A cross section through the up-graded dike is given in Figure 10.15. As may be seen, the lower half of the downstream slope was trimmed to a 1.8:1 slope and a trench was excavated along the toe to a substantial depth into the foundation. Non-woven filter fabric was placed on the prepared surfaces of the slope and trench to prevent soil movement into a 12-in. thick layer of 3/8 in. to 1 in. crushed stone that feeds seepage to the bottom of the trench where a 12 in. dia. perforated pipe collects the seepage and conducts it to exits. In the repair, some 1.5 million square yards of the non-woven fabrics was used, making it the largest single use of this fabric on record. Some of the engineers on the review panel had thought of using a blend of sand and gravel comparable to concrete sand to feed the water to the pipe; however, the dis-

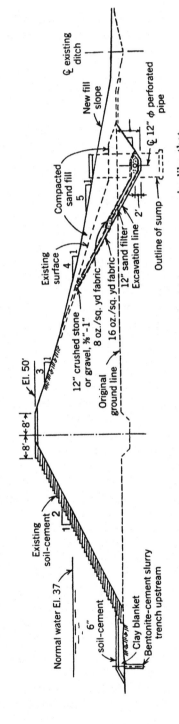

FIG. 10.15 New seepage control measures for Florida storage reservoir dike that failed by piping. Reprinted from *Civil Engineering*, January, 1981, with permission, ASCE.

charge capability of such a layer was open to question. The system selected makes use of the protective features of good fabrics, and the good discharge capabilities of the open-graded crushed rock. This general type of drainage system (fabrics for filters and open-graded aggregate for discharge) has many uses (see Chapter 5, Filter and Drain Design), and is far superior to systems that try to use one layer of blended aggregate to obtain filter protection and discharge capability. Drains of this general design are also shown in Figure 5.12c and Figure 6.19.

Reservoir Linings

Many reservoir sites are covered with relatively impervious soils and are adequately watertight without the use of special linings. In such cases the natural soil is usually compacted in place with rubbertired rollers or sheepsfoot rollers after moistening, to improve its watertightness. A layer of gravel may be placed on the surface to prevent erosion and to keep the water clean.

When water losses must be kept to very small amounts (as in many reservoirs in arid countries), thorough studies are required to eliminate errors in predicting rates. Aisenstein et al. (1960) report that measured seepage losses from Zohar Reservoir in southern Israel were several times the estimated losses of a few millimeters per day. Seepage losses through the heavy clay in the bottom of the reservoir were about 2 cm. These authors conclude that standard types of permeability test may be misleading. Discrepancies in permeability can be caused by testing in an unsaturated state, for the saturated permeability may be several times that of the unsaturated value. Also, the dispersion effects of salts in semiarid soils may result in higher permeabilities of aggregated clays in reservoir bottoms than in tests. In the testing for the Zohar Reservoir, permeabilities in the order of 10^{-8} cm/sec were obtained from the normal permeability tests, whereas the true permeability is believed to be in the order of 10^{-5} cm/sec. This wide variation accounts for the measured discrepancy in water losses.

Many reservoir sites, particularly those in dry or semiarid climates, have little or no natural soil cover and often are composed of pervious formations in which the water table may be low. In these locations special linings are placed on the reservoir bottom and sides to reduce the losses of costly water, to protect adjacent properties, and to keep the water clean.

The type of lining that is most suitable for a given reservoir depends largely on the cost of the water, the watertightness needed for safety considerations, and the availability and cost of materials. Several of the more widely used linings are described in this section.

Asphalt Concrete and Asphalt Plank. Specially designed asphalt concrete mixes have been used to line a substantial number of important dams and storage reservoirs. The Ludington Pumped-Storage Project in Michigan made use of such a lining with a special drainage system. This lining was required to be impervious to water, deformable, resistant to ice erosion, and free from cracking and fissuring because of heating or aging.

Whitehead and Ruotolo (1973), in describing this project, say that the drainage system under the asphalt concrete lining is an important safety feature of the reservoir. Any water leaking through the lining will collect at the bottom of the sloping face drainage layer and be pumped back to the surface by submersible pumps spaced at 150-ft intervals around its 6-mile circumference. Water-level recorders at points all around the reservoir will chart water levels inside the sloping drain and thus forewarn operators of any leakage as soon as it develops.

Lehnert (1973) describes the construction work carried out since 1968 in Germany, Spain, Italy, and Rumania, in tropical and subtropical areas, and Japan. He says that bituminous linings for pumped storage projects, in which water-loading conditions change once or twice a day, must withstand large stresses. Drainage of the bottom of the pavement (lining) is important, and the use of a thicker, one-layer asphalt concrete course is recommended to avoid trapping water between successive layers. He also describes the construction of linings of pumped storage schemes in Belgium, the United Kingdom, and the United States. (See also "Thin Sloping Membranes" in Section 6.2).

When asphalt plank is required, preconstructed sheets of various dimensions are used in thicknesses of $\frac{1}{8}$ to $\frac{1}{4}$ in. and in sizes that vary from 5 by 10 ft to 10 by 40 ft. Joints are lapped or butted and cemented, and the lining is laid on the smoothed soil, care being taken to avoid puncturing it during construction. With close control over its placement and care in making the joints watertight, many successful linings have been constructed with asphalt plank. Figures 10.13a and 10.13c show cross sections through reservoirs lined with asphalt plank or other impermeable material and drainage systems for the control of leakage through the linings. Permeable drainage blankets can either discharge the collected seepage through outfall pipes (Fig. 10.13a) or into the drawndown reservoir through one-way valves (Fig. 10.13c), or be evacuated with submersible pumps.

Butyl Rubber and Plastic Sheeting. Reservoir linings may be constructed of heavy rubber or plastic sheeting placed carefully on the smoothed, compacted subgrade to avoid puncturing. If all joints are made according to manufacturers recommendations, these linings can be highly watertight. They must be protected against livestock and people because they are easily damaged. Punctures can be repaired, although the water level must be drawn down to the level required to permit the repairs to be made in the dry. Although the length of dependable life of these linings has not been fully demonstrated, some manufacturers are offering to guarantee installations for 10 years. The general designs in Figure 10.13a and 10.13c apply also to butyl and plastic linings. Great care must be taken to avoid the placement of thin linings on formations containing shallow, hidden openings that can lead to the rupture of these linings.

Thin linings have been used in the construction of reservoirs of substantial sizes. Chuck (1970) describes the 1.5-billion-gallon Kualapuu Reservoir in Hawaii in which an earth dam embankment of compacted earth rises 60 ft (18.3 m). Altogether 4,800,000 sq ft (446,000 sq m) of nylon reinforced butyl rubber lining 1/32 in. thick was installed at a cost of 23 cents/sq ft. The choice of rubber was made after considering several alternatives, according to Chuck.

Sprayed Asphalt. Sprayed asphaltic emulsions, reinforced with filter fabrics, line some reservoirs. These fabrics are destroyed by hot asphalts, but they can be used as reinforcement for cold asphaltic emulsions. Relatively inexpensive nonwoven fabrics have been used for reinforcement, but the cost of spreading, sewing, and handling the fabrics adds substantially to the overall cost of an installation. Special catalytically blown asphalt membranes have been constructed for reservoirs and canals and when done carefully have been said to be highly watertight. Catalytically blown asphalts are highly resistant to flow or creep under the small weight of the thin earth cover placed over these linings after they have cured. Approximately 1.5 gal/sq yd (total amount) is generally placed in three or more nearly equal applications, with joints staggered.

Before any thin membrane is placed, the soil should be thoroughly sterilized to discourage the growth of grass, weeds, and plants that would rupture the lining. The subgrade should be smoothed and compacted before the membrane is placed and a protective soil or fine gravel cover carefully placed by shoving with light spreaders that will not "ball up" the lining with the cover or tear or damage the lining. If thin membranes are placed on foundations that contain natural groundwater or become saturated from leakage, bulges can occur when it becomes necessary to drain the reservoir, in which case repairs will be needed before the reservoir is refilled.

Compacted Earth Lining. If adequate quantities of good quality impervious soils are available in the vicinities of reservoirs, they usually make the most satisfactory linings, for soils do not deteriorate. Compacted impervious earth linings are usually at least 5 to 10 ft thick (Fig. 10.13*b*), although even 2 or 3 ft of highly impervious clay can reduce leakage to small amounts. Permeability tests should be made of representative samples of available materials compacted to densities expected after a long period of use. Losses by seepage through linings can be estimated with Darcy's law or flow nets, and if drainage is required behind these linings drains can be designed by using Eq. 10.1 or 10.2.

When clay linings are depended on for watertightness, it is important that possible increases in water losses from deflocculation (see Sec. 5.2) or increased permeability from the character of the water (see Sec. 6.1) be considered. Also, after all known factors have been taken into consideration a reasonable factor of safety (5 to 10) should be applied to allow for the fact that earth linings are almost never so watertight as predicted by small-scale permeability tests.

If compacted earth linings or thin membrane linings are to be placed on the sides and bottoms of reservoirs in gravel formations or in rocks showing evidence of jointing, faulting, open seams, or coarse conglomerate layers, special precautions should be taken to protect the linings against piping into these formations. The drainage layer in Figure 10.13b can be adapted to these conditions by placing several feet of cohesionless sand and gravel adjacent to the foundation, an overlying coarser layer for fast removal of seepage, and a topping of filter material to protect the earth lining.

In all locations in which ruptures in reservoir linings could lead to detrimental seepage, piezometers should be installed at frequent intervals around the reservoir and water pressures should be read periodically. Also, the effluent from drainage outlets should be observed regularly. If either the pressures or the seepage quantities should rise appreciably for no apparent reason, the reservoir should be observed closely for possible trouble.

10.6   OVERFLOW WEIRS AND SPILLWAY CHUTES

This section describes drainage facilities for overflow weirs and dams on soil foundations and for spillway chutes, which are two types of structure that are highly susceptible to seepage failures unless thoroughly protected.

A number of cases of confined flow under hydraulic structures are analyzed in this chapter by flow nets.

Overflow Weirs and Dams on Soil Foundations

General. High masonry dams must rest on strong rock foundations, for foundation weaknesses may lead to total failure. Low diverting weirs and dams, on the other hand, are built across rivers at locations in which rock is deep. Sand and gravel make suitable foundations for these structures if adequate precautions are taken to prevent excessive settlement and failures due to (1) washing out at the downstream toe by the overpouring water or (2) piping of the foundation by seepage under the structure.

Washing out by scour at the downstream toe usually can be avoided by the use of concrete or stone aprons of adequate size and shape and by maintaining an adequate pool of water to cushion the action of the overflowing water.

Piping failures can be prevented by controlling seepage velocities and uplift pressures by the use of cutoffs and aprons and by protecting seepage exits with graded filters.

The type of structure developed for a given site depends on the flowing:

1. The head to which the structure will be subjected.
2. The quantity of water that will flow over it.
3. The nature of the earth foundation.

Cutoffs may be of concrete (usually in bouldery foundations), wood piling (if permanently submerged), steel sheet piling, or chemical grout. Cutoffs are most effective if they penetrate to impervious strata, for partial cutoffs in deep pervious strata produce only limited reductions in uplift pressures and seepage quantities. (See Figure 6.5 for the relation between the seepage quantity and the depth of a cutoff trench beneath an earth dam and Figure 10.12 for a study of grout curtains under masonry dams.) When used in deep alluvial foundations, sheet pile cutoffs improve seepage conditions under weirs and dams by forcing flow lines down into foundations, thereby reducing escape gradients (Fig. 10.18b).

Downstream cutoffs protect foundations against progressive scour in the event protective riprap is washed out. They increase uplift under aprons if not used with properly located drains.

For general lengthening of the seepage path, upstream aprons are preferable to downstream aprons because they do not create uplift problems. The are inefficient if used without vertical cutoffs and drains.

Downstream aprons provide for dissipation of the energy of the overflowing water. They are susceptible to uplift if not adequately drained.

Filtered drains properly located should be provided under all overflow weirs and dams on alluvial foundations because they control uplift pressures and allow seepage to escape harmlessly (Fig. 10.18c). They also trap foundation soil which may start to move from beneath a hydraulic structure due to "roofing" or other inconsistencies. Good drainage is essential for the protection of weirs and dams on soil foundations.

Many overflow weirs and dams have been constructed in the United States and in other countries by combining methods for the control of underseepage. Borovoi, Razin, and Eristov (1963) describe the spillway dam for the Volga River Hydroelectric Station (Fig. 10.16), which has a head of 27 m and a length of 725 m. This spillway dam, which is expected to pass a flow of 37,600 cu m/sec, was constructed on 6 to 10 m of alluvium. Seepage control is obtained by two rows of steel sheeting driven through the alluvium into the sand-aleurite rock, which has a permeability one-seventieth that of the sand. Two rows of relief wells relieve uplift head under the dam and stilling basin.

Casagrande (1935, 1937, 1961) was a long-time advocate of the use of rational methods for the design of structures with seepage and for the control over seepage forces and pressures with drainage. In modern design, dams and overflow weirs on earth foundations should always be developed with the aid of rational methods and planned seepage control measures.

Mechanics of Piping Due to Heave. Nonuniformities in the deposition of soils and vertical holes made by burrowing animals, rotted roots, *unfilled drill holes,* abandoned water wells, and the like often permit seepage to concentrate and emerge in the form of *boils* at the landward toe of dams and other hydraulic structures. Sherard et al. (1963) describe numerous causes of localized piping failures in dams. Piping failures caused by heave can be expected to occur at the downstream side of a hydraulic structure when the uplift forces of seep-

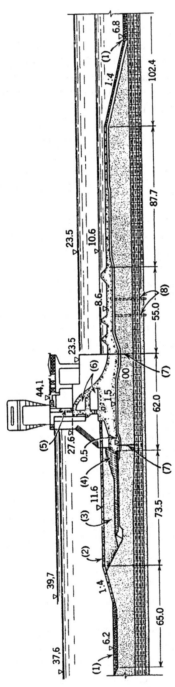

FIG. 10.16 Spillway dam for Volga Hydroelectric Station, USSR. (After Borovoi, Razin, and Eristov.) (1) rockfill, (2) concrete plates, (3) sand, (4) compacted clay, (5) main gate, (6) hollows, (7) steel sheeting, (8) relief wells. Dimensions are given in meters. (See "Some Large Dams of Hydro-projects in the U.S.S.R.," contribution to "Topmost Dams of the World," The Japan Dam Association, Tokyo, October 1963, pp. 224–225.)

age exceed the downward forces due to the submerged weight of the soil (Sec. 3.5).

The method described by Terzaghi and Peck (1948a) for determining the factor of safety against heave is illustrated with reference to a row of sheet piles in sand (Fig. 10.17). The principles developed here apply equally to the soil at the landward sides of weirs or dams on permeable earth foundations. First, a flow net is drawn (Fig. 10.17a) from which the excess hydrostatic pressure on a horizontal plane such as Ox at a depth D can be determined (Sec. 3.5). For a head h_1 on the left side of the sheet pile wall the uplift pressure (excess hydrostatic pressure) can be represented by the ordinates of curve C using line Ox as a reference line. It is shown that the uplift pressure is greatest just to the right of the wall; hence the greatest danger of uplift exists near the wall. By tests with models Terzaghi found that when the upward forces of seepage on a portion of Ox near the wall become equal to the downward forces exerted by the submerged soil the surface of the sand rises (Fig. 10.17a). This heave occurs simultaneously with an expansion of the volume of the sand, which causes its permeability to increase. Additional seepage causes the sand to boil, which accelerates the flow of water and leads to complete failure. Terzaghi's model tests demonstrated that heave occurs within a distance of about $D/2$ from the sheet piles. To calculate a factor of safety against failure, forces are determined on the prism $efaO$ (Fig. 10.17b) which has a depth D and a width $D/2$.

The average excess hydrostatic pressure on the base of prism $efaO$ is equal to $\gamma_w h_a$, and the uplift force U is equal to $\gamma_w h_a D/2$. Piping failure occurs when U becomes equal to the submerged weight of the sand which is its volume $D^2/2$ times its unit submerged weight γ' or $W' = 1/2D^2\gamma'$.

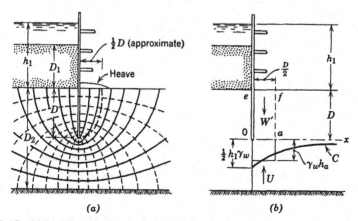

(a)   (b)

FIG. 10.17 Use of flow net to determine factor of safety of row of sheet piles in sand with respect to piping. (After Terzaghi and Peck 1948.) (Fig. 104, p. 231, *Soil Mechanics in Engineering Practice*, K. Terzaghi, and R. B. Peck, Wiley, New York, 1948.)

The factor of safety with respect to piping can therefore be expressed as

$$G_s = \frac{W'}{U} = \frac{D\gamma'}{h_a\gamma_w} \qquad (10.3)$$

If it is not economical to drive the sheet piles deep enough to prevent heave, the factor of safety can be increased by placing a weighted filter over prism *efaO*. If the weight of this filter is *W*, the total downward force within a distance *D/2* of the pile wall is *W* + *W'*, and the factor of safety is increased to

$$G_s' = \frac{W + W'}{U} \qquad (10.4)$$

The upward force *U* exerted by the seeping water can also be determined with hydraulic gradients obtained from the flow net, for the seepage force is equal to the average hydraulic gradient in prism *efaO* multiplied by its volume and the unit weight of water (Sec. 3.5). Thus

$$F = \gamma_w i(\text{vol}) = 62.5i(V) \qquad (3.21)$$

When the hydraulic gradient becomes equal to 1.0, the uplift exerted on a cubic foot of submerged soil is 62.5 lb. Under this gradient a soil with a unit weight double that of water (125 lb/cu ft) has a unit submerged weight of 62.5 lb/cu ft; hence with an uplift gradient of 1.0 the effective downward force is zero. This must be, because the *body force* (Sec. 3.5) must be zero. For this state of stress the frictional resistance at the base of prism *efaO* is zero. When these conditions exist, a state of *flotation* exists in the soil that leads to heave and to boiling and piping.

Hydraulic gradients that produce flotation and heave vary with the unit weight of the soil. Extremely lightweight soils are lifted by small gradients, whereas heavy soils such as most sands and gravels require an uplift gradient of about 1.0 or somewhat greater.

In the design of masonry dams and weirs on earth foundations flow nets should be used to study seepage patterns and Eq. 10.3 or 10.4 for calculating factors of safety with respect to heave. Equation 3.21 determines the uplift forces at discharge exists.

As with other structures, the design of overflow weirs is largely based on experience; however, analytical seepage methods can be of great value in pointing up the good and bad features of alternate seepage control measures and in establishing sound design principles.

Although empirical methods used in the past for designing overflow weirs (see Lane, 1935) recognized that inclined or vertical contacts between structures and foundations offer greater resistance to seepage than horizontal contacts, these methods generally failed to evaluate details that can have a major influence on the security of hydraulic structures; for example, the position of a cutoff greatly influences the magnitude of escape gradients and the degree of safety against piping. Thus Figure 10.18a shows a flow net for an overflow weir with a sheet pile cutoff under the *upstream edge,* and Figure 10.18b shows a flow net with a cutoff of the same dimensions under the *downstream edge.* Except for the direction of flow, these two flow systems are identical, and the lines of creep, or weighted lines of creep (shortest flow paths), are identical. Nevertheless the factor of safety against piping is vastly different. The escape gradient for the section in Figure 10.18a approaches infinity at point x and the factor of safety against heaving approaches zero; the maximum escape gradient in Figure 10.18b with the sheet pile wall at the downstream edge is about 0.2, which is a reasonably safe level. A still more satisfactory design (Fig. 10.18c) utilizes a sheet pile cutoff at the upstream toe in combination with a drain under the downstream toe of the weir.

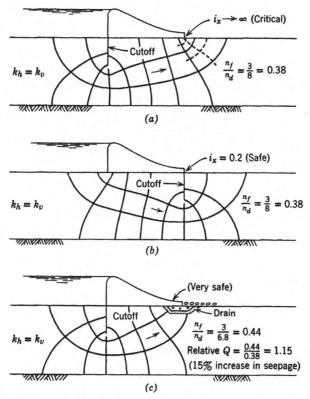

FIG. 10.18   Flow-net study of seepage control measures versus safety against piping ($k_h = k_v$). (*a*) Cutoff wall under upstream toe, no drain. (*b*) Cutoff wall under downstream toe, no drain. (*c*) Cutoff wall under upstream toe and drain at downstream toe.

In the design of overflow weirs and small masonry dams on alluvial foundations *uplift pressures* must be kept below safe levels or aprons can be lifted and damaged.

Spillway Chutes

Overflow weirs on pervious foundations are designed to pass the surplus flows of rivers; hence they are *spillways.* Masonry dams on rock foundations are frequently designed with spillway sections that pass the surplus flows of rivers over these structures. These dams (Sec. 10.4) are made safe against underseepage by grouting and drainage.

Almost all earth dams and some masonry dams are designed with spillways that are cut into rock formations weathered to various degrees. Unless the rock is fresh and nonerodible, at least the upper portion of a spillway must be lined with concrete. If seepage and uplift pressures under the lining are not controlled, spillways may be severely damaged during critical river stages. In severe cases failures of dams have been initiated by spillway failures; consequently the control of seepage and uplift pressures under spillway linings is of extreme importance.

Whenever a rigid structure is built on erodible or compressible weathered foundations, the slightest erosion or settlement of the foundation opens channels along which seepage concentrates and causes further channeling. Unless the cycle is prevented, complete collapse of the structure can occur. Many of the failures of earth dams are caused by piping along outlets, under spillways, or along other rigid appurtenant facilities.

A number of measures are commonly used for the prevention of piping and the control of uplift pressures under spillways.

1. Grout curtains at the upper ends of spillways.
2. Cutoffs several feet into foundations at the upstream ends and at periodic intervals under lined discharge chutes to inhibit the free flow of seepage from the upstream sides to the downstream sides.
3. Concrete or other impervious aprons extending upstream from the overflow elements of spillways.
4. Cutoff walls of reinforced concrete extending into fills placed at the sides of spillway chutes.
5. Drainage facilities to collect and remove underseepage while preventing the movement of soil fines, thereby also preventing erosion and excessive uplift pressures.

Chute spillways may be short and flat or long and steep, depending on the topography and the amount of head to be dissipated. Ungated spillways usually have rather small acting heads, whereas gated spillways are often subjected to substantial heads. If a large amount of head exists at a spillway or if large uplift pressures can develop under its chute, measures should be taken to protect the spillway and the chute from underseepage damage. Grout curtains and upstream aprons increase the seepage path and reduce seepage quantities and help to control uplift pressures. Drainage blankets or line drains often are highly effective in removing seepage from beneath the structure thus ensuring a high degree of protection.

Details of a comprehensive drainage system for a large, wide chute spillway designed by the U.S. Bureau of Reclamation are given in Figure 10.19. As seen in the plan in Figure 10.19a, this chute has a width of 170 ft and a length of more than 400. Seepage control is provided by a grout curtain near the upstream edge of the structure, by porous concrete drains, and by a network of pipe drains laid in filter gravel (see Fig. 10.19b). Channeling beneath the structure is further controlled by several cutoffs under the spillway structure and a cutoff to a minimum depth of 10 ft at the downstream end of the chute.

The U.S. Army Corps of Engineers and other experienced builders of major dams require extremely careful excavation and clean up of foundations for spillways and similar care in all important phases of the construction. Minor imperfections in this work can lead to trouble; hence careless design and construction cannot be permitted in important hydraulic structures.

Even minute-appearing details of design have been developed by experience after other designs have been found deficient. A typical illustration of an important detail is the construction joint shown in Figure 10.20. Although several types of joint are in use, many engineers prefer one of this general type because it anchors the upper ends of individual slabs into the rock, thus resisting downhill sliding. Also, the upper slab extends over the lower slab, preventing high velocity water from eroding soft rock out from under the slabs and undermining them. A transverse drain is placed near the joint, as shown in Figure 10.20 to remove seepage and control uplift pressures.

The details of drainage systems for chute spillways must be adapted to the foundation conditions at individual sites. If distinct jointing exists, which will feed water to narrow line drains, a system of the type shown in Figure 10.19 offers an economical solution. On the other hand, if the formations cannot be adequately drained in this manner, a blanket drain with pipe collectors may be required under all or a substantial part of the lining. The seepage from line or blanket drains can be discharged through pipe outlets or it can be bled up through weep holes in the lining. If weeps are used, they must be protected from clogging.

Wherever drains are installed for the control of underseepage, the seepage paths are shortened, gradients are steepened, and seepage quantities are increased. Water losses are not usually important if erodible rocks and soils are held permanently in place and prevented from migrating by good construction practices and the use of *graded filters*. In developing designs for spillways and other hydraulic structures, uplift pressures, seepage gradients, and seepage quantities should be analyzed with flow nets. Flow nets in Figure 10.21 for the spillway in Figure 10.19 show that without drainage (Fig. 10.21a) the spillway

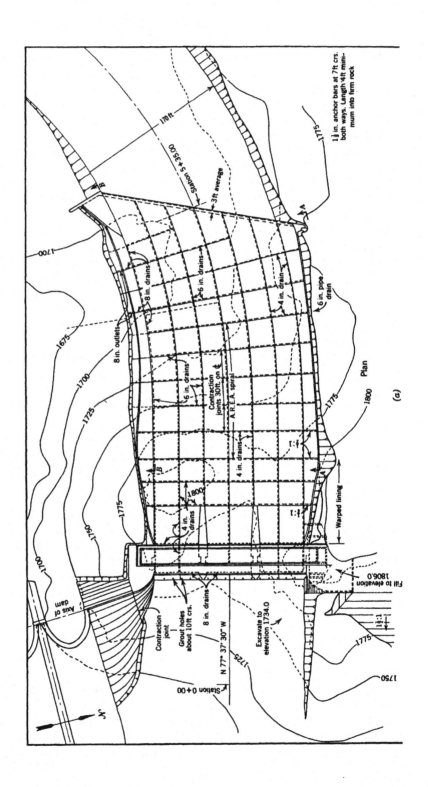

(a)

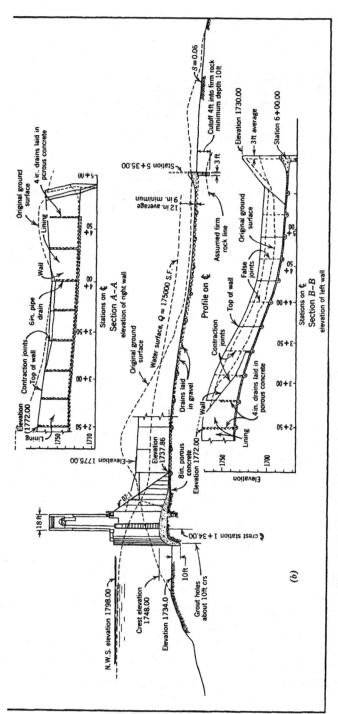

FIG. 10.19 Design of a large spillway. (From Fig. 12, par. 12.37, USBR "Treatise on Dams," Des. Sup. No. 2 to part 2, Engr. Des., of Vol. 10, Des., and Constr., Recl. Manual, Chap. 12, Spillways, Washington, D.C., October 20, 1950.) (*a*) Plan, (*b*) sections and profile.

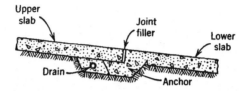

FIG. 10.20   Spillway construction joint detail.

lining would be subjected to excessive uplift pressures, whereas with the drainage system in operation (Fig. 10.21*b*) uplift pressures are held to a safe level. Without any knowledge of the permeability of a foundation, the *relative seepage quantities* for alternative designs can be estimated by Eq. 3.18: $q = kh(n_f/n_d)$, because for one design

$$q_1 = kh \frac{n_{f-1}}{n_{d-1}}$$

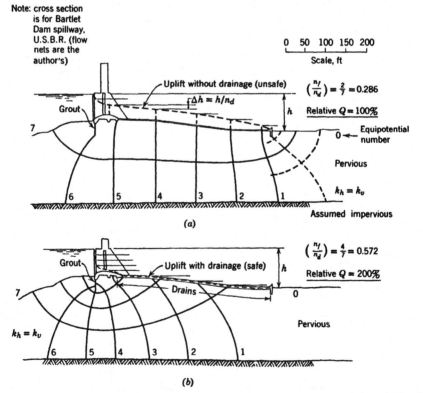

FIG. 10.21   Flow-net study of seepage under a spillway lining (from Fig. 10.19). (*a*) No drainage under lining. (*b*) Effective system of drains under lining.

and for an alternate design

$$q_2 = kh \frac{n_{f-2}}{n_{d-2}}$$

and

$$\frac{q_1}{q_2} = \frac{kh \dfrac{n_{f-1}}{n_{d-1}}}{kh \dfrac{n_{f-2}}{n_{d-2}}} = \frac{\dfrac{n_{f-1}}{n_{d-1}}}{\dfrac{n_{f-2}}{n_{d-2}}} \qquad (10.5)$$

Hence for alternate designs of a hydraulic structure on a given foundation and *for a given amount of head* seepage quantities are simply proportional to the *shape factors* obtained from flow nets.

For the spillway illustrated in Figure 10.19 the design with the drainage system will pass about two times the quantity that would flow without the drains. If the actual volume of seepage is large, it may be necessary to conduct it through suitable ditches or pipes to natural drainage courses or to sumps where it can be lifted back into the reservoir.

REFERENCES

Abbett, Robert W., E. E. Halmos, et al. (1956), *American Civil Engineering Practice,* Robert W. Abbett (Ed.), Wiley, New York, Vol. 2, Section 21, p. 142; (1956a) p. 145.

Aisenstein, B., G. Korlath, and A. Yevnin (1960), "A Seepage Problem in a Clay Reservoir," *Proceedings,* First Asian Regional Conference, New Delhi, February 1960 (Int. Soc. of S.M. and Foundation Engineering), Paper I (c) (i).

Anderegg, M. (1972), "The Use of Electro-Osmosis Against Rising Damp in Walls," Schweizerische Bauzeitung, Zurich, Vol. 90, No. 39, pp. 954–958.

ASCE Subcommittee on Uplift in Masonry Dams (1952), *Transactions,* A.S.C.E., Vol. 117, p. 1237.

Barrett, Robert J. (1966), "Use of Plastic Filters in Coastal Structures," *Proceedings,* 10th International Conference on Coastal Engineering, Tokyo, September 1966, pp. 1048–1067.

Borovoi, A. A., N. V. Razin, and V. S. Eristov (1963), "Some Large Dams of Hydro-projects in the U.S.S.R.," Department of Power Designing, Ministry of Power and Electrification of the U.S.S.R.; a contribution to "Topmost Dams of the World, The Japan Dam Association, Tokyo, October 1963, pp. 224–225.

Brandl, H. (1987), "Retaining Walls and Other Retaining Structures," Chap. 47, *Ground Engineer's Reference Book,* Edited by F. G. Bell; Butterworth and Co. (Pub.) Ltd., London, p. 47/15.

Carlson, R. W. (1957), "Permeability, Pore Pressure, and Uplift in Gravity Dams," *Transactions,* A.S.C.E., Vol. 122, pp. 587-613.

Casagrande, A. (1935), "Discussion of 'Security from Under-Seepage—Masonry Dams on Earth Foundations,' by E. W. Lane," *Transactions,* A.S.C.E., Vol. 100 p. 1289.

Casagrande, A. (1937), "Seepage Through Dams," Harvard University Publication 209, reprinted from *Journal of the New England Water Works Association,* June 1937.

Casagrande, A. (1961), "Control of Seepage Through Foundations and Abutments of Dams," *Geotechnique,* Vol. XI, September 1961, pp. 161-181.

Casagrande, A., S. D. Wilson, and E. D. Schwantes (1972), "The Baldwin Hills Reservoir Failure in Retrospect," A.S.C.E. Specialty Conference "Performance of Earth and Earth-Supported Structures," Purdue University, Lafayette, Ind., June 1972, Vol. 1, Part 1, pp. 551-588.

Chuck, R. T. (1970), "Largest Butyl Rubber Lined Reservoir," *Civil Engineering,* A.S.C.E., Vol. 40, No. 5, pp. 44-47.

Civil Engineering, ASCE, Aug., 1987, pp. 46-48.

Feld, Jacob (1957), *American Civil Engineering Practice,* Robert W. Abbett (Ed.), Wiley, New York, Vol. 3, Section 26, pp. 50-52.

Fukuoka, M. and Y. Imamura (1982), "Fabric Retaining Walls," Proc. 2nd Int. Conf. on Geotextiles, Las Vegas, Nevada, Aug. 1-6, 1982, Vol. III, pp. 575-580.

Gilkey, Herbert J. (1959), *Civil Engineering Handbook,* Editor-in-Chief, L. C. Urquhart, McGraw-Hill, New York, Section 7, p. 175.

Horsky, O. (1969), "Seepage from the Dam Foix in Calcareous Karst," Vodni hospodàrstvi, Prague, Vol. 7, pp. 201-203.

Jones, C. J. F. P. (1982), "Practical Construction Techniques for Retaining Structures Using Fabrics and Geogrids," Proc., 2nd Int. Conf. on Geotextiles, Las Vegas, Nevada, Aug. 1-6, 1982, Vol. III, pp. 581-585.

Keener, K. B. (1951), "Uplift Pressures in Concrete Dams," *Transactions,* A.S.C.E., Vol. 116, pp. 1218-1237.

King, John C., and Edward G. W. Bush (1963), "Symposium on Grouting: Grouting of Granular Materials," *Transactions,* A.S.C.E., Vol. 128, Part I, pp. 1279-1317.

Lane, E W. (1935), "Security from Underseepage—Masonry Dams on Earth Foundations," *Transactions,* A.S.C.E., Vol. 100, p. 1235.

Lehnert, J. (1973), "Bituminous Linings for Dams and Pumped Storage Schemes," Schriftenreihe der Strabag-Bau-AG, Köln, 9 Sequenz, No. 1, pp. 7-60.

Levinton, Zusse, and N. H. Feldstein (1957), *American Civil Engineering Practice,* Robert W. Abbett (Ed.), Wiley, New York, Vol. 3, Section 27, pp. 1-25.

Londe, Pierre (1972), "The Mechanics of Rock Slopes and Foundations," Rock Mechanics Research Report No. 17, Imperial College of Science and Technology, University of London, April, 1972.

Merriman, Thaddeus, and Thos. H. Wiggin (1946), *Civil Engineer's Handbook,* Wiley, New York, p. 1842.

Schroeder, W. L. (1987), "Caissons and Cofferdams," Chapter 4, *Ground Engineer's Reference Book,* Edited by F. G. Bell; Butterworth and Co. (Pub.), Ltd. London, p. 40/13.

Sherard, J. L., R. J. Woodward, S. F. Gizienski, and W. A. Clevenger (1963), *Earth and Earth-Rock Dams,* Wiley, New York, pp. 115-127.

Tate, T. N. (1961), "World's Largest Drydock," *Civil Engineering,* December 1961, pp. 33-37.

Terzaghi, K. (1943), *Theoretical Soil Mechanics,* Wiley, New York, pp. 247-251.

Terzaghi, K., and R. B. Peck (1948), *Soil Mechanics in Engineering Practice,* Wiley, New York, pp. 506-507; (1948a) p. 231.

U.S. Department of the Navy, Naval Facilities Engineering Command, "Design Manual, Soil Mechanics, Foundations, and Earth Structures," NAVFAC DM-7.2, "Foundations and Earth Structures," May, 1982.

Wegmann, Edward (1922), *The Design and Construction of Dams,* Wiley, New York, 7th ed., p. 221 (1922a) p. 222.

Whitehead, Carl F., and Donn Ruotolo (1973), "Ludington Pumped-Storage Project Wins 1973 Outstanding CE Achievement Award," *Civil Engineering* A.S.C.E., June 1973, pp. 64-68.

Zierke, K. R. (1971), "Trends in Development in Underground Railway Construction," Strasse-Brucke-Tunnel, Vol. 23, No. 1, pp. 18-20.

Zola, S. P., and P. M. Boothe (1960), "Design and Construction of Navy's Largest Drydock," *Journal,* Waterways and Harbors Division, A.S.C.E., Vol. 86, No. WW1, March 1960, Part I pp. 53-84.

CHAPTER ELEVEN

WASTE DISPOSAL FACILITIES AND INFILTRATION PONDS

This chapter deals mainly with matters pertaining to the preservation of the world's groundwater resources from serious contamination. Section 11.1 discusses problems of deterioration of groundwaters by industrial and domestic wastes including (1) toxic and nuclear wastes, (2) industrial plant wastes, and (3) mining wastes. Section 11.2 discusses methods for replenishing diminishing groundwaters by the use of infiltration ponds.

Section 11.1 also discusses the physical hazards created by the amassing of large volumes of semi-liquid mining sludges behind "dams" that often are of questionable stability.

Throughout this book, examples are given of the way application of basic seepage fundamentals such as Darcy's law can enable engineers to understand the *nature* of a seepage or drainage problem and develop realistic solutions. Here, simple Darcy's law calculations explain why serious groundwater contamination caused by thousands of toxic sources went on so long without being noticed. Also, methods of reducing or preventing future contamination are described.

11.1 WASTE DISPOSAL FACILITIES

The Basic Problem

Throughout the developed parts of the world, human activities produce contaminated wastes in such great volumes as to pose serious threats to public safety and health. The *need* for proper disposal of the liquid sewage wastes and solid wastes of municipalities has been recognized for a long time (i.e., Metcalf & Eddy, 1930). Cholera epidemics in Europe led to the construction of elaborate sewerage systems in Paris and London in the early 1800's. In America, sewers were built in cities such as New York as early as 1805. But not until more outbreaks of cholera and typhoid fever and the discovery of the bacillus of typhoid fever in 1880 by Eberth in Germany (Prescott and Horwood, 1935) was the need for disinfecting water with chlorine (starting in 1935) or other disinfectants recognized.

Even with this long background of knowledge of the needs for treating and purifying sewage and other contaminated wastes, discharges of untreated sewage caused severe contamination of rivers, lakes, ocean borders, and groundwater supplies around the world in the past several decades. Because of the growing public concern over the quality of water supplies, numerous public agencies have been formed to set more stringent rules and regulations over the amount of contamination that can be tolerated in wastes before they can be released into bodies of water that can be harmed by unregulated discharges. Many types of wastes are put in retaining facilities of various kinds that hold the wastes until they have been purified to the point where they no longer represent any threat to water supplies and other bodies of water. Some of the more common types of waste-retaining facilities are discussed here.

Toxic Wastes and Nuclear Wastes

The rapid explosion of technology has led to the production of large quantities of highly toxic wastes and nuclear wastes, which in many areas are finding their way into groundwater supplies used for commercial, industrial, and residential purposes, threatening the well being of millions of people. In contrast with the long-time knowledge that sewage discharges contaminate water supplies unless adequately treated, the groundwater contamination has been slow in being discovered. Why is this so? The biggest reason is the vast difference between the rate of flow of surface waters and groundwaters. While flows above the ground can move several feet a second or more, groundwater movements rarely exceed a few feet a *day*. Thus, for contaminants to spread even a few hundred feet from a source can require several years. If the horizontal coefficient of permeability of the most permeable layers in an aquifer is 10 ft/day, the slope of the water table is 0.002, and the effective porosity is 30%, the seepage velocity (Eq. 3.1) would be (10 ft/day x 0.002 / 0.3) = 0.06 ft/day or 22 ft/year. Many groundwaters will flow much slower than this rate, but some will be much faster.

How do contaminants get into groundwater systems? There are literally infinite ways. Before the seriousness of the contamination problem was realized, little effort was generally made to keep toxic chemicals out of the groundwater systems; potent cleaning agents, degreasing chemicals, solvents, and other toxic substances used in industrial and military activities were often put into unlined retaining ponds, in "dry wells", or allowed to discharge into drainage channels or onto the adjacent land areas, where they soaked into the ground. Then, during rainy periods, such contaminants gradually worked their way into underlying or surrounding groundwaters.

Industrial and military activities aren't the only sources of pollutants that enter water supplies, as innumerable other things add to the problems. Leaky fuel storage tanks are considered by some investigators to be severe polluters. And landfills used for disposal of garbage are other major sources of contaminants. Every household in America is contributing to the problem each time a toxic substance is flushed down a drain or placed in a garbage can to be hauled to the local dump (landfill). Major contaminants are wood preservatives, insecticides, fungicides, and a variety of weed and brush killers such as herbicides 2,4,5-T and 2,4,5-TP. Other troublesome wastes are discarded batteries, used motor oil, and old paint.

Even the fertilizers and other chemicals used on lawns, trees, etc. can find their way into the underlying groundwaters or into sewer systems of every municipality in our land, thus compounding the contamination problems. In spite of the regulations that have been imposed on the disposal of toxic substances, "accidental" releases of damaging substances into drainage ditches, streams, rivers, etc., are common topics reported by the public media.

After the contamination problem was widely recognized, most groundwater regions formed authorities or agencies that monitor the spreading of contaminants by regularly taking water samples from all available wells, and analyzing for impurities. If insufficient numbers of water supply wells are available for sampling, additional monitoring wells are often installed. In addition to the taking of samples for laboratory testing for the kinds and quantities of contaminants, the water levels in unpumped wells ought to be measured, so that groundwater contour maps can be developed in an area under study. A groundwater level contour is an "equipotential" line (see Chap. 3), and water flows at right angles to an equipotential, so groundwater contours enable investigators to determine the *directions* in which contamination is spreading. Though great efforts are now being made to protect groundwater supplies from contamination, so much detrimental material is already in the ground at many places that little more can be done than to watch its spread, and to try to limit the entry of additional toxic materials into the soil-groundwater systems. On August 24, 1988 the Environmental Protection Agency proposed stringent new rules to reduce the amount of toxics getting into U.S. groundwaters from garbage dumps. Covering abandoned dumps with waterproof membranes, cleaning up all existing dumps leaking contaminants into groundwaters, and installing extensive pollution-detection equipment are among the requirements proposed.

In some areas with serious contamination, municipalities are installing expensive water treatment systems, using granular activated carbon or other filters to remove contaminants. Also, the blending of contaminated water with good quality outside water supplies has been used to produce water with contaminant levels within safe levels set by states or other agencies. To prevent future contamination, stiff regulations have been widely imposed that require treatment or retention of waste products in special, lined reservoirs or by other safe disposal procedures. To try to reduce the amounts of toxic chemicals reaching garbage dumps or "landfills," many local agencies are developing "recycling and collection" programs to encourage the public to bring their half-empty bottles and cans of noxious materials to collection sites, from which they are hauled to safe, controlled disposal area.

Programs for cleanup of groundwater aquifers are underway in thousands of locations in the U.S. alone. One of the biggest and most difficult is in the San Gabriel Basin in southern California (see *Engineering News-Record*, 1986). Here, the groundwater supply for more than one million people has suffered widespread pollution with high levels of trichloroethylene (up to 1800 parts per billion [ppb] in one water supply well), carbon tetrachloride, and perchloroethylene. These compounds and others used as solvents and metal degreasers are far above state action levels of 1 to 16 ppb, as well as federal recommended levels. Twenty landfills and more than 230 industrial operations are believed to be contributing to the contamination problem in that basin. Though the contamination must have been going on for years, the first evidence was not found until 1979. Engineers and groundwater experts studying the problem say that groundwater flowing at velocities of 500 to 1000 ft/year in these relatively permeable formations could spread the contamination to the adjoining Central Basin. Cleaning up this groundwater system may be so costly and time consuming that the people living in this valley just east of Los Angeles face the prospect of losing their water to contamination (Foreman and Ziemba, 1987). The Environmental Protection Agency (see *ENR*, March 10, 1988) said that "depending on the cleanup option selected, the project could take as long as 50 years and cost $800 million." Devising a cleanup program alone "will cost more than $50 million and take up to a decade."

Because of the enormity of the task of cleaning up contaminated groundwater supplies, efforts are being made to keep contaminants out of groundwater aquifers. Storing hazardous wastes in special, lined impoundment reservoirs has taken on great importance as a way to protect water supplies from contamination. In general, the principles outlined in Sec. 10.5, "Storage Reservoirs," can be used for designing containments for dangerous waste products, but higher levels of restriction on leakage are needed than for clean water reservoirs.

Asphalt plank, asphalt concrete, butyl rubber, and plastic sheeting are some of the types of thin liners to be considered. Compacted clay is another good possibility. Daniel (1985) suggests that well compacted clay liners may offer a good solution for containing some kinds of hazardous wastes, but says clay liners may be ineffective if adequate precautions are not taken to insure watertightness. He warns that leachites of some organic and inorganic chemicals can attack clay liners; therefore, tests are needed to prove the compatibility of a given clay liner material and the waste liquid to be stored. Also, he says, it is necessary to avoid exposing the liner to concentrated acids, bases, and organic chemicals. He also comments that leakage may be greater than predicted, and one should not rely solely on laboratory tests to predict the permeability of a full scale clay liner.

In order to prevent losses of contaminants into groundwaters, double liners are often used, with a water-tight type of liner underlaid by a permeable drainage layer feeding seepage to pipes which collect the water and lead it to sumps where it can be pumped back into the impoundment. Daniel (1987) cautions readers by saying "all liner materials are subject to attack by certain chemicals." Obviously, there is no single solution to all impoundments of hazardous wastes, and great care is needed in designing systems for their storage. Each one must be treated as an individual problem.

Daniel (1987a) says that it is safer and cheaper to keep hazardous wastes out of the groundwater than to remove them, and better monitoring is necessary to minimize the threat to man and the environment. According to Daniel, a study of 39 hazardous waste sites by R. Allen Freeze of British Columbia showed that they leaked after an average storage time of 14 years. Further, when repairs were attempted, the efforts were totally successful only 16% of the time, and 43% of the repairs failed altogether. In addition to monitoring wells from which water samples can be taken to check the spread of contaminants he suggests various sensors, including types that can be implanted *beneath* containments before they are built, to check the movement of contaminants into underlying saturated zones.

In my own experience, I have found that several lines of observation wells in different directions around leaking impoundments offer a good way to study the buildup and spreading of saturation mounds produced by the leaking fluid. By drawing *groundwater contours* at various intervals of time, I could get a good understanding of the amount the groundwater level was rising and how fast it was spreading into adjacent areas where contamination could cause problems. At one location that I studied, I recommended that a comprehensive system of deep wells be put in and pumped when necessary to prevent the developing mound from rising to harmful levels and the spreading of contaminants into a nearby residential and industrial area.

Proposals have been made to reduce the permeability of aquifiers to slow down the movement of polluted groundwater and to allow more time to treat the problem. Chemical treatment is one of the ideas that has been proposed [see *ENR* (formerly *Engineering News-Record*), April 2, 1987, p. 16]. Other methods are being studied, including groundwater crystalization using high-voltage electricity to glassify the soil surrounding hazardous material and chemicals rather than electricity to solidify the earth.

Wastes from nuclear plants are mixtures of industrial chemicals and radioactive materials. Their containment without leakage is of paramount importance. Exhaustive studies have been made of the problems with nuclear wastes from nuclear bomb-making sites and the best ways of keeping them out of groundwaters (see *Engineering News-Record,* Jan. 30, 1986, pp. 28–31).

How to dispose of radioactive wastes that have been stored in underground steel tanks has been a major problem. One solutioin is a process developed by a Department of Energy contractor to solidify 600,000 gallons of highly radioactive waste into glass logs (see *Engineering News-Record,* Feb. 9, 1984, p. 24).

Clearly, the problems of maintaining water supplies fit for human use is one of the most serious challenges facing engineers today. The seepage fundamentals presented in this book provide a sound basis for understanding the nature of groundwater flow phenomena, why the contamination problems have proceeded undetected for so long, and how to design retaining facilities that can provide a high degree of protection against pollution.

Industrial Plant Wastes

Many kinds of industrial plants use furnaces, steam turbines, etc. that produce exhausts (smoke) that in the past was allowed to discharge its contaminants into the atmosphere with little or no treatment or control. Not until serious disasters occurred were major efforts started to prevent the pollution of the atmosphere by these plants. In Dec. 1930, for example, more than 60 people died in the Meuse Valley in Belgium from severe air pollution caused by steel mills, power plants, glass works, lime kilns, zinc refining plants, a coking plant, a sulfuric acid plant, a fertilizer plant, and smoke from homes and buildings heated with coal (see *Encyclopaedia Britannica,* Vol. 18, 1971, p. 185). And, between Oct. 27 and 31, 1948 20 people died and 6,000 became ill at Donora, Pennsylvania (near Pittsburgh), from severe pollution caused by air stagnation while large amounts of contaminants were being discharged into the atmosphere. The largest disaster of all occurred in London after highly severe contamination created extremely unhealthy conditions from Dec. 5 to 9, 1952. In the one-month period that followed, more than 3,500 people died. In the late 1950's, serious steps were taken to prevent industrial plant exhausts from entering the atmosphere. Cities that had intolerable conditions began again to have clean air to breathe.

Now, nearly all industrial plant exhausts are being treated to prevent the pollutants from entering the atmosphere. Various methods are used, but some of the most common involve the cleaning or "scrubbing" of exhausts by complex facilities that collect the contaminants into liquid slurries that are pumped into specially constructed reservoirs. Fly ash, for example, is often placed behind earth dams or levees constructed on clay foundations that hold the level of leakage down to levels that are considered permissible without severely contaminating adjacent groundwaters. Lining systems such as are shown in Figures 10.13 and 10.14 may be suitable for many installations where the leakage must be held down to very low levels. In some cases, however, agencies have told an industry that no leakage at all can be tolerated and very expensive retaining systems have been constructed.

In southeastern Montana, for example, the owners of a major coal-burning power plant constructed a concrete panel wall three miles long and 50 to 90 feet deep around a huge disposal basin which will contain about 100,000 tons a year of scrubber waste slurry that will be piped to it for retention (see *Engineering News-Record,* March 1, 1984, p. 29). To meet the high quality of the stack discharges required in the permit, a new flue-gas scrubber system capable of removing upwards of 96% of the sulfur dioxide in the exhaust was designed and constructed. The disposal basin constructed there is capable of holding the amount of slurry expected over a 22-year life. The concrete panel was designed to provide a water-tight cutoff to prevent the spreading of contaminants into the surrounding groundwater systems.

Wherever industrial plant wastes are stored in reservoirs, monitoring wells should be installed around outside edges and carefully checked for any rising of groundwater levels caused by seepage out of the reservoir, and samples tested periodically for any evidence of pollutants entering the groundwater. If any potentially dangerous condition starts to develop, consideration should be given to the installation of a system of pumped wells that can be used to lower the water levels around the lower edges of the reservoir. If the water levels are kept lower than the general groundwater levels beyond a reservoir, the spreading of seepage can be prevented. The pumped water can be fed back into the reservoir, or disposed of in another safe method.

Mining Wastes

Mine wastes constitute the largest tonnage of materials handled by industry. The refining of coal, copper, and other minerals often involves crushing of excavated materials high in impurities to 1/50 in. or smaller, and the addition of water and oil or other agent to allow the valuable minerals to be separated from the dirt and other impurities by floatation or sedimentation processes. In the past, large "gob" piles of coarse waste were constructed along the edges of valleys, and the fine waste materials generally seeped to and contaminated local streams, rivers, or lakes. After the passage by Congress of the Clean Water Restoration Act of 1966, the sludges had to be held in retaining ponds until they stabilized by evaporation or purified by oxidation or treatment. Other major classes of nearly liquid wastes having to be retained behind dikes or "dams" or in lined reservoirs are dredging wastes and fly ash from power plant chimneys. The destabilizing effects of excess water can create safety hazards to the people living below these facilities, if they are not properly constructed; and the liquid contaminants can harm surface or groundwaters surrounding the waste ponds, if they are allowed to escape.

Figure 11.1 shows sludge being pumped into a pond of a typical Appalachian coal mining waste dump. Another view of a waste dump pond into which sludge has been pumped is shown in Figure 11.2. The culvert pipe protruding from the new dirt fill (in Fig. 11.2) was installed to provide a spillway to try to prevent overtopping in the event of heavy rainstorms filling the pond to overflowing. The cracks in the sludge show that it has dried slightly at the top; however, it is still very liquid below the thin top crust.

Refuse piles, dumps, or "dams" can reach heights of several hundred feet, as shown in Figure 11.3. The outer fringes of these piles are constructed of coarse wastes that are relatively stable, but the sludges in back are often highly saturated, loose, weak, and potentially susceptible to failure—particularly in

FIG. 11.1 Coal mining sludge being pumped into a retaining pond in Appalachia.

FIG. 11.2 Coal mining sludge behind a retaining dike in Appalachia.

FIG. 11.3   Looking up the downstream slope of a 400-ft high coal mining waste dump in Appalachia. Very wet sludge similar to that in Fig. 11.2 is in back of this dump.

rainy regions where water tables are high and the drying out process is very slow, as seen in Figure 11.2.

Thousands of refuse piles or "dams" exist in mining regions of the United States alone, and the U.S. Department of the Interior, Bureau of Mines, Washington, D.C. conducts an extensive program of inspection and evaluation of the safety of these structures. This program was greatly accelerated by the sudden collapse of a coal refuse dam on February 26, 1972, near Saunders, West Virginia (*Engineering News-Record,* 1972), which flooded Buffalo Creek Valley and caused the deaths of 118 people; 7 others were reported missing, (see Fig. 11.4). More than 500 homes were lost and extensive property damage sustained. In a small valley above Buffalo Creek a refuse dam of coarse waste (retaining a sludge pond) was being built on a thick layer of wet, loose sludge, highly susceptible to piping and liquefaction failures. It collapsed suddenly

FIG. 11.4 Buffalo Creek Dam (near Saunders, W. Va.) after its failure on February 26, 1972. Looking diagonally upstream over a large spoil pile that was eroded by the wave of mud and water released by the failure (Courtesy, West Virginia Dept. of Highways).

during a rainstorm which filled its pond and released several hundred acre-feet of sludge and water into the valley below, causing the devastation already noted. To avoid more failures great efforts were made to identify waste piles that had questionable stability and to require changes needed to make them safe.

The high water content of sludges often causes pore pressures that reduce the factor of safety of the waste "dams." Stojadinivić and Pantelić (1969), for example, tell about the collapse of part of a stock pile of flotation waste products caused by its saturation by water that had been prevented from flowing away when the dike was filled by the hydraulic filling method. Other regularly drained areas were stable. Improvement of the drainage of the sludges, together with the retaining structures, is essential to the safety of these structures.

If failures of waste disposal structures are to be avoided, the same fundamental seepage principles used in the design of earth dams, levees, and storage reservoirs must be employed.

Fortunately, many mine tailings dams are now being built in this manner. The Buffalo Creek dam failure in 1972 drew attention to the need for stricter design and construction standards for coal mine waste dams. The Mine Safety and Health Administration (MSHA) got very active in the engineering and permitting aspects of these structures, according to Joseph J. Ellam, Chief of the Pennsylvania Division of Dam Safety and President of the Association of State Dam Officials (see *Engineering News-Record,* August 8, 1985, p. 11). According to Ellam, because of MSHA's influence, and concern of other dam building and safety organizations, coal-tailings dams are now fully engineered structures.

In arid regions where water tables are low, stability problems are much less severe, because of two factors: (1) the sludges dry out much faster than in humid regions, and (2) the low water tables allow *vertically downward,* or *bottom* drainage of the excess water, which is the best kind of drainage, as pointed out elsewhere in this book. In many mining operations in arid regions, the sludges dry out sufficiently within one or two weeks so that they can be excavated to build additional height of retaining dike or dam to provide additional storage for new sludges (see photo in Fig. 11.5). Here, a dragline is excavating dried sludge from a moderate distance within the outside edge of a pond, and dumping it at the outside edge. After it is compacted the pipe line feeding sludge to the pond will be raised to the top of the new dike, where additional sludge will be fed into the pond from spigets placed at frequent intervals.

Because of the great importance of tailings impoundments, improvements in the construction and monitoring of these structures have been made world-

FIG. 11.5   After two weeks of draining and drying, this sludge in a mine waste area in Arizona can be excavated to form a new dike along its outside edge.

wide, Mackenchnie (1981) says that waste structures for the mining industry in Zimbabwe are now being engineered, and construction control and monitoring are paying big dividends. Roth, et al. (1980) say that safety inspections of coal refuse sites have been aided by rapid monitoring systems using convergent and vertical photography from conventional fixed-wing aircraft.

In situations where leached acids or other contaminants could harm the environment, extensive measures are often taken to retain the pollutants. A roller-compacted concrete dam was built 150 miles southeast of Phoenix, Arizona, to prevent an acidic copper-laden processing solution from being washed into the downstream San Francisco River during major storm runoffs (see *ENR*, September 10, 1987, p. 24). In a novel process, an acid solution in water is sprayed over a large mine tailings dump from which most of the copper has been removed but salvageable amounts still remain. The acid solution, rich in copper, flows out of the bottom of the pile into a pond from which the copper can be recovered. Though the acid leachate may never escape into downstream waters, the new dam provides storage to prevent any possible downstream contamination during a flash rainstorm. Large amounts of money were spent here to try to protect the environment from contamination.

To reduce the hazards created by mine waste impoundment structures, these "dams" ought to be designed and constructed using standards comparable to those used for regular earth dams. Foundations need to be cleared of all debris or pervious materials that could induce piping failures in foundations. Embankment materials should be placed in lifts of moderate, specified thickness and compacted to specified density and moisture content to ensure stable fills. To avoid failures such as occurred at Buffalo Creek, no fill should be placed on unconsolidated or poorly consolidated sludge or other weak material. To protect against sloughing caused by seepage exiting on the downstream slope, properly designed and constructed toe drains should be provided (see Chapter 5 for criteria and methods for designing good drains).

Adequate spillways should be provided for mine tailings dams to prevent their overtopping in the event of heavy runoff from rainstorms. Also, as previously noted, all tailings dams of importance should be carefully monitored to make sure that no unsafe condition can develop.

Mine tailings dams have taken on such great importance that they have attracted the attention of many investigators, often leading to valuable contributions in this field of engineering. Al-Hussaini et al. (1981), for example, performed centrifuge model experiments with varying geometrics and properties of coal waste material, and confirmed the fact that positive seepage control can be beneficial to tailings dams. They report that a toe drain controlled seepage on the downstream slope and prevented sloughing at the toe.

Jeyapalan et al. (1983) concluded that when mine tailings dams fail, the tailings tend to liquefy and flow over considerable distances, often causing great losses in property and human life. They conducted flume tests to check the validity of analytical procedures developed to analyze the flow movements and obtained good agreement with the theory. Among the case histories they studied was the failure of a gypsum tailings impoundment in East Texas in 1966; the Aberfan Coal Waste Tip in Wales in 1965 that moved nearly a half mile, covered a school and other buildings, and killed 120 people in its path; and the Buffalo Creek Dam failure in West Virginia in 1972, flowing 40 miles and causing 118 deaths and considerable property damage. They feel that these experiments and the analytical methods will be of value to those desiring to evaluate the potential for damage in the event of the failure of other mine tailings dams of similar properties and construction.

Mining operations often require the stripping of large amounts of soil and non-ore-bearing materials to reach productive layers. The stripped materials are placed by end dumping or other methods in waste embankments that can become very high and very large. Since no water or sludge is impounded behind these structures, they normally represent relatively minor hazards, but their failures can create nuisances, clean-up problems, and encroachment on properties beyond mining companies' easements or rights-of-way. Vandre (see U.S. Department of Agriculture Forest Service, 1980) presents criteria for the design and construction of these non water-impounding waste embankments. He recommends that adequate subsurface explorations be made to obtain geologic information and groundwater levels, identification of the engineering characteristics of foundations and embankment materials be achieved by appropriate tests, and that appropriate stability calculations be used for the design of these structures. He advises that weak material should be removed from foundations wherever required to prevent failures because of such materials.

Vandre says drains "to control water levels in a positive manner," should be provided "where questions exist as to the probable behavior of water." To minimize surface water runoff from areas adjacent to and above an embankment, interceptor drains should be provided to prevent water from flowing onto the fill, both during and after operations. The top surface should be graded, both during and after waste embankment construction, to prevent surface water from flowing onto the embankment slopes. When internal drains are provided to lower saturation levels and reduce seepage forces near toes, the coarse layer should be protected against intrusion of fines by using filter layers designed according to accepted filter criteria (see Chapter 5), or by filter cloth. By the use of such design measures and precautions, he concludes that the failures of these structures can be greatly minimized, even though they cannot all be prevented.

As with all of the other kinds of facilities discussed in this book, well drained waste embankments are inherently more stable than their undrained counterparts. These structures can be extremely variable in the materials used and the way they are placed. Each needs to be treated as an individual situation, but many can be greatly improved by drainage. Toe drains can relieve tendencies for sloughing. And a blanket drain of the kind shown in Figure 7.11, if placed under the major part of an embankment including outer edges, can be beneficial in several ways, such as:

1. By producing a general lowering of saturation levels in an embankment subject to water infiltration.

2. Through "bottom drainage", inducing vertically downward flow in any embankment subject to large amounts of rainfall or other saturating inflows, and thus nullifying horizontally directed seepage and the destabilizing effects it produces.

3. By removing groundwater from "springs" or other sources of trapped water under an embankment, thus preventing the destabilizing effects of the buildup of uplift pressures, or the rise of saturation levels into the embankment.

This represents a different kind of usage for this very versatile type of drain, which consists of highly permeable, open-graded aggregate enveloped within fabric filters.

11.2 INFILTRATION PONDS

Basic Concepts

Throughout the world, wherever groundwater supplies are diminishing from overpumping, there is great need to conserve every possible useable drop of water. Infiltration ponds are being built in many areas to add to the replenishment of groundwater and to purify the supply that is returned to adjacent streams. In California's Central Valley extensive pond systems are being used for these purposes. Stormwater runoff is being captured by infiltration facilities designed as part of many highway and airfield pavement systems. Flood flows in rivers in dry valleys are being ponded by small retaining dikes or dams until the water can soak into the ground.

When water is being returned to groundwater systems, it is important that contamination be avoided. Also, in systems design it is important to be sure that no legal water rights are being violated.

Designing infiltration pond systems requires careful estimation of the quantities of water that will be put into a given system and the applicaton of seepage principles to determine the size and details of a facilty needed for disposing of the water. Prevention of clogging of the surface by silt, mud, and other matter carried by water is a major problem in any storm water or treated sewerage water collection and infiltration system and good maintenance programs are essential to their continued success. Also, good monitoring programs for checking groundwater levels and quality around ponds are essential to their effective, safe operation.

Infiltration ponds cannot be effectively used unless soil and groundwater conditions are favorable to their use. Designers should make detailed investigations at sites and use Darcy's law and other seepage principles to verify the following:

1. The capability of the bottoms of ponds to infilter water on a long-term operational basis.
2. The capability of the underlying soil to discharge the inflowing water into the surrounding groundwater system.
3. The capability of the surrounding groundwater system to accept water.

Accumulations of colloids and other fine materials that settle out of the water in infiltration ponds can gradually reduce the rate of infiltration. One very successful system in California had approximately 1,200 acres of ponds. About one-third of the ponds were being allowed to dry out on a regular basis, so that the bottoms could be scarified or scraped as necessary to increase the permeability and bring the infiltration rates up to satisfactory levels. With this maintenance practice, these ponds were kept in very effective operating condition. A number of lines of monitoring wells were installed around this infiltration system shortly after it was first put into operation, and water level readings were made regularly to determine if any rising trends could impair the discharge rates into surrounding areas. After a series of unusually wet years, the water levels under the ponds rose to levels that were seen as approaching conditions that could greatly reduce the effectiveness of the system. Accordingly, a number of deep, pumped wells were installed, and pumped for a few months to lower the water levels to adequate depths. This water was piped away from the site to another disposal area. These temporary measures helped to keep this system in good condition.

Illustrative Examples

Figure 11.6 is a simplified illustration of flow from infiltration ponds. A 100-acre plot (Fig. 11.6a) had been proposed for an infiltration pond in a city required to dispose of 20 million gallons a day of treated sewerage. The site was covered with 20 ft of sandy soil with a vertical permeability $k_v = 1$ ft/day and a horizontal permeability $k_h = 5$ ft/day. The water table stood at a 10-ft depth, and an impermeable clay layer appeared at 20 ft.

The capability of the site for downward percolation was calculated by Darcy's law ($Q = kiA$) by using the *entire plan area* and a downward hydraulic gradient of 1.0 (Fig. 11.6b) as

$$Q_v = kiA = 1.0 \text{ ft/day}(1.0) \ (2640 \times 1650) = 4,400,000 \text{ cu ft/day or}$$
$$33,000,000 \text{ gal/day}$$

This was more than adequate to meet the city's needs; however, this rate of infiltration (1.0 cu ft/sq ft/day) would fill the 10-ft column of soil above the water table in 3.5 days (assuming a porosity of 0.35). The flow would then suddenly change from vertically downward to horizontal (Fig. 11.6c) and the ability of the site to discharge water would become $Q_h = kiA$, in which both

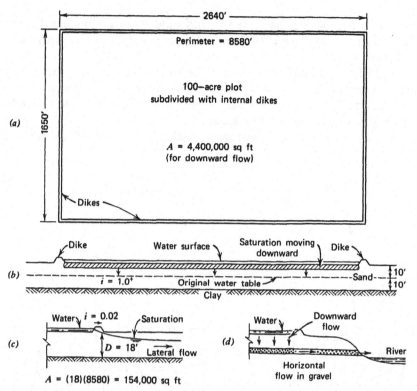

FIG. 11.6 Illustration of flow from infiltration ponds. (*a*) Plan of 100-acre plot. (*b*) Cross section showing initial condition with *downward* flow. (*c*) Edge section showing permanent condition with *horizontal* (lateral) flow. (*d*) Another site with underlying permeable layers and much better drainage.

i and *A* would be sharply reduced. A much smaller hydraulic gradient would apply, and *A* would become the perimeter length times the depth of saturated soil discharging water outward. By using the values for *i* and *A* shown in Figure 11.6*c*

$$Q_h = kiA = 5.0 \text{ ft/day}(0.02) \ (154,000 \text{ sq ft})$$
$$= 15,400 \text{ cu ft/day or } 115,000 \text{ gal/day}$$
$$\text{(less than 1\% of the required rate)}$$

Even though the designer had originally recommended the site (he had calculated only Q_v), it had to be discarded.

For several years a nearby city had been disposing of 5 million gal/day of treated sewerage on a 20-acre plot, which had led the designer to think that the 100-acre site could handle 20 million gal/day readily. The 20-acre plot, however, is near a river bank and is underlaid by highly permeable gravels (Fig. 11.6*d*), which provide fast underdrainage and allow permanent downward flow.

Obviously the capabilities of sites to remove infiltration can vary substantially and depend not only on the depth to water but also on subsurface conditions. Thorough studies are needed if reasonable estimates are to be made of possible discharge rates for individual sites.

Figure 11.7 illustrates the way regular readings of water levels in monitoring

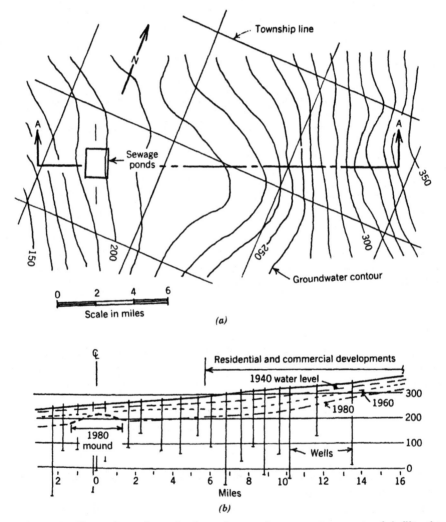

(a)

(b)

FIG. 11.7 Illustration of monitoring of groundwater systems around infiltration ponds. (*a*) Plan showing groundwater contours between some ponds and an adjacent city. (*b*) Sectional view showing ground level and several water levels over a 40-year period. (Sec. A-A).

wells around infiltration ponds and the development of groundwater contour maps can keep infiltration system managers abreast of the behavior of the groundwater systems being used for the disposal of their water. A city located in the eastern half of the area shown in Figure 11.7a is piping its treated wastewater through large trunk sewers to an 800-acre pond system, west of the developed areas. The lands surrounding the ponds are devoted to agricultural uses. Since the project is in a warm, sunny area, part of the water is eliminated by evaporation, but a large portion soaks through the bottoms of the ponds to reach the groundwater below. This water builds up a saturation mound of sufficient height and size to disperse the water into the surrounding groundwater aquifers. The system was first put in operation about 1900. In recent years, heavy well pumping throughout the inland valley in which this city is located has been progressively lowering the water in the entire area. Disposal of treated sewage into the ground helps to slow down the fall of the groundwater in this area, but it is still going down at a fairly rapid rate, as may be seen by the water level profiles in Sec. A-A, Figure 11.7b, for 1940, 1960, and 1980. Pumping from city wells is contributing heavily to the lowering of the water table to the east of the sewage ponds.

In 1980, the top of the groundwater mound under the ponds was at about the same elevation as the groundwater level under the city, nine miles to the east of the ponds, but the water table had a positive slope from the city toward the west, which made it impossible for water from the ponds to reach the city's water supply wells. Since the sewage plant effluent is treated, it is highly unlikely that it could harm the city's water supply even if it could reach some of the city's wells. Nevertheless, to totally eliminate any possibility of degradation of the city's water supply by the teated water, the plant managers have installed wells around and under the ponds, and have periodically operated the wells to prevent spreading of the infiltered water toward the city. Thus, with good monitoring and sound management practices, substantial amounts of water are being returned to the region's aquifers for a partial reversal of the downward trend in the groundwater.

REFERENCES

Al-Hussaini, Mosaid M., Deborah J. Goodings, and Andrew N. Schofield (1981), "Centrifuge Modeling of Coal Waste Embankments," *Journal of the Geotechnical Engineering Division,* American Society of Civil Engineers, V. 107, No. 4, April, 1981, pp. 481–499.

Daniel, David E. (1985), "Can Clay Liners Work?," *Civil Engineering,* ASCE, April, 1985, pp. 48–49.

Daniel, David E. (1987), "Waste Disposal and Underground Storage," Chap. 43, *Ground Engineer's Reference Book,* Edited by F. G. Bell; Butterworth and Co. (Pub.) Ltd., London, p. 43/9.

Daniel, David E. (1987a), "Monitoring for Hazardous Waste Leaks," *Civil Engineering,* ASCE, Feb., 1987, pp. 48–50.

Engineering News-Record (1972), "Failure of Coal Waste Embankment Blamed on Lack of Engineering," March 9, 1972, p. 11.

Engineering News-Record (1986), "Massive Groundwater Fix Studied," Nov. 20, 1986, pp. 28–29.

Foreman, Terry L. and Neil L. Ziemba (1987), "Cleanup on a Large Scale," *Civil Engineering,* ASCE, August, 1987, pp. 46–48.

Mackenchnie, W. R. (1981), "Impoundment of Tailings by the Mining Industry in Zimbabwe," Proc. Int. Conf. on Soil Mechanics and Foundation Engineering—10th, Stockholm, Sweden, June 15-19, 1981, Vol. 2, pp. 361–368.

Metcalf, Leonard, and Harrison P. Eddy (1930), *Sewerage and Sewage Disposal,* Mc-Graw-Hill Book Co., Inc., New York and London, pp. 1–15.

Jeyapalan, Jey K., J. Michael Duncan, and H. Bolton Seed (1983), "Investigation of Flow Failures of Tailings Dams," *Journal of the Geotechnical Engineering Division,* American Society of Civil Engineers, V. 109, No. 2, Feb. 1983, pp. 172–189.

Prescott, Samuel C., and Murray P. Horwood (1935), *Sedgwick's Principles of Sanitary Science and Public Health,* The Macmillan Company, N. Y. p. 39; p. 182.

Roth, Lawrence H., Joseph A. Cesare, and George S. Allison (1980), "Rapid Monitoring of Coal Refuse Embankments," Annual Meeting, Society of Mining Engineers, AIME, Las Vegas, Nevada, Feb. 24-28, 1980.

Stojadinvić, R., and I. Pantelić (1969), "An Analysis of Slope Stability in a Stockpile of Flotation Waste Products," *Izgrandnja,* Vol. 23, No. 3, pp. 37–43.

U.S. Department of Agriculture, Forest Service, Intermountain Region (1980), "Stability of Non Water Impounding Mine Waste Embankments," (Prepared by Bruce Vandre), *Tentative Engineering Guide,* Ogden, Utah, March, 1980.

AUTHOR INDEX

441

SUBJECT INDEX